T0206964

Communications in Computer and Information Science 1514

More information about this series at https://link.springer.com/bookseries/7899

Nicholas N. Olenev · Yuri G. Evtushenko ·
Milojica Jaćimović · Michael Khachay ·
Vlasta Malkova (Eds.)

Advances in Optimization and Applications

12th International Conference, OPTIMA 2021
Petrovac, Montenegro, September 27 – October 1, 2021
Revised Selected Papers

Editors
Nicholas N. Olenev
Dorodnicyn Computing Centre
FRC CSC RAS
Moscow, Russia

Yuri G. Evtushenko
Dorodnicyn Computing Centre
FRC CSC RAS
Moscow, Russia

Milojica Jaćimović
University of Montenegro
Podgorica, Montenegro

Michael Khachay
Institute of Mathematics and Mechanics
Ekaterinburg, Russia

Vlasta Malkova
Dorodnicyn Computing Centre
FRC CSC RAS
Moscow, Russia

ISSN 1865-0929 ISSN 1865-0937 (electronic)
Communications in Computer and Information Science
ISBN 978-3-030-92710-3 ISBN 978-3-030-92711-0 (eBook)
https://doi.org/10.1007/978-3-030-92711-0

This Springer imprint is published by the registered company Springer Nature Switzerland AG
The registered company address is: Gewerbestrasse 11, 6330 Cham, Switzerland

Preface

This volume contains the second part of the refereed proceedings of the XII International Conference on Optimization and Applications (OPTIMA 2021)[1].

Organized annually since 2009, the conference has attracted a significant number of researchers, academics, and specialists in many fields of optimization, operations research, optimal control, game theory, and their numerous applications in practical problems of operations research, data analysis, and software development.

The broad scope of OPTIMA has made it an event where researchers involved in different domains of optimization theory and numerical methods, investigating continuous and discrete extremal problems, designing heuristics and algorithms with theoretical bounds, developing optimization software, and applying optimization techniques to highly relevant practical problems, can meet together and discuss their approaches and results. We strongly believe that this facilitates collaboration between researchers working in modern optimization theory, methods, and applications and those employing them to resolve valuable practical problems.

The conference was held during September 27 – October 1, 2021, in Petrovac, Montenegro, in the Budvanian riviera on the azure Adriatic coast. Due to the COVID-19 pandemic situation the Program Committee (PC) decided to organize online sessions for those who were not able to come to Montenegro this year. The main organizers of the conference were the Montenegrin Academy of Sciences and Arts, Montenegro, the Dorodnicyn Computing Centre, FRC CSC RAS, Russia, the Moscow Institute of Physics and Technology, Russia, the Lomonosov Moscow State University, Russia, and the University of Évora, Portugal. This year, the key topics of OPTIMA were grouped into seven tracks:

I. Mathematical programming
II. Global optimization
III. Stochastic optimization
IV. Optimal control
V. Mathematical economics
VI. Optimization in data analysis
VII. Applications.

The Program Committee (PC) and the reviewers of the conference included more than one hundred well-known experts in continuous and discrete optimization, optimal control and game theory, data analysis, mathematical economy, and related areas from leading institutions of 25 countries including Argentina, Australia, Austria, Belgium, China, Finland, France, Germany, Greece, India, Israel, Italy, Lithuania, Kazakhstan, Mexico, Montenegro, The Netherlands, Poland, Portugal, Russia, Serbia, Sweden, Taiwan, Ukraine, the UK, and the USA. This year we received 98 submissions mostly

[1] http://agora.guru.ru/display.php?conf=OPTIMA-2021.

from Russia but also from Azerbaijan, Belarus, China, Finland, France, Germany, Kazakhstan, Moldova, Montenegro, Poland, Saudi Arabia, Serbia, UAE, and Ukraine. Each submission was reviewed by at least three PC members or invited reviewers, experts in their fields, to supply detailed and helpful comments. Out of 63 qualified submissions, the Program Committee decided to accept 22 full and 3 short papers to the first volume of the proceedings, published in LNCS volume 13078. Thus the acceptance rate for the LNCS volume was about 40%.

In addition, after a short presentation of the candidate submissions, discussion at the conference, and subsequent revision, the Program Committee proposed 19 papers out the remaining 38 papers to be included in this, second, volume of the proceedings.

The conference featured five invited lecturers, and several plenary and keynote talks. The invited lectures included:

- Anton Bondarev, International Business School Suzhou, Xi'an Jiaotong-Liverpool University, China, "Optimality of sliding dynamics in hybrid control systems"
- Nenad Mladenovic, Khalifa University of Science and Technology, Abu Dhabi, UAE, "Formulation Space Search Metaheuristic"
- Yurii Nesterov, CORE/INMA, Université Catholique de Louvain, Belgium, "Inexact high-order proximal-point methods with auxiliary search procedure"
- Panos M. Pardalos, University of Florida, USA, "Artificial Intelligence, Data Sciences, and Optimization in Economics and Finance"
- Alexey Tret'yakov, Siedlce University of Natural Sciences and Humanities, Poland, "Exit from singularity. New optimization methods and the p-regularity theory applications"

We would like to thank all the authors for submitting their papers and the members of the PC for their efforts in providing exhaustive reviews. We would also like to express special gratitude to all the invited lecturers and plenary speakers.

November 2021

Nicholas N. Olenev
Yuri G. Evtushenko
Milojica Jaćimović
Michael Khachay
Vlasta Malkova

Organization

Program Committee Chairs

Milojica Jaćimović	Montenegrin Academy of Sciences and Arts, Montenegro
Yuri G. Evtushenko	Dorodnicyn Computing Centre, FRC CSC RAS, Russia
Igor G. Pospelov	Dorodnicyn Computing Centre, FRC CSC RAS, Russia

Program Committee

Majid Abbasov	St. Petersburg State University, Russia
Samir Adly	University of Limoges, France
Kamil Aida-Zade	Institute of Control Systems of ANAS, Azerbaijan
Alla Albu	Dorodnicyn Computing Centre, FRC CSC RAS, Russia
Alexander P. Afanasiev	Institute for Information Transmission Problems, RAS, Russia
Yedilkhan Amirgaliyev	Suleyman Demirel University, Kazakhstan
Anatoly S. Antipin	Dorodnicyn Computing Centre, FRC CSC RAS, Russia
Adil Bagirov	Federation University, Australia
Artem Baklanov	International Institute for Applied Systems Analysis, Austria
Evripidis Bampis	LIP6 UPMC, France
Olga Battaïa	ISAE-SUPAERO, France
Armen Beklaryan	National Research University Higher School of Economics, Russia
Vladimir Beresnev	Sobolev Institute of Mathematics, Russia
Anton Bondarev	Xi'an Jiaotong-Liverpool University, China
Sergiy Butenko	Texas A&M University, USA
Vladimir Bushenkov	University of Évora, Portugal
Igor A. Bykadorov	Sobolev Institute of Mathematics, Russia
Alexey Chernov	Moscow Institute of Physics and Technology, Russia
Duc-Cuong Dang	INESC TEC, Portugal
Tatjana Davidovic	Mathematical Institute, Serbian Academy of Sciences and Arts, Serbia
Stephan Dempe	TU Bergakademie Freiberg, Germany
Askhat Diveev	FRC CSC RAS and RUDN University, Russia
Alexandre Dolgui	IMT Atlantique, LS2N, CNRS, France
Olga Druzhinina	FRC CSC RAS, Russia
Anton Eremeev	Omsk Division of Sobolev Institute of Mathematics, SB RAS, Russia
Adil Erzin	Novosibirsk State University, Russia
Francisco Facchinei	Sapienza University of Rome, Italy

Vladimir Garanzha	Dorodnicyn Computing Centre, FRC CSC RAS, Russia
Alexander V. Gasnikov	Moscow Institute of Physics and Technology, Russia
Manlio Gaudioso	Universita della Calabria, Italy
Alexander I. Golikov	Dorodnicyn Computing Centre, FRC CSC RAS, Russia
Alexander Yu. Gornov	Institute System Dynamics and Control Theory, SB RAS, Russia
Edward Kh. Gimadi	Sobolev Institute of Mathematics, SB RAS, Russia
Andrei Gorchakov	Dorodnicyn Computing Centre, FRC CSC RAS, Russia
Alexander Grigoriev	Maastricht University, The Netherlands
Mikhail Gusev	N.N. Krasovskii Institute of Mathematics and Mechanics, Russia
Vladimir Jaćimović	University of Montenegro, Montenegro
Vyacheslav Kalashnikov	ITESM, Monterrey, Mexico
Maksat Kalimoldayev	Institute of Information and Computational Technologies, Kazakhstan
Valeriy Kalyagin	Higher School of Economics, Russia
Igor E. Kaporin	Dorodnicyn Computing Centre, FRC CSC RAS, Russia
Alexander Kazakov	Institute for System Dynamics and Control Theory, SB RAS, Russia
Michael Khachay	Krasovsky Institute of Mathematics and Mechanics, Russia
Oleg V. Khamisov	L. A. Melentiev Energy Systems Institute, Russia
Andrey Kibzun	Moscow Aviation Institute, Russia
Donghyun Kim	Kennesaw State University, USA
Roman Kolpakov	Moscow State University, Russia
Alexander Kononov	Sobolev Institute of Mathematics, Russia
Igor Konnov	Kazan Federal University, Russia
Vera Kovacevic-Vujcic	University of Belgrade, Serbia
Yury A. Kochetov	Sobolev Institute of Mathematics, Russia
Pavlo A. Krokhmal	University of Arizona, USA
Ilya Kurochkin	Institute for Information Transmission Problems, RAS, Russia
Dmitri E. Kvasov	University of Calabria, Italy
Alexander A. Lazarev	V.A. Trapeznikov Institute of Control Sciences, Russia
Vadim Levit	Ariel University, Israel
Bertrand M. T. Lin	National Chiao Tung University, Taiwan
Alexander V. Lotov	Dorodnicyn Computing Centre, FRC CSC RAS, Russia
Olga Masina	Yelets State University, Russia
Vladimir Mazalov	Institute of Applied Mathematical Research, Karelian Research Center, Russia
Nevena Mijajlović	University of Montenegro, Montenegro
Nenad Mladenovic	Mathematical Institute, Serbian Academy of Sciences and Arts, Serbia
Mikhail Myagkov	University of Oregon, USA
Angelia Nedich	University of Illinois at Urbana Champaign, USA
Yuri Nesterov	CORE, Université Catholique de Louvain, Belgium

Yuri Nikulin	University of Turku, Finland
Evgeni Nurminski	Far Eastern Federal University, Russia
Nicholas N. Olenev	Dorodnicyn Computing Centre, FRC CSC RAS, Russia
Panos Pardalos	University of Florida, USA
Alexander V. Pesterev	V.A. Trapeznikov Institute of Control Sciences, Russia
Alexander Petunin	Ural Federal University, Russia
Stefan Pickl	Universität der Bundeswehr München, Germany
Boris T. Polyak	V.A. Trapeznikov Institute of Control Sciences, Russia
Yury S. Popkov	Institute for Systems Analysis, FRC CSC RAS, Russia
Leonid Popov	IMM UB RAS, Russia
Mikhail A. Posypkin	Dorodnicyn Computing Centre, FRC CSC RAS, Russia
Alexander N. Prokopenya	Warsaw University of Life Sciences, Poland
Oleg Prokopyev	University of Pittsburgh, USA
Artem Pyatkin	Novosibirsk State University and Sobolev Institute of Mathematics, Russia
Ioan Bot Radu	University of Vienna, Austria
Soumyendu Raha	Indian Institute of Science, India
Leonidas Sakalauskas	Institute of Mathematics and Informatics, Lithuania
Eugene Semenkin	Siberian State Aerospace University, Russia
Yaroslav D. Sergeyev	University of Calabria, Italy
Natalia Shakhlevich	University of Leeds, UK
Alexander A. Shananin	Moscow Institute of Physics and Technology, Russia
Angelo Sifaleras	University of Macedonia, Greece
Mathias Staudigl	Maastricht University, The Netherlands
Petro Stetsyuk	V.M. Glushkov Institute of Cybernetics, Ukraine
Fedor Stonyakin	V. I. Vernadsky Crimean Federal University, Russia
Alexander Strekalovskiy	Institute for System Dynamics and Control Theory, SB RAS, Russia
Vitaly Strusevich	University of Greenwich, UK
Michel Thera	University of Limoges, France
Tatiana Tchemisova	University of Aveiro, Portugal
Anna Tatarczak	Maria Curie-Skłodowska University, Poland
Alexey A. Tretyakov	Dorodnicyn Computing Centre, FRC CSC RAS, Russia
Stan Uryasev	University of Florida, USA
Frank Werner	Otto von Guericke University Magdeburg, Germany
Adrian Will	National Technological University, Argentina
Vitaly G. Zhadan	Dorodnicyn Computing Centre, FRC CSC RAS, Russia
Anatoly A. Zhigljavsky	Cardiff University, UK
Julius Žilinskas	Vilnius University, Lithuania
Yakov Zinder	University of Technology, Australia
Tatiana V. Zolotova	Financial University under the Government of the Russian Federation, Russia
Vladimir I. Zubov	Dorodnicyn Computing Centre, FRC CSC RAS, Russia
Anna V. Zykina	Omsk State Technical University, Russia

Organizing Committee Chairs

Organizing Committee

Contents

Mathematical Economics

Optimization in Data Analysis

Applications

Mathematical Programming

On Numerical Estimates of Errors in Solving Convex Optimization Problems

A. Birjukov and A. Chernov(✉)

Moscow Institute of Physics and Technology, Moscow, Russian Federation
http://www.mipt.ru

Abstract. In this article we consider schemes of the error estimates development for numerical methods solving optimization problems. Suggested error estimates are formulated as functions depended on various values generated by numerical methods (i.e. points, function values, gradient values, etc.). These functions do not depend on such problem's parameters like Lipshitz constant, strong convexity constant and etc. The numerical value of the error estimates is calculated after receiving accurate enough problem's solution. Error estimates were received both for target function value and argument value.

Keywords: Convex optimization problems · Estimates of errors of the values of the target function and its argument · Error estimates correction coefficients · Theoretical and numerical error estimates · Smooth optimization · Non-smooth optimization · Strictly decreasing sequence · Upper and lower estimates

1 Introduction

The article deals with the actual problem of building error estimates of solving convex optimization problems (1) via numerical method.

$$f(x) \to \min_{x \in G \subset R^n} \tag{1}$$

The set G is convex and the function $f : G \to R^1$ is convex and can be either smooth or non-smooth. We assume that the solution of the problem (1) exists and denote it by f^* and x^* for the function and the point.

Most numerical methods are represented by some iteration process with starting point which builds the sequence of the points $\{x^k\}_{k=0,1,2,\ldots}$. This iteration process can be written in the form:

$$x^{k+1} = x^k + \lambda_k \cdot p^k, \quad k = 0, 1, 2, \ldots \tag{2}$$

In (2) p^k is the descending direction of the function $f(x)$ and λ_k is the size of the step over direction p^k. We assume that the sequence $\{x^k\}$ from (2) converges to the point x^*: $x^* = \lim_{k \to \infty} x^k$.

The research was supported by Russian Science Foundation (project No. 21-71-30005).

N. N. Olenev et al. (Eds.): OPTIMA 2021, CCIS 1514, pp. 3–18, 2021.
https://doi.org/10.1007/978-3-030-92711-0_1

Definition 1. *Let $\varepsilon > 0$ be some small enough number. If for the point x inequalities* (3) *are valid then x is ε-solution for the problem* (1).

$$f(x) - f(x^*) \leq \varepsilon \quad or \quad ||x - x^*|| \leq \varepsilon. \tag{3}$$

The error estimates in the form (3) are not always available in practice. That is why some empirical rules [1] are used instead. Some examples of this rules are listed below:

- gradient norm (for unconstrained optimization): $||\nabla f(x^k)|| \leq \varepsilon$;
- distance between points: $||x^{k+1} - x^k|| \leq \varepsilon$ or $|f(x^k) - f(x^{k+1})| \leq \varepsilon$. (4)

In the last 20–30 years, the theory of the numerical methods for convex optimization problems, where error estimates have the form (5), has been widely used.

$$f(x^k) - f(x^*) \leq \varphi(L, R, \mu, k) \leq \varepsilon. \tag{5}$$

In (5) L (Lipshitz constant for $\nabla f(x)$), R (estimate of $||x^0 - x^*||$), μ (constant of the strong convexity) are parameters that characterize the class of the objective function , k is iteration number, $\varepsilon > 0$ is required accuracy and small enough [2,3]. Ideas on how to build functions $\varphi(\cdot)$ were suggested in [4] and later extended in [5,6] and in the other articles. But, in general, the problem of idenitification of L, R, μ constants is more difficult then initial optimization problem. Thus, error estimates (5) are difficult to use in practice.

The simplest examples of the error estimates (5) are error estimates for the unconstrained optimization problem with μ-strongly convex objective function [11]:

$$f(x^k) - f(x^*) \leq \frac{||\nabla f(x^k)||^2}{2\mu} \leq \varepsilon; \quad ||x^k - x^*|| \leq \frac{||\nabla f(x^k)||}{\mu} \leq \varepsilon.$$

Usage of the error estimates (3) in practice requires reliable calculations, which can be done in multi-precision arithmetic. In [7,8] the method of the error estimate of the calculations was suggested for the configurable mantissa of the machine representation of the real number. This approach is used in the numerical experiment presented in this paper.

In this section we build the error estimates in the smooth case using the parameters available while iteration process execution in the forms (6) and (7). In the Sect. 3 results obtained will be extended in the non-smooth case.

$$f(x^k) - f(x^*) \leq \varphi_1(x^k, x^N, \nabla f(x^k)) \leq \varepsilon, \tag{6}$$

$$||x^k - x^*||^2 \leq \varphi_2(x^k, x^N, \nabla f(x^k)) \leq \varepsilon, \tag{7}$$

where N is the final iteration's number, functions φ_1 and φ_2 depend on the specified arguments and some other computable parameters.

The error estimates specified in (3), (5), (6), (7) are not the only ones. Some classification of the error estimates usefull for this article is presented on the Fig. 1 with the following description:

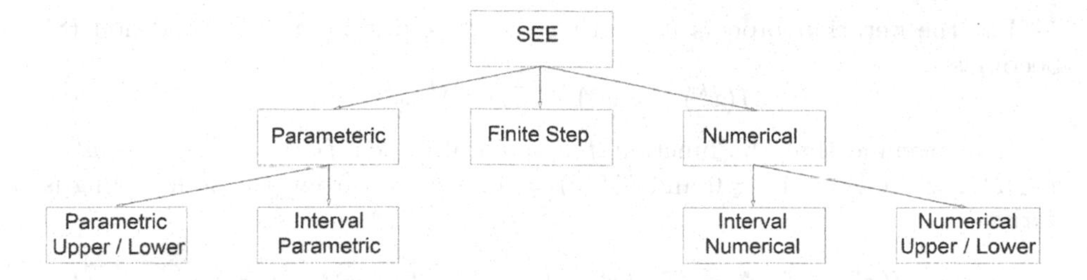

Fig. 1. Classification of the solution error estimates

- SEE - solution error estimate.
- A parametric estimate of the solution error can be written in the form (5) with known function φ and unknown numerical values of the parameters L, R, μ. These estimates represent the result of the theoretical analysis of the convergence rate of the various numerical methods.
- A numerical estimate of solution error can be written in the forms (6), (7) with known functions φ_1 and φ_2, as well as the values of its argument.
- A finite step estimate of solution error estimate is estimation of the solution error for finite-step numerical methods, which in exact arithmetic calculate the exact solution of the problem in a finite number of steps.

One can find examples of the error estimates in the finite step, interval parametric, parametric upper/lower forms however numerical error estimates forms in optimization theory are unknown for authors.

Definition 2. *Parametric error estimate of the solution is the error estimate written in the form* (5).

Numerical error estimate of the solution is the error estimate written in the form (6) *or* (7).

So-called duality gap (difference between values of the objective functions from prime and dual problems) can be considered as an example of the estimate (6). It is used often in practice but for constrained optimization problems [9,10]. In order to use it one has to formulate the dual problem.

2 Numerical Estimates of Errors for Convex Optimization Problems

The first-order criteria of the function f convexity on the set $G \subset R^n$ have the following form:

$$\begin{array}{ll} smoothcase & f(y) - f(x) \le \langle \nabla f(y), y - x \rangle, \quad \forall x, y \in G, \\ non - smoothcase & f(y) - f(x) \le \langle a, y - x \rangle, \end{array} \tag{8}$$

where $a \in \partial f(y)$, $\partial f(y)$ is subdifferential of the function f in the point $y \in G$.

For the iteration process (2) with $y := x^k$ and $x := x^*$ the condition (8) becomes:

$$0 \leq f(x^k) - f(x^*) \leq \nabla f(x^k)^T(x^k - x^*). \tag{9}$$

If we assume that the function f is quadratic, i.e.: $f(x) = x^T Ax/2 - b^T x$, $x \in R^n$, where $A = A^T \succeq 0$ and $\nabla f(x) = Ax - b$ then from (8) the following is derived:

$$f(x^k) - f(x^*) \leq \langle \nabla f(x^k), x^k - x^* \rangle = 2(f(x^k) - f(x^*)). \tag{10}$$

Thus for the quadratic function inequality (10) is strict.

Definition 3. *Let N be the last executed step of the iteration process* (2). *Then the point x^N and the value $f(x^N)$ are called controlling solution of the optimization problem* (1).

Definition 4. *The number γ_k defined in* (11) *is called accuracy coefficient for numerical error estimate in the form* (6)

$$\gamma_k = \frac{\langle \nabla f(x^k), x^k - x^* \rangle}{f(x^k) - f(x^*)} \tag{11}$$

Remark 1. In the same way we can introduce the number δ_k that can be called accuracy coefficient for numerical error estimate in the form (7). But due to the form of the function $||x^k - x^*||^2$ the following equality (12) is valid:

$$\delta_k = \frac{\langle \nabla_x ||x^k - x^*||^2, x^k - x^* \rangle}{||x^k - x^*||^2} = 2 \tag{12}$$

If one will replace in (11) and (12) x^* by the controlling solution x^N, $f(x^N)$ except the $\nabla_x ||x^k - x^*||^2$ in (12), then sequences will be defined:

$$\gamma_k^N = \frac{\langle \nabla f(x^k), x^k - x^N \rangle}{f(x^k) - f(x^N)}, \quad \delta_k^N = \frac{\langle \nabla_x ||x^k - x^*||^2, x^k - x^N \rangle}{||x^k - x^*||^2 - ||x^N - x^*||^2} \tag{13}$$

Accuracy coefficients introduced in Definition 4 are important, since they characterize reliability of the inequalities (6) and (7).

Remark 2. Reliability and quality of the numerical error estimate are opposing concepts: reliability improvement (i.e. increasing of γ_k and δ_k) leads to quality decreasing and vice versa.

Also reliability improvement can be based on reducing of $f(x^N) - f(x^*)$ and $||x^N - x^*||$ or via increasing of $\langle \nabla f(x^k), x^k - x^N \rangle$ and $||x^k - x^N||$.

Remark 3. For linear function $f(x) = \langle a, x \rangle$, where a is problem's parameter, the introduced accuracy coefficients will be:

$$\gamma_k = \frac{\langle a, x^k - x^* \rangle}{\langle a, x^k - x^* \rangle} = 1; \quad \gamma_k^N = \frac{\langle a, x^k - x^N \rangle}{\langle a, x^k - x^N \rangle} = 1$$

Remark 4. Let objective function $f(x)$, $x \in G$ be a μ-strongly convex. Then

$$f(x^k) - f(x^N) \leq \langle \nabla f(x^k), x^k - x^N \rangle - \mu \|x^k - x^N\|^2 < \langle \nabla f(x^k), x^k - x^N \rangle$$

Definition 5. *The point x^N is called M_f-function solution of the optimization problem* (1) *if for all $k \leq M_f$ $\gamma_k^N \geq \gamma > 1$ and inequality below is valid.*

$$f(x^k) - f(x^N) > (\gamma - 1)^{-1}(f(x^N) - f(x^*))$$

If $f(x) = \|x - x^*\|^2$ then the point x^N will be M_x-function solution if $\delta_k^N \geq \delta > 1$ for all $k \leq M_x$ and inequality below is valid:

$$\|x^k - x^*\|^2 \geq \frac{\delta}{\delta - 1}\|x^N - x^*\|^2.$$

In this case the point x^N will be called M_x-argument solution.

Calculation methodique of the values γ, δ, M_f and M_x are presented in this article at the the Sect. 4.

Theorem 1. *Let:*

- *the point x^* be the solution of the convex problem* (1)*;*
- *$\{x^k\}$, $k = 0, 1, 2, \ldots$ be the sequence that converges to the point x^*;*
- *$\forall k$ $f(x^k) > f(x^{k+1})$;*
- *x^N is M_f-function solution.*

Then for any $k \leq M_f$ inequalities below are valid:

$$f(x^k) - f(x^*) \leq \langle \nabla f(x^k), x^k - x^N \rangle; \tag{14}$$

Proof. Let us consider the number $k \leq N$.

Following the definition of γ_k^N in (13) we can write:

$$\begin{aligned}\langle \nabla f(x^k), x^k - x^N \rangle &= \gamma_k^N (f(x^k) - f(x^N)) \\ &= \gamma_k^N (f(x^k) - f(x^*)) - \gamma_k^N (f(x^N) - f(x^*))\end{aligned}$$

Thus:

$$\begin{aligned}\langle \nabla f(x^k), x^k - x^N \rangle &= (f(x^k) - f(x^*)) + A_k^N; \\ A_k^N &= (\gamma_k^N - 1)(f(x^k) - f(x^*)) - \gamma_k^N (f(x^N) - f(x^*)).\end{aligned}$$

Due to above the equality below is valid:

$$A_k^N = (\gamma_k^N - 1)(f(x^k) - f(x^N)) - (f(x^N) - f(x^*)).$$

The value of A_k^N has to be greater than 0 in order to inequality (14) be valid. But:

$$\begin{aligned}A_k^N &= (\gamma_k^N - 1)(f(x^k) - f(x^N)) - (f(x^N) - f(x^*)) \\ &\geq (\gamma - 1)(f(x^k) - f(x^N)) - (f(x^N) - f(x^*)) > 0\end{aligned}$$

The last inequality is valid since x^N is M_f-function solution of the optimization problem (1) considered.

Theorem 2. *Let:*

- *the point x^* be the solution of the convex problem* (1);
- $\{x^k\}$, $k = 0, 1, 2, \dots$ *be the sequence that converges to the point* x^*;
- $\forall k\ f(x^k) > f(x^{k+1})$;
- x^N *is* M_x*-argument solution.*

Then for any $k \leq M_x$ inequalities below are valid:

$$\|x^k - x^*\| \leq 2\|x^k - x^N\|, \quad \|x^k - x^*\| \leq (\delta + \sqrt{\delta(\delta - 1)})\|x^k - x^N\| \tag{15}$$

Proof. Let us introduce $\overline{f}(x) = \|x - x^*\|^2$. Then $\overline{f}(x^*) = 0$ and due to Theorem 1 inequality (14) will have form

$$\|x^k - x^*\|^2 \leq 2\left\langle x^k - x^*, x^k - x^N \right\rangle^2 \leq 2\|x^k - x^*\|\|x^k - x^N\|.$$

Thus: $\|x^k - x^*\| \leq 2\|x^k - x^N\|$. Now, let us proof the second inequality in (15):

$$\|x^k - x^*\| = \|x^k - x^N + x^N - x^*\| \leq \|x^k - x^N\| + \|x^N - x^*\|.$$

Since x^N is M_x-argument solution we can further improve inequality:

$$\|x^k - x^*\| \leq \|x^k - x^N\| + \sqrt{(\delta - 1)/\delta}\|x^k - x^*\|$$

$$\|x^k - x^*\| \leq \left(1 - \sqrt{(\delta - 1)/\delta}\right)^{-1} \|x^k - x^N\| = (\delta + \sqrt{\delta(\delta - 1)})\|x^k - x^N\|.$$

The last statement corresponds to the second inequality in (15).

Remark 5. If $\delta = 1,33$ both inequalities in (15) are the same. The multiplyer 2 in (15) corresponds to numerical estimate while the multiplyer $(\delta + \sqrt{\delta(\delta - 1)}$ corresponds to parametrical one.

Remark 6. 1. Result of the Theorem 1 is valid if and only if for any $k \leq M_f$ with γ_k^N from (13):

$$\gamma_k^N \frac{f(x^k) - f(x^N)}{f(x^k) - f(x^*)} \geq 1 \quad or \quad f(x^k) - f(x^*) \leq \gamma_k^N (f(x^k) - f(x^N)) \tag{16}$$

2. Result of the Theorem 2 is valid if and only if there exists constant $c \geq 1$:

$$c\frac{\|x^k - x^N\|}{\|x^k - x^*\|} \geq 1 \quad or \quad \|x^k - x^*\| \leq c\|x^k - x^N\|. \tag{17}$$

Remark 7. 1. The practical meaning of the Theorem 1 is: accuracy coefficient γ_k^N is calculated explicitly while the value of $(f(x^k) - f(x^N))/(f(x^k) - f(x^*))$ is less than 1 and close enough to 1 for big enough N. This estimate is reliable for $\gamma_k^N \geq 2$ that is observed in practice.

2. It is recommended to specify $c = 2$ or $c = 3$ for the Theorem 2. According to this estimate we can say that $\|x^k - x^*\| \leq c\|x^k - x^N\| \leq \varepsilon$. Condition of the strict relaxation is a good signal for convergence $x^k \to x^*$. It is evident that if one will increase value of parameter c then reliability of inequality is increasing, while the accuracy is decreasing.

Let introduce values ε_1 and ε_2 in order to calculate estimates (16) and (17) in the form (18) below:

$$\begin{aligned} \varepsilon_1 = c_f(f(x^k) - f(x^N)), \text{ where } c_f = max(2, \gamma_k^N) \\ \varepsilon_2 = c_x\|x^k - x^N\|, \text{ where } c_x = 2, 3, \ldots \text{ or } c_x = \delta + \sqrt{\delta(\delta - 1)}. \end{aligned} \tag{18}$$

The algorithm to calculate solution error estimates consists of two parts: the first part calculates the test solution of the problem (1) $(x^N, f(x^N))$; the second part calculates values ε_1, ε_2 accuracy estimates:

Algorithm 1. *1. set* $k := 0$ *and* $\varepsilon > 0$ *is a small enough;*
2. compute and store p^k, λ_k, x^{k+1}, $\nabla f(x^k)$ *for each step* $k = 0, 1, 2, \ldots$*;*
3. test inequality (4). If it is not valid then $k := k + 1$ *and go to algorithm step 2 otherwise we go to step 4;*
4. calculate ε_1, ε_2 *according to (18).*

Remark 8. Inequalities (16), (17) for smooth and non-smooth (see Sect. 3 below) objective functions represent the **M**ethod to **E**stimate the **E**rror of the **S**olution (MEES) for the convex optimization problems (CoMEES).

3 Numerical Estimates of Errors in Solving Convex Non-smooth Optimization Problems

Assume that the problem (1) solution (values x^* and $f(x^*)$) exists, but now we will not require that function $f(x)$ is smooth. We also assume that subdifferential set $\partial f(y)$ [11] is not empty. Let $a_k \in \partial f(x^k)$ then:

$$0 \leq f(x^k) - f(x^*) \leq \langle a_k, x^k - x^* \rangle, \tag{19}$$

Introduce subgradient $\overline{a}^k$ as a direction of the fastest grow rate of the function f at the point x^k [11]:

$$\overline{a}^k = \pi_{\partial f(x^k)}(0_n) \equiv \arg \min_{a \in \partial f(x^k)} \langle a, a \rangle . \tag{20}$$

In order to simplify further statements we denote $p^k = \overline{a}^k$, $x^{k+1} = x^k - \lambda_k \overline{a}^k$ and $\lambda_k = \arg\min_{\lambda > 0} f(x^k - \lambda \overline{a}^k)$ for $k = 0, 1, 2, \ldots$ and assume $x^k - \lambda_k \overline{a}^k \in G$.

Assume that for function f and iteration process of the method following is valid: $f(x^{k+1}) < f(x^k)$ and $\lim_{k \to \infty} x^k = x^*$, $\lim_{k \to \infty} f(x^k) = f(x^*)$. Note that inequality $f(x^{k+1}) < f(x^k)$ is valid if $a_k = \overline{a}^k$. If step size is not equal to λ_k defined above and converges to 0 for $k \to \infty$ then related iteration process can be non-decreasing [11].

For a non-smooth convex optimization problem, introduce the following coefficients:

$$\begin{aligned} \gamma_k^N &= \frac{\langle a^k, x^k - x^N \rangle}{f(x^k) - f(x^N)}, \; \delta_k^N = 2\frac{\langle x^k - x^*, x^k - x^N \rangle}{\|x^k - x^*\|^2 - \|x^N - x^*\|^2}, \\ \gamma_k &= \frac{\langle a^k, x^k - x^* \rangle}{f(x^k) - f(x^*)}, \quad \delta_k = \frac{2\|x^k - x^*\|^2}{\|x^k - x^*\|^2} = 2. \end{aligned} \tag{21}$$

Theorem 3. *Let*

- *the point x^* be the solution of the convex problem (1);*
- *$\{x^k\}$, $k = 0, 1, 2, \ldots$ be the sequence that converges to the point x^*;*
- *be valid $\forall k$ inequality $f(x^k) > f(x^{k+1})$;*
- *x^N be M_f-function solution of the considered problem.*

Then for any $k \leq M_f$:

$$f(x^k) - f(x^*) \leq \langle \overline{a}^k, x^k - x^N \rangle \tag{22}$$

Theorem 4. *Let*

- *the point x^* be the solution of the convex problem (1);*
- *$\{x^k\}$, $k = 0, 1, 2, \ldots$ be the sequence that converges to the point x^*;*
- *be valid $\forall k$ inequality $f(x^k) > f(x^{k+1})$;*
- *x^N be M_x-argument solution of the considered problem.*

Then for any $k \leq M_x$:

$$\|x^k - x^*\| \leq 2\|x^k - x^N\| \quad or \quad \|x^k - x^*\| \leq (\delta + \sqrt{\delta(\delta - 1)})\|x^k - x^N\|. \tag{23}$$

Proof. Direction $\overline{a}^k$ is analogue of the function gradient in non-smooth case: it characterizes the maximum rate of increase of the objective function $f(x)$ at the point x^k. Thus the proofs of the Theorems 3 and 4 are similar to the proofs of the Theorems 1 and 2.

Remark 9. 1. Result of the Theorem 3 is valid if and only if for any $k \leq M_f$ with γ_k^N from (21)

$$\gamma_k^N \frac{f(x^k) - f(x^N)}{f(x^k) - f(x^*)} \geq 1 \quad or \quad f(x^k) - f(x^*) \leq \gamma_k^N (f(x^k) - f(x^N));$$

2. Result of the Theorem 4 is valid if and only if there exists constant $c \geq 1$ (one can also select $c = \delta + \sqrt{\delta(\delta - 1)}$ if δ is known):

$$c \frac{\|x^k - x^N\|}{\|x^k - x^*\|} \geq 1 \quad or \quad \|x^k - x^*\| \leq c\|x^k - x^N\| \quad \forall k \leq M_x.$$

Introduce new quality parameters:

$$\begin{aligned} \varepsilon_3 &= \gamma_k^N (f(x^k) - f(x^N)); \\ \varepsilon_4 &= c\|x^k - x^N\|. \end{aligned} \tag{24}$$

Let $\varepsilon > 0$ be a small number, ε_3 and ε_4 be required accuracy of the solution then we define the following algorithm:

Algorithm 2. *1. let* $k := 0$*;*
2. define $\partial f(x^k)$*;*
3. if $0 \in \partial f(x^k)$ *then stop, we find an exact solution (for unconstrained optimization problem).*
4. calculate $\overline{a}^k$*,* λ_k*, and* x^{k+1}*;*
5. calculate $\|x^k - x^{k+1}\|$ *and* $f(x^k) - f(x^{k+1})$*;*
6. test inequality (4). If it is not valid then $k := k+1$ *and go to step 2, otherwise go to step 7.*
7. calculate ε_3*,* ε_4*.*

4 Rounding Method to Estimate the Error of the Solution (RoMEES)

The theory described above allows building some structure of the solution error estimates for all points provided by some method rather than for a single point. Using the item 4 in Algorithm 1 (or item 7 in Algorithm 2) values γ_k^N are calculated for the all $k \leq N-1$. On the next step this values are used to check Theorem 2 in case of $f(x^N) \approx f(x^*)$ and $x^N \approx x^*$. Thus for some k estimates (16) and (17) will be valid. These conditions can be invalid for some k close enough to the value N. Usage of another approach for solution error estimates called "rule of the first digits matches" in the mantissa of the $f(x^k)$ and $f(x^N)$ for the some $k \leq N-1$ allows to solve this problem. This rule to estimate the error of the solution for strictly decreasing sequence of the function values was suggested in the article [8].

Denote $f^* = f(x^*)$, $f^k = f(x^k)$. The values f^k and f^* are represented in decimal number system by the sequence $a_m \in [0, 9]$, and the integer value t (the order of the value) where m is the length of the mantissa according to the machine number definition and $a_0 \neq 0$:

$$f^k = a_0^k, a_1^k \dots a_m^k \cdot 10^t.$$
$$f^* = a_0^*, a_1^* \dots a_m^* \cdot 10^t.$$

Since the sequence f^k converges then for some large enough N the value $f(x^N)$ is "close enough" to f^*, i.e. for the some m_k the following is valid:

$$a_r^k = a_r^N = a_r^* \quad \forall 0 \leq r \leq m_k \leq m_{N-1} = m_N \leq m+1. \tag{25}$$

Note that the sequence m_k is the monotonically increase sequence. Thus we can obtain the lower bound $\underline{f_k^*}$ and upper bound $\overline{f_k^*}$ estimates for the value f^*:

$$\begin{array}{r} \text{for } f^* > 0: \underline{f_k^*} = a_0, a_1 a_2 \dots a_{m_k} \cdot 10^t;\ \overline{f_k^*} = a_0, a_1 a_2 \dots (a_{m_k}+1) \cdot 10^t \\ \text{for } f^* < 0: \underline{f_k^*} = -a_0, a_1 a_2 \dots (a_{m_k}+1) \cdot 10^t;\ \overline{f_k^*} = -a_0, a_1 a_2 \dots a_{m_k} \cdot 10^t \end{array} \tag{26}$$

Thus the error estimates can be presented in the following form:

$$\begin{array}{c} f^k - f^* \leq f^k - \underline{f_k^*} \leq 10^{t-m_k};\quad \overline{f_k^*} - \underline{f_k^*} \leq 10^{t-m_k}; \\ f^k - \overline{f_N^*} \leq f^k - f^* \leq f^k - \underline{f_N^*} \leq 10^{t-m_k}. \end{array} \tag{27}$$

The relative error for $f^k - f^*$ is:

$$\text{for } f^* > 0: \frac{f^k - f^*}{f^*} \le \frac{f^k - f^*}{\underline{f_k^*}} < 10^{-m_k}. \quad (28)$$

$$\text{for } f^* < 0: \frac{f^k - f^*}{|f^*|} \le \frac{f^k - f^*}{|\underline{f_k^*}|} < 10^{-m_k}. \quad (29)$$

Thus it is possible to build the table of the f^k values where $k \le N$ and select for each k related m_k number where $a_r^k = a_r^N$ for all $r \le m_k$. Thus one can calculate solution error estimates using the inequalities above. In the other case ($m_k = 0$) it is necessary to use (16) and (17).

The result theorem of this section is formulated below:

Theorem 5. *Let us consider the solution of the optimization problem* (1) *where:*

1. $\{f(x^k)\}_{k=0,1,2,\dots}$ *is a strictly decreasing convergent sequence:* $\lim_{k\to\infty} x^k = x^*$ *and* $\lim_{k\to\infty} f(x^k) = f^*$.
2. $k = N$ *is the finally calculated number and* (25) *is valid for some* k;
3. *conditions in* (25) *are valid.*

Then error estimates (27), (28) *and* (29) *are valid.*

Remark 10. If $f^* = 0$ then error estimate for any k is the value $f(x^k)$.

Definition 6. *Method based on the error estimates* (27), (28) *and* (29) *above we will call rounding method to estimate the error of the solution (RoMEES).*

Consider A_k^N from the proof of the Theorems 1 and 3:

$$\begin{aligned} A_k^N &= (\gamma_k^N - 1)(f(x^k) - f(x^N)) - (f(x^N) - f(x^*)) \\ &= \left\langle \nabla f(x^k), x^k - x^N \right\rangle - (f(x^k) - f^*) \ge 0. \end{aligned}$$

Thus one can obtain the following estimate for the values M_f and γ:

$$M_f = \max_{k \in J_N^f} k,$$
$$J_N^f = \left\{ i \in [0, N-1] : \forall k \le i \ \left\langle \nabla f(x^k), x^k - x^N \right\rangle \ge f(x^k) - \underline{f_N^*} \right\},$$
$$\gamma = \min_{k \le M_f} \frac{\left\langle \nabla f(x^k), x^k - x^N \right\rangle}{f(x^k) - f(x^N)}$$

Let $\|x^N - x^*\| \le (c-1)\|x^k - x^N\|$ then $\|x^k - x^*\| \le \|x^k - x^N\| + \|x^N - x^*\| < c\|x^k - x^N\|$. Thus we can estimate the values M_x and δ as follows:

$$M_x = \max_{k \in J_N^x} k,$$
$$J_N^x = \left\{ i \in [0, N-1] : \forall k \le i \ (c-1)\|x^k - x^N\| \ge \|x^N - x^*\| \right\},$$
$$\delta = \min_{k \le M_x} \frac{2\left\langle x^k - x^*, x^k - x^N \right\rangle}{\|x^k - x^*\|^2 - \|x^N - x^*\|^2}.$$

This estimates can be used if one will do substitutions: $x^* \rightarrow x^N$ and $x^N \longrightarrow x^{N_0}$.

If $f(x^*)$ and $f(x^N)$ are close enough then according to (27) then the estimate of the high quality can be calculated:

$$f(x^k) - f(x^*) < f(x^k) - \underline{f_N^*}.$$

Remark 11. Usage of the (27) and (28) provides another view point on remarks 6 and 9. In order to do this one should use the best estimate $\underline{f_k^*}$ from (26). On the next step one should compare the values $\nabla f(x^k)^T(x^k - x^N)$ and 10^{t-m_k}.

Remark 12. RoMEES estimates are valid in case of the specified assumptions. But this estimates can seriously exceed the real error estimate. For example, if for some k: $f(x^k) = 1.570000001756$, $f(x^{k+1}) = 1.569999999756$ and $f(x^N) = 1.569999998545$ then $m_k = 1$, $m_{k+1} = 8$, $\Delta f(x^k) - f(x^N) = 10^{-8}$. In this case one can use CoMEES.

On the other hand, for large enough values N and k, the error estimate provided by CoMEES can be less than $f^k - f^*$. In this case one can use RoMEES.

Thus these methods complement each other and thus improve the error estimation quality.

Remark 13. For non-convex optimization problems and related strictly decreasing sequences constructed by any method when solving the problem, one can use RoMEES. If the sequence converges but is not monotonic, one can choose the necessary strictly decreasing convergent subsequence and use RoMEES.

Remark 14. To prove that the point x^N, $f(x^N)$ is the approximation of the solution optimization problem with known error, one should use corresponding sufficient conditions of the optimality for this problem.

5 Experiments

Experiments were completed on the PC with RAM 16 GB and CPU Intel Core I7 2.8 GHz with OS Windows 10 (64) in PyCharm IDE (Python 3.8). Bold numbers in tables in each row illustrate the matching of the corresponding numbers in $f(x^k)$ and $f(x^N)$ for RoMEES.

5.1 Smooth Convex Function

Let us consider the problem (1) where $G = R^n$ with the following strongly convex objective function:

$$f(x) = \sum_{i=1}^{n} \alpha_i (x_i - \beta_i)^2 + \exp^{c^T x} + \gamma \exp^{-c^T x}, \tag{30}$$

where $\alpha_i = i^2$, $\beta_i = 1$, $c_i = 0.1$ for all $i = 1, \ldots n$, $\gamma = 2$, $n = 40$.

To solve the problem Polak-Ribiere-Polyak method [11] with golden search as a one-dimensional search is used (initial point $x^0 = 1_n$, the mantissa is 50, the required accuracy of the solution is 10^{-8} for the norm of the objective function gradient (stop condition of the iteration process), one dimensional search accuracy is 10^{-11}). Results of the experiment are presented in the Table 1.

Table 1. Smooth unconstrained convex optimization problem output ($\varepsilon_1 = \nabla f(x_k)^T (x^k - x^N)$, $\varepsilon_2 = 10^{-m_k}$, m_k is the amount of the same numbers after the comma in $f(x^k)$ and $f(x^N)$, $\varepsilon_3 = \sqrt{2}||x^k - x^N||$, $N = 103$).

#	$f(x^k)$	ε_1	ε_2	γ_k^N	ε_3
0	627387.116318387452895008643484	$1.00 \cdot 10^6$	10^6	1.9637	324.2905
1	478900.083088491681976382588882	$7.28 \cdot 10^5$	10^6	2.0150	311.7882
2	243348.130698388336052959732978	$2.45 \cdot 10^5$	10^6	1.9494	247.6078
3	216169.692279384052760407350101	$1.97 \cdot 10^5$	10^6	2.0013	239.7694
8	**1**31996.906021836025384060445899	$2.90 \cdot 10^4$	10^5	2.0001	130.1494
13	**1**20121.261640740892823384176226	$5.26 \cdot 10^3$	10^5	2.0000	56.4248
18	**117**854.159033531029297795044290	$7.23 \cdot 10^2$	10^3	2.0001	18.8308
23	**117**530.925085920908178025845055	$7.71 \cdot 10^1$	10^3	1.9979	5.3423
28	**11749**6.629560200623172382736530	$8.57 \cdot 10^0$	10^1	2.0009	1.6442
33	**117492**.624757779447610055971921	$5.58 \cdot 10^{-1}$	10^0	2.0000	0.4656
38	**117492.3**78753366524838502751313	$6.60 \cdot 10^{-2}$	10^{-1}	2.0000	0.2119
43	**117492.3**56031599605433167243964	$2.06 \cdot 10^{-2}$	10^{-1}	2.0000	0.1163
48	**117492.345**996952543060912913084	$4.89 \cdot 10^{-3}$	10^{-3}	2.0000	0.0100
53	**117492.3457**62464305993232808333	$2.05 \cdot 10^{-4}$	10^{-4}	2.0000	0.0021
58	**117492.345752325**708989538086367	$1.86 \cdot 10^{-7}$	10^{-6}	2.0000	$1.97 \cdot 10^{-4}$
63	**117492.345752235**074160763575294	$4.63 \cdot 10^{-9}$	10^{-8}	2.0000	$3.85 \cdot 10^{-5}$
68	**117492.345752232**852670052957178	$1.90 \cdot 10^{-10}$	10^{-9}	2.0000	$5.93 \cdot 10^{-6}$
73	**117492.34575223275**8708228546886	$1.63 \cdot 10^{-12}$	10^{-11}	2.0000	$7.03 \cdot 10^{-7}$
78	**117492.345752232757**950445777728	$1.13 \cdot 10^{-13}$	10^{-12}	2.0000	$2.22 \cdot 10^{-7}$
83	**117492.34575223275789**7164252487	$6.19 \cdot 10^{-15}$	10^{-14}	2.0000	$6.38 \cdot 10^{-8}$
88	**117492.345752232757894**104723483	$7.59 \cdot 10^{-17}$	10^{-15}	2.0000	$3.64 \cdot 10^{-9}$
93	**117492.34575223275789406**7167474	$8.28 \cdot 10^{-19}$	10^{-17}	2.0000	$4.89 \cdot 10^{-10}$
98	**117492.3457522327578940667**68423	$3.01 \cdot 10^{-20}$	10^{-19}	2.0000	$8.59 \cdot 10^{-11}$
101	**117492.345752232757894066753**815	$8.96 \cdot 10^{-22}$	10^{-20}	2.0000	$9.26 \cdot 10^{-12}$
102	**117492.345752232757894066753**746	$7.57 \cdot 10^{-22}$	10^{-21}	2.0000	$7.99 \cdot 10^{-12}$
103	**117492.345752232757894066753**367	0.0	10^{-21}	Not defined	0.0

5.2 Non-smooth Convex Unconstrained Optimization Problem

In this section we provide results of the experiment for the problem (1) for $G = R^n$ with continuous non-smooth convex objective function:

$$f(x) = \sum_{j=1}^{2} d_j \left| c_j^T x - b_j \right| + \sum_{i=1}^{n} \alpha_i (x_i - \beta_i)^2, \tag{31}$$

where $\alpha_i = i^2$, $\beta_i = 1/2$, $i = 1, \ldots, n$, $d_1 = 2$, $d_2 = 1$, $c_1 = 1_n$, $c_{2,j} = (-1)^{j+1}, j = 1, \ldots, n$, $b_j = c_j^T \beta$, $j = 1, 2$[1], $n = 20$.

[1] Parameters of the objective function allows explicitly find problem solution $x^* = \beta$, $f(x^*) = 0$.

Table 2. Non-smooth unconstrained convex optimization problem output ($\varepsilon_1 = a_k^T(x^k - x^N)$, $\varepsilon_2 = 10^{-m_k}$, m_k is the amount of the same numbers after the comma in $f(x^k)$ and $f(x^N)$, $\varepsilon_3 = \sqrt{2}||x^k - x^N||$, $N = 656$).

#	$f(x^k)$	ε_1	ε_2	γ_k^N	ε_3
1	123.28934998833217774	$2.36 \cdot 10^2$	10^3	1.9142	$2.09 \cdot 10^0$
2	52.09959143810754034	$9.59 \cdot 10^1$	10^3	1.8402	$1.71 \cdot 10^0$
3	30.72335415840225609	$5.49 \cdot 10^1$	10^3	1.7866	$1.53 \cdot 10^0$
38	0.36905197041285442	$6.59 \cdot 10^{-1}$	10^0	1.7860	$3.10 \cdot 10^{-1}$
73	0.02787459012696062	$5.57 \cdot 10^{-2}$	10^{-1}	2.0000	$1.10 \cdot 10^{-1}$
108	0.00471345433445760	$9.43 \cdot 10^{-3}$	10^{-2}	2.0000	$4.70 \cdot 10^{-2}$
143	0.00089439467000347	$1.79 \cdot 10^{-3}$	10^{-3}	2.0000	$2.05 \cdot 10^{-2}$
178	0.00017343988402011	$3.47 \cdot 10^{-4}$	10^{-3}	2.0000	$9.09 \cdot 10^{-3}$
213	0.00003376259407778	$6.75 \cdot 10^{-5}$	10^{-4}	1.9999	$3.99 \cdot 10^{-3}$
248	0.00000657677448323	$1.31 \cdot 10^{-5}$	10^{-5}	1.9999	$1.77 \cdot 10^{-3}$
283	0.00000128126955778	$2.56 \cdot 10^{-6}$	10^{-5}	1.9997	$7.77 \cdot 10^{-4}$
318	0.00000024961853096	$4.99 \cdot 10^{-7}$	10^{-6}	1.9993	$3.45 \cdot 10^{-4}$
353	0.00000004863116722	$9.72 \cdot 10^{-8}$	10^{-7}	1.9984	$1.51 \cdot 10^{-4}$
388	0.00000000947442832	$1.89 \cdot 10^{-8}$	10^{-8}	1.9962	$6.71 \cdot 10^{-5}$
423	0.00000000184583272	$3.68 \cdot 10^{-9}$	10^{-8}	1.9916	$2.93 \cdot 10^{-5}$
458	0.00000000035961401	$7.12 \cdot 10^{-10}$	10^{-9}	1.9806	$1.30 \cdot 10^{-5}$
493	0.00000000007006581	$1.37 \cdot 10^{-10}$	10^{-10}	1.9576	$5.62 \cdot 10^{-6}$
528	0.00000000001365543	$2.59 \cdot 10^{-11}$	10^{-10}	1.9043	$2.42 \cdot 10^{-6}$
563	0.00000000000266545	$4.72 \cdot 10^{-12}$	10^{-11}	1.8009	$9.91 \cdot 10^{-7}$
598	0.00000000000052436	$7.69 \cdot 10^{-13}$	10^{-12}	1.5900	$3.69 \cdot 10^{-7}$
633	0.00000000000010722	$8.66 \cdot 10^{-14}$	10^{-12}	1.3034	$9.00 \cdot 10^{-8}$
653	0.00000000000004593	$7.07 \cdot 10^{-15}$	10^{-13}	1.3642	$9.03 \cdot 10^{-9}$
654	0.00000000000004412	$3.45 \cdot 10^{-15}$	10^{-13}	1.0233	$6.13 \cdot 10^{-9}$
655	0.00000000000004239	$3.30 \cdot 10^{-15}$	10^{-13}	2.0000	$3.71 \cdot 10^{-9}$
656	0.00000000000004075	0.0	10^{-13}	Not defined	0.0

Subdifferential of the objective function (31) can be written exactly:

$$\partial f(x) = \sum_{j=1}^{2} d_j \delta_j(x) c_j + \alpha .\!* (x - \beta),$$
$$\delta_j(x) = \begin{cases} \text{sign}(c_j^T x - b_j), & c_j x - b_j \neq 0; \\ [-1; 1], & c_j x - b_j = 0. \end{cases} \tag{32}$$

where binary operator .* means component wise multiplication of the two vectors (i.e. $(u. * v)_i = u_i \cdot v_i$).

To solve this problem we use the subgradient method [11] in the form below:

$$\begin{aligned} x^{k+1} &= x^k - \gamma_k a^k, \\ \gamma_k &= \arg\min\nolimits_{\gamma \geq 0} f(x^k + \gamma a^k), \\ a^k &= \arg \min_{a \in \partial f(x^k)} a^T a. \end{aligned} \tag{33}$$

Usage of (32) and (33) allows to exactly calculate the values a^k while each step.

5.3 Smooth Convex Constrained Optimization Problem

In this section we consider the following constrained convex optimization problem:

$$\begin{aligned} f(x) &= \sum_{i=1}^{n} \alpha_i (x_i - \beta_i)^2 + \exp^{c^T x} + \gamma \exp^{-c^T x} \to \min_{x \in G} \\ G &= \left\{ x \in R^n : x_i \geq 0, i = 1, \ldots, n-1; \sum_{i=1}^{n-1} i x_i^2 + x_n \leq n \right\} \end{aligned} \tag{34}$$

In the problem (34) we use parameters $\alpha_i = i^2$, $\beta_i = 1$, $c_i = 0.1$ for all $i = 1, \ldots n$, $\gamma = 2$, $n = 40$.

To solve the problem (34) we used the method of modified Lagrange function [11] (MMLF) where Polak-Ribiere-Polyak method is used to solve unconstrained optimization problem with accuracy $\|\nabla M(x, \lambda_k)\| \leq 10^{-10}$ where $M(x, \lambda)$ is modified Lagrange function.

The results of the experiment are presented in the Table 3. It is important to mention that each point x^k provided by MMLF is not belong to G and was projected easily to the point $\overline{x}^k \in G$.

Remark 15. Objective functions (30), (31), (34) are strongly convex, the solutions of the corresponding optimization problems are exist. Sufficient optimality conditions for are: $\nabla f(x^*) = 0$ for the problem (30); $0 \in \partial f(x^*)$ for the problem (31); $\nabla f(x^*) + \nabla \varphi(x^*)^T \lambda^* = 0$ for the problem (34).

For the problem (30) $\|\nabla f(x^N)\| = 7.85 \cdot 10^{-11}$, for the problem (31) $\|a^N\| = 9.55 \cdot 10^{-7}$ where $a^N \in \partial f(x^N)$, for the problem (34): $\|\nabla f(x^N) + \nabla \varphi(x^N)^T \lambda^N\| = 8.58 \cdot 10^{-11}$ and $\|\nabla f(\overline{x}^N) + \nabla \varphi(\overline{x}^N)^T \lambda^N\| = 6.97 \cdot 10^{-5}$.

Table 3. Smooth constrained convex optimization problem output ($\varepsilon_1 = \nabla f(x_k)^T (x^k - x^N)$, $\varepsilon_2 = 10^{-m_k}$, m_k is the amount of the same numbers after the comma in $f(x^k)$ and $f(x^N)$, $\varepsilon_3 = \sqrt{2}||x^k - x^N||$, $N = 733$).

#	$f(x^k)$	ε_1	ε_2	γ_k^N	ε_3
1	575692.051375784582	$1.8972 \cdot 10^4$	10^5	1.8459	20.8345
2	**566**816.389333221359	$2.505 \cdot 10^3$	10^4	1.7862	7.4170
3	**565**802.333142664310	$6.83 \cdot 10^2$	10^3	1.7586	3.8336
17	**565413.8**85095520080	$1.1 \cdot 10^{-2}$	10^{-1}	1.7162	$1.5 \cdot 10^{-2}$
44	**565413.8**81832801886	$3.2 \cdot 10^{-3}$	10^{-1}	$1 + 4.3 \cdot 10^{-7}$	$8.3 \cdot 10^{-6}$
74	**565413.8**80132263156	$1.5 \cdot 10^{-3}$	10^{-1}	$1 + 4.8 \cdot 10^{-7}$	$6.0 \cdot 10^{-6}$
104	**565413.87**9304121492	$6.4 \cdot 10^{-4}$	10^{-2}	$1 + 5.4 \cdot 10^{-7}$	$4.2 \cdot 10^{-6}$
134	**565413.878**952784211	$2.8 \cdot 10^{-4}$	10^{-3}	$1 + 5.5 \cdot 10^{-7}$	$2.8 \cdot 10^{-6}$
164	**565413.878**800598179	$1.3 \cdot 10^{-4}$	10^{-3}	$1 + 5.1 \cdot 10^{-7}$	$1.8 \cdot 10^{-6}$
194	**565413.878**732762156	$6.4 \cdot 10^{-5}$	10^{-3}	$1 + 4.5 \cdot 10^{-7}$	$1.2 \cdot 10^{-6}$
224	**565413.878**701378429	$3.3 \cdot 10^{-5}$	10^{-3}	$1 + 3.6 \cdot 10^{-7}$	$7.8 \cdot 10^{-7}$
254	**565413.8786**86192141	$1.8 \cdot 10^{-5}$	10^{-4}	$1 + 2.8 \cdot 10^{-7}$	$5.0 \cdot 10^{-7}$
284	**565413.8786**78469752	$1.0 \cdot 10^{-5}$	10^{-4}	$1 + 2.1 \cdot 10^{-7}$	$3.2 \cdot 10^{-7}$
314	**565413.8786**74342364	$5.8 \cdot 10^{-6}$	10^{-4}	$1 + 1.5 \cdot 10^{-7}$	$2.1 \cdot 10^{-7}$
344	**565413.8786**72034077	$3.4 \cdot 10^{-6}$	10^{-4}	$1 + 1.0 \cdot 10^{-7}$	$1.4 \cdot 10^{-7}$
374	**565413.8786**70693408	$2.1 \cdot 10^{-6}$	10^{-4}	$1 + 6.7 \cdot 10^{-8}$	$8.4 \cdot 10^{-8}$
404	**565413.87866**9891634	$1.3 \cdot 10^{-6}$	10^{-5}	$1 + 4.4 \cdot 10^{-8}$	$5.4 \cdot 10^{-8}$
434	**565413.87866**9401755	$8.2 \cdot 10^{-7}$	10^{-5}	$1 + 2.9 \cdot 10^{-8}$	$3.4 \cdot 10^{-8}$
464	**565413.87866**9097907	$5.1 \cdot 10^{-7}$	10^{-5}	$1 + 1.8 \cdot 10^{-8}$	$2.2 \cdot 10^{-8}$
494	**565413.878668**907492	$3.2 \cdot 10^{-7}$	10^{-6}	$1 + 1.2 \cdot 10^{-8}$	$1.4 \cdot 10^{-8}$
524	**565413.878668**787320	$2.0 \cdot 10^{-7}$	10^{-6}	$1 + 7.5 \cdot 10^{-9}$	$8.7 \cdot 10^{-9}$
554	**565413.878668**711112	$1.2 \cdot 10^{-7}$	10^{-6}	$1 + 4.7 \cdot 10^{-9}$	$5.5 \cdot 10^{-9}$
584	**565413.878668**662621	$7.7 \cdot 10^{-8}$	10^{-6}	$1 + 2.9 \cdot 10^{-9}$	$3.4 \cdot 10^{-9}$
614	**565413.878668**631684	$4.6 \cdot 10^{-8}$	10^{-6}	$1 + 1.8 \cdot 10^{-9}$	$2.0 \cdot 10^{-9}$
644	**565413.878668**611907	$2.6 \cdot 10^{-8}$	10^{-6}	$1 + 1.0 \cdot 10^{-9}$	$1.1 \cdot 10^{-9}$
674	**565413.8786685**99242	$1.3 \cdot 10^{-8}$	10^{-7}	$1 + 5.2 \cdot 10^{-10}$	$5.9 \cdot 10^{-10}$
704	**565413.8786685**91124	$5.1 \cdot 10^{-9}$	10^{-7}	$1 + 2.0 \cdot 10^{-10}$	$2.3 \cdot 10^{-10}$
730	**565413.878668586**458	$4.3 \cdot 10^{-10}$	10^{-9}	$1 + 1.8 \cdot 10^{-11}$	$1.9 \cdot 10^{-11}$
731	**565413.878668586**311	$2.8 \cdot 10^{-10}$	10^{-9}	$1 + 1.2 \cdot 10^{-11}$	$1.2 \cdot 10^{-11}$
732	**565413.878668586**169	$1.4 \cdot 10^{-10}$	10^{-9}	$1 + 5.8 \cdot 10^{-12}$	$6.3 \cdot 10^{-12}$
733	**565413.878668586**029	0.0	10^{-9}	Not defined	0.0

6 Conclusion

In this article we suggest two methods to estimate the error of the solution of convex optimization problems (CoMEES, RoMEES) and non-convex smooth optimization problems (RoMEES). The high accuracy of the results in the experiments was achieved because of the selected mantissa value 50.

In the Tables 1, 2 and 3 we demonstrate the results of the experiments, which allows estimating the error of the solution for each point k ($k = \overline{1, N}$) of the related iteration process. We assume that suggested methods to estimate the error of the solution are actual for usage in practice and can be expanded in further research works.

References

1. Gill, P.E., Murray, W., Wright, M.H.: Practical Optimization. Academic Press, London (1981)
2. Gasnikov, A.V.: Modern Numerical Optimization Methods, 2nd edn. MIPT, Moscow (2018)
3. Nesterov, Y.E.: Introduction to Convex Optimization. MTsNMO, Moscow (2018)
4. Nemirovski, A.S., Iudin, D.B.: Problem Complexity and Method Efficiency in Optimization. Nauka, Moscow (1979)
5. Bubeck, S.: Convex optimization: algorithms and complexity. Found. Trends Mach. Learn. **8**(3–4), 231–357 (2015)
6. Gasnikov, A.V., et al.: Accelerated meta-algorithm for convex optimization problems. Comput. Math. Math. Phys. **61**(1), 17–28 (2021). https://doi.org/10.1134/S096554252101005X
7. Birjukov, A.G., Grinevich, A.I.: Multiprecision arithmetic as a guarantee of the required accuracy of numerical calculation results. TRUDY MIPT **4**(3), 171–180 (2012)
8. Birjukov, A.G., Grinevich, A.I.: Methods for rounding errors estimation in the solution of numerical problems using arbitrary precision floating point calculations. TRUDY MIPT **5**(2), 160–174 (2013)
9. Chernov, A., Dvurechensky, P., Gasnikov, A.: Fast primal-dual gradient method for strongly convex minimization problems with linear constraints. In: Kochetov, Y., Khachay, M., Beresnev, V., Nurminski, E., Pardalos, P. (eds.) DOOR 2016. LNCS, vol. 9869, pp. 391–403. Springer, Cham (2016). https://doi.org/10.1007/978-3-319-44914-2_31
10. Anikin, A.S., Gasnikov, A.V., Dvurechensky, P.E., et al.: Dual approaches to the minimization of strongly convex functionals with a simple structure under affine constraints. Comput. Math. and Math. Phys. **57**(8), 1270–1284 (2017)
11. Polyak, B.T.: Introduction to Optimization, 2nd edn., corr. and ext. LENAND, Moscow (2014)

Sequential Subspace Optimization for Quasar-Convex Optimization Problems with Inexact Gradient

Ilya A. Kuruzov[1,3](✉) and Fedor S. Stonyakin[1,2]

[1] Moscow Institute of Physics and Technology, Moscow, Russia
kuruzov.ia@phystech.edu
[2] V. I. Vernadsky Crimean Federal University, Simferopol, Russia
[3] Institute for Information Transmission Problems RAS, Moscow, Russia

Abstract. It is well-known that accelerated first-order gradient methods possess optimal complexity estimates for the class of convex smooth minimization problems. In many practical situations it makes sense to work with inexact gradient information. However, this can lead to an accumulation of corresponding inexactness in the theoretical estimates of the rate of convergence. We propose one modification of the Sequential Subspace Optimization Method (SESOP) for minimization problems with γ-quasar-convex functions with inexact gradient. A theoretical result is obtained indicating the absence of accumulation of gradient inexactness. A numerical implementation of the proposed version of the SESOP method and its comparison with the known Similar Triangle Method with an inexact gradient is carried out.

Keywords: Subspace optimization method · Inexact gradient · Quasar-convex functions

1 Introduction

It is well-known that accelerated gradient-type methods possess optimal complexity estimates [9] for the class of convex smooth minimization problems. In many practical situations it makes sense to work with inexact gradient information (see e.g. [1,2,10,11]). For example, this is relevant for gradient-free optimization methods (when estimating the gradient by finite differences) in infinite dimensional spaces for inverse problems (see, e.g. [7]).

However, this can lead to an accumulation of corresponding inexactness in the theoretical estimates of the rate of convergence. Let us consider minimization problems of convex and L-smooth function f ($\|\cdot\|$ is a usual Euclidean norm)

$$\|\nabla f(x) - \nabla f(y)\| \leqslant L\|x - y\| \quad \forall x, y \in \mathbb{R}^n \tag{1}$$

The research was supported by Russian Science Foundation (project No. 21-71-30005).

N. N. Olenev et al. (Eds.): OPTIMA 2021, CCIS 1514, pp. 19–33, 2021.
https://doi.org/10.1007/978-3-030-92711-0_2

with an inexact gradient $g : \mathbb{R}^n \to \mathbb{R}^n$:

$$\|g(x) - \nabla f(x)\| \leqslant \delta, \tag{2}$$

where $L > 0$ and $\delta > 0$. For the considered class of problems, the following estimate for accelerated gradient-type methods:

$$f(x_N) - \min_{x \in \mathbb{R}^n} f(x) = O\left(L\|x_0 - x^*\|^2 N^{-2} + \delta \max_{k \leq N} \|x_k - x^*\|\right)$$

is known [3,11] for each $x^* : f(x^*) = \min_{x \in \mathbb{R}^n} f(x)$. It is clear that the quantity $\max_{k \leq N} \|x_k - x^*\|$ can be not small enough. In this paper, we propose one modification of the Sequential Subspace Optimization Method [8] with an δ-additive noise in gradient (2) and prove for $\|x_0 - x^*\| >> 1$ the following estimate:

$$f(x_N) - \min_{x \in \mathbb{R}^n} f(x) = O\left(L\|x_0 - x^*\|^2 N^{-2} + \delta \|x_0 - x^*\|\right),$$

where x_0 is the starting point of the algorithm. Thus, a certain solution to the problem of accumulating the gradient inexactness is proposed for a special accelerated gradient method. It is important that we also consider some type of non-convex problems [5,6].

The article consists of an introduction, 3 main sections and a conclusion.

In the first main Sect. 2, we propose and analyze a new modification of Subspace Optimization Method (Algorithm 1) for minimization problems with γ-quasar-convex functions with an inexact gradient. The use of such a specific method made it possible to obtain a significant result on the non-accumulation of the additive gradient inexactness in the estimate of the convergence rate (Theorem 1).

However, this result for Algorithm 1 is essentially tied to the structure of this method, which is associated with auxiliary low-dimensional minimization problems. Therefore, it is important to investigate the influence of errors in solving such problems on the final estimate of the rate of convergence. Section 3 is devoted to this question and Theorem 2 is obtained.

The last main Sect. 4 is devoted to numerical illustration of the obtained theoretical results for one example of a quadratic function minimization problem. Firstly, we show that the convergence may be significantly better than the theoretical estimates for Algorithm 1. Secondly, we compare Algorithm 1 with another known accelerated Similar Triangles Method (STM) for the case of additive gradient noise [11]. The STM was chosen for comparison with Algorithm 1 for the following reasons:

- in the case of exact gradient information ($\delta = 0$ in (2)) both the STM and the SESOP possess the optimal rate of convergence $O(N^{-2})$;
- for the STM with an inexact gradient (2), a theoretical estimate of the quality of the solution is known ([11], Theorem 1 and Remark 3).

Let us introduce some auxiliary notations and definitions.

Throughout this paper $\langle \mathbf{x}, \mathbf{y} \rangle$ $\|\cdot\|$ means the inner product of vectors $\mathbf{x} = (x_1, x_2, ..., x_n)$, $\mathbf{y} = (y_1, y_2, ..., y_n) \in \mathbb{R}^n$ and is given by the formula $\langle \mathbf{x}, \mathbf{y} \rangle = \sum\limits_{k=1}^{n} x_k y_k$.

It turns out that it is possible to formulate the main results of the work for a certain class of not necessarily convex problems. Let us recall the definition of the class of γ-quasar-convex functions (see [5,6]).

Definition 1. *Assume that $\gamma \in (0, 1]$ and let $\mathbf{x}^*$ be a minimizer of the differentiable function $f : \mathbb{R}^n \to \mathbb{R}$. The function f is γ-quasar-convex with respect to $\mathbf{x}^*$ if for all $\mathbf{x} \in \mathbb{R}^n$,*

$$f(\mathbf{x}^*) \geq f(\mathbf{x}) + \frac{1}{\gamma} \langle \nabla f(\mathbf{x}), \mathbf{x}^* - \mathbf{x} \rangle. \tag{3}$$

For example, a non-convex function $f(x) = |x|(1 - e^{-|x|})$ is a 1-quasar-convex [4,6]. The class of γ-quasar-convex functions is also called γ-weakly quasi-convex functions (see [4]). Clearly, each convex function is also 1-quasar-convex. So, all results of this paper are applicable to convex optimization problems with an inexact gradient information.

2 Subspace Optimization Method with Inexact Gradient

In this section we present some variant of the SESOP (Sequential Subspace Optimization) method [8] for γ-quasar-convex functions with an inexact gradient. We generalize the results [4] for the SESOP method with an inexact gradient on the class of γ-quasar-convex functions. In other words, our modifications of the SESOP method work with some approximation $g(x)$ of gradient $\nabla f(x)$ at each point $x \in \mathbb{R}^n$.

Similarly to [4,8] we start with a description of the investigated algorithm. Let $D_k = \|\mathbf{d}_k^0 \, \mathbf{d}_k^1 \, \mathbf{d}_k^2\|$ be an $n \times 3$ matrix, the columns of which are the following vector:

$$\mathbf{d}_k^0 = g(\mathbf{x}_k), \quad \mathbf{d}_k^1 = \mathbf{x}_k - \mathbf{x}_0, \quad \mathbf{d}_k^2 = \sum_{i=0}^{k} \omega_i g(\mathbf{x}_i),$$

where $\omega_0 = 1, \omega_i = \frac{1}{2} + \sqrt{\frac{1}{4} + \omega_{i-1}^2}$.

For all $k \geq 1$ we have $\omega_k = \frac{1}{2} + \sqrt{\frac{1}{4} + \omega_{k-1}^2} \geq \frac{1}{2} + \omega_{k-1}$ and $\omega_k = \frac{1}{2} + \sqrt{\frac{1}{4} + \omega_{k-1}^2} \leq 1 + \omega_{k-1}$.

So, it holds that

$$k + 1 \geq \omega_k \geq \frac{k+1}{2}. \tag{4}$$

The matrices D_k will generate the subspaces over which we will minimize our objective function. With D_k defined this way, the proposed algorithm takes the following form:

Algorithm 1. A modification of the SESOP method with an inexact gradient

Require: objective function f with an inexact gradient g, initial point $\mathbf{x}_0$, number of iterations T.

1: **for** $k = 0, \ldots, T-1$ **do**

2: Find the optimal step

$$\tau_k \leftarrow \arg\min_{\tau \in \mathbb{R}^3} f(\mathbf{x}_k + D_k \tau) \tag{5}$$

3: $\mathbf{x}_{k+1} \leftarrow \mathbf{x}_k + D_k \tau_k$

4: **end for**

5: **return** $\mathbf{x}_T$

Let us show that the main advantage of the SESOP method with an inexact gradient is the absence of the term $\max_k \|x_k - x^*\|$ in the theoretical estimate. Before proving this result, we need to estimate the error accumulation in $\mathbf{d}_k^3$ vector.

We will use the inequality for all $a, b \geq 0$ that:

$$\sqrt{a+b} \leq \sqrt{a} + \sqrt{b}. \tag{6}$$

Lemma 1. *Let the objective function f be L-smooth and γ-quasar-convex with respect to $\mathbf{x}^*$. Suppose also for the inexact gradient $g: \mathbb{R}^n \to \mathbb{R}^n$ of function f there is some constant $\delta_1 \geq 0$ such that for all $\mathbf{x} \in \mathbb{R}^n$*

$$\|g(\mathbf{x}) - \nabla f(\mathbf{x})\| \leq \delta_1. \tag{7}$$

Let $\{\mathbf{x}_j\}_j$ be a sequence of points generated by Algorithm 1. Then the following inequality holds:

$$\left\| \sum_{k=0}^{T} \omega_k g(\mathbf{x}_k) \right\|^2 \leq 2 \sum_{j=0}^{T} \omega_j^2 \|g(\mathbf{x}_j)\|^2 + 72 T^4 \delta_1^2 \tag{8}$$

for each $T \in \mathbb{N}$.

Proof. Let us define $W_T = \left\| \sum_{k=0}^{T} \omega_k g(\mathbf{x}_k) \right\|$. Note that $\mathbf{d}_j^2 = \sum_{k=0}^{j} \omega_k g(\mathbf{x}_k)$ and $\|d_j^2\| \leq W_j$ for each j. For W_T we have the following equality for this value:

$$W_T^2 = \omega_T^2 \|g(\mathbf{x}_T)\|^2 + 2 \left\langle w_T g(\mathbf{x}_T), \mathbf{d}_{T-1}^2 \right\rangle + W_{T-1}^2 \tag{9}$$

and we have that

$$\nabla f(\mathbf{x}_T) \perp \mathbf{d}_{T-1}^2,$$

because of optimizing on the subspace $\mathbf{x}_{T-1} + \mathbf{d}_{j-1}^2$. Therefore, we have the following inequality $\left(\langle g(\mathbf{x}_j),\ \mathbf{d}_{j-1}^2 \rangle = 0\right)$:

$$|\langle g(\mathbf{x}_j),\ \mathbf{d}_{j-1}^2 \rangle| = |\langle \nabla f(\mathbf{x}_j), \mathbf{d}_{j-1}^2 \rangle + \langle g(\mathbf{x}_j) - \nabla f(\mathbf{x}_j), \mathbf{d}_{j-1}^2 \rangle| \leq \delta_1 W_{j-1} \tag{10}$$

for all $j \geq 1$. So, we have the following correlations for W_T:

$$W_T^2 = \sum_{j=0}^{T} \omega_j^2 \|g(\mathbf{x}_j)\|^2 + 2\sum_{j=1}^{T} w_j \left\langle g(\mathbf{x}_j), \mathbf{d}_{j-1}^2 \right\rangle,$$

and

$$W_T^2 \leq \sum_{j=0}^{T} \omega_j^2 \|g(\mathbf{x}_j)\|^2 + 2\delta_1 \sum_{j=1}^{T} w_j W_{j-1}. \tag{11}$$

On the other hand, we can estimate the inner product by the Cauchy-Schwarz inequality for $T = j$ in (9) and to get the estimate $W_j^2 \geq -2w_j\delta_1 W_{j-1} + W_{j-1}^2$.

Solving this inequality on W_j, we have ($\sqrt{a+b} \leq \sqrt{a} + \sqrt{b}$ for all $a, b \geq 0$)

$$W_{j-1} \leq w_j\delta_1 + \sqrt{w_j^2\delta_1^2 + W_j^2} \leq 2w_j\delta_1 + W_j.$$

By induction we have the following estimate:

$$W_j \leq 2\delta_1 \sum_{k=j+1}^{T} w_k + W_T \text{ for } j = \overline{0, T-1}.$$

From $\sum\limits_{k=j+1}^{T} w_k \leq \frac{1}{2}T(T+3)$ we have that

$$W_j \leq \delta_1 T(T+3) + W_T \text{ for } j = \overline{0, T-1}.$$

The last inequality and (11) mean that

$$W_T^2 \leq \sum_{j=0}^{T} \omega_j^2 \|g(\mathbf{x}_j)\|^2 + 2\delta_1^2 T(T+3) + 2\delta_1 W_T \sum_{j=1}^{T} w_j, \tag{12}$$

$$W_T^2 \leq \sum_{j=0}^{T} \omega_j^2 \|g(\mathbf{x}_j)\|^2 + 2\delta_1^2 T(T+3) + T(T+3)\delta_1 W_T. \tag{13}$$

The value $W_T \geq 0$ by definition. One of the roots of the previous quadratic inequality is always negative. So, for W_T to meet this inequality, its value must be not more than the largest root of the corresponding quadratic function:

$$W_T \leq \frac{1}{2}(T^2+3T)\delta_1 + \frac{1}{2}\sqrt{(T^2+3T)^2\delta_1^2 + 4\sum_{j=0}^{T} \omega_j^2 \|g(\mathbf{x}_j\|^2 + 8\delta_1^2 T(T+3)}.$$

Taking into account $T \geq 1$, we have

$$W_T \leq 2T^2\delta_1 + \frac{1}{2}\sqrt{48T^4\delta_1^2 + 4\sum_{j=0}^{T} \omega_j^2 \|g(\mathbf{x}_j)\|^2},$$

Due to the inequality (6) we have

$$W_T \leq 6T^2\delta_1 + \sqrt{\sum_{j=0}^{T} \omega_j^2 \|g(\mathbf{x}_j)\|^2}.$$

Further, the inequality $(a+b)^2 \leq 2a^2 + 2b^2$ (for all $a, b \in \mathbb{R}$) means that

$$W_T^2 \leq 72T^4\delta_1^2 + 2\sum_{j=0}^{T} \omega_j^2 \|g(\mathbf{x}_j)\|^2.$$

Using Lemma 1, we can prove the following main result of this section.

Theorem 1. *Let the objective function f be L-smooth and γ-quasar-convex with respect to $\mathbf{x}^*$. Also, for the inexact gradient $g : \mathbb{R}^n \to \mathbb{R}^n$ there is some constant $\delta_1 \geq 0$ such that for all $\mathbf{x} \in \mathbb{R}^n$*

$$\|g(\mathbf{x}) - \nabla f(\mathbf{x})\| \leq \delta_1. \tag{14}$$

Then the sequence $\{\mathbf{x}_k\}$ generated by Algorithm 1 satisfies

$$f(\mathbf{x}_k) - f^* \leq \frac{8LR^2}{\gamma^2 k^2} + 4\left(\frac{R}{\gamma} + 17\right)\delta_1, \tag{15}$$

where $R = \|\mathbf{x}^* - \mathbf{x}_0\|$.

Proof. By the construction of $\mathbf{x}_{k+1}$ we have the following inequality:

$$f(\mathbf{x}_{k+1}) = \min_{\mathbf{s}\in\mathbb{R}^3} f\left(\mathbf{x}_k + \sum_{i=0}^{2} s_i d_k^i\right) \leq f\left(\mathbf{x}_k + s_0 g(\mathbf{x}_k)\right). \tag{16}$$

On the other hand, from L-smoothness we have that

$$f(\mathbf{y}) \leq f(\mathbf{x}) + \langle \nabla f(\mathbf{x}), \mathbf{y} - \mathbf{x} \rangle + \frac{L}{2}\|\mathbf{x} - \mathbf{y}\|^2$$

for all $\mathbf{x}, \mathbf{y} \in \mathbb{R}^n$. Further, from (14) follows the corresponding inequality for an inexact gradient:

$$f(\mathbf{y}) \leq f(\mathbf{x}) + \langle g(\mathbf{x}), \mathbf{y} - \mathbf{x} \rangle + \frac{L}{2}\|\mathbf{x} - \mathbf{y}\|^2 + \delta_1\|\mathbf{x} - \mathbf{y}\| \quad \forall \mathbf{x}, \mathbf{y} \in \mathbb{R}^n.$$

Using the inequality $\delta_1\|\mathbf{x}-\mathbf{y}\| = \left(\sqrt{\frac{1}{L}}\delta_1\right)\left(\sqrt{\frac{L}{1}}\|\mathbf{x}-\mathbf{y}\|\right) \leq \frac{\delta_1^2}{2L} + \frac{L}{2}\|\mathbf{x}-\mathbf{y}\|^2$, we have

$$f(\mathbf{y}) \leq f(\mathbf{x}) + \langle g(\mathbf{x}), \mathbf{y} - \mathbf{x} \rangle + L\|\mathbf{x} - \mathbf{y}\|^2 + \frac{\delta_1^2}{2L} \quad \forall \mathbf{x}, \mathbf{y} \in \mathbb{R}^n. \tag{17}$$

On the base of the last inequality and the right part of (16) for $\mathbf{y} := \mathbf{x}_k + s_0 g(\mathbf{x}_k)$ and $\mathbf{x} = \mathbf{x}_k$ we can conclude that

$$f(\mathbf{x}_{k+1}) \leq f(\mathbf{x}_k) + \left(s_0 + s_0^2 L\right) \|g(\mathbf{x}_k)\|^2 + \frac{1}{2L}\delta_1^2 \tag{18}$$

for each $s_0 \in \mathbb{R}$. Further,

$$-\left(s_0 + s_0^2 L\right) \|g(\mathbf{x}_k)\|^2 \leq f(\mathbf{x}_k) - f(\mathbf{x}_{k+1}) + \frac{1}{2L}\delta_1^2. \tag{19}$$

Maximizing the left part of (19) by s_0, we have the following estimate for the inexact gradient norm:

$$\|g(\mathbf{x}_k)\|^2 \leq 4L(f(\mathbf{x}_k) - f(\mathbf{x}_{k+1})) + 2\delta_1^2. \tag{20}$$

So, $\nabla f(\mathbf{x}_k) \perp \mathbf{x}_k - \mathbf{x}_0$ for all $k > 0$ because $\mathbf{x}_k$ is a minimizer of f on the subspace containing the directions $\mathbf{x}_k - \mathbf{x}_{k-1}$ and $\mathbf{x}_{k-1} - \mathbf{x}_0$. Because of this, we can write

$$\langle \nabla f(\mathbf{x}_k), \mathbf{x}_k - \mathbf{x}^* \rangle = \langle \nabla f(\mathbf{x}_k), \mathbf{x}_0 - \mathbf{x}^* \rangle.$$

From this equality and (14) we have

$$f(\mathbf{x}_k) - f(\mathbf{x}^*) \leq \frac{1}{\gamma}\langle g(\mathbf{x}_k), \mathbf{x}_0 - \mathbf{x}^* \rangle + \delta_1 \frac{R}{\gamma}.$$

Similarly to the proof of Theorem 3.1 from [4] we have the following chain of correlations:

$$\begin{aligned}\sum_{k=0}^{T-1} \omega_k (f(\mathbf{x}_k) - f^*) &\leq \frac{1}{\gamma}\left\langle \sum_{k=0}^{T-1} \omega_k g(\mathbf{x}_k), \mathbf{x}_0 - \mathbf{x}^* \right\rangle + \delta_1 \frac{R}{\gamma}\sum_{k=0}^{T-1}\omega_k \\ &\leq \frac{1}{\gamma}\left\| \sum_{k=0}^{T-1} \omega_k g(\mathbf{x}_k) \right\| R + \delta_1 \frac{R}{\gamma}\sum_{k=0}^{T-1}\omega_k.\end{aligned} \tag{21}$$

Now we can estimate the multiplier $\left\| \sum_{k=0}^{T-1} \omega_k g(\mathbf{x}_k) \right\|$. According to Lemma 1 and (4), we have the following estimates:

$$\begin{aligned}\left\| \sum_{k=0}^{T-1} \omega_k g(\mathbf{x}_k) \right\|^2 &\leq 2\sum_{j=0}^{T-1} \omega_j^2 \|g(\mathbf{x}_j)\|^2 + 256 T^4 \delta_1^2 \\ &\leq 8L \sum_{k=0}^{T-1} \omega_k^2 (f(\mathbf{x}_k) - f(\mathbf{x}_{k+1})) + 260 T^4 \delta_1^2.\end{aligned}$$

Note that the choice of ω_k is equivalent to choosing the largest ω_k satisfying

$$\omega_k = \begin{cases} 1, \text{if } k = 0, \\ \omega_k^2 - \omega_{k-1}^2, \text{otherwise.} \end{cases}$$

Now we can estimate the left part of (21) denoting $\varepsilon_k = f(\mathbf{x}_k) - f^*$ and using inequality (6) in the following way:

$$\begin{aligned}
S &= \sum_{k=0}^{T-1} \omega_k \varepsilon_k \\
&\leq \frac{1}{\gamma} \left\| \sum_{k=0}^{T-1} \omega_k g(\mathbf{x}_k) \right\| R + \delta_1 \frac{R}{\gamma} \sum_{k=0}^{T-1} \omega_k \\
&\leq \left(\frac{8LR^2}{\gamma^2} \sum_{k=0}^{T-1} \omega_k^2 (\varepsilon_k - \varepsilon_{k+1}) + 260 T^4 \delta_1^2 \right)^{\frac{1}{2}} + \delta_1 \frac{R}{\gamma} \sum_{k=0}^{T-1} \omega_k \qquad (22) \\
&\leq \left(\frac{8LR^2}{\gamma^2} \sum_{k=0}^{T-1} \omega_k^2 (\varepsilon_k - \varepsilon_{k+1}) \right)^{\frac{1}{2}} + 17 T^2 \delta_1 + \delta_1 \frac{R}{\gamma} \sum_{k=0}^{T-1} \omega_k \\
&= \sqrt{\frac{8LR^2}{\gamma^2}} \sqrt{S - \varepsilon_T \omega_{T-1}^2} + 17 T^2 \delta_1 + \delta_1 \frac{R}{\gamma} \sum_{k=0}^{T-1} \omega_k .
\end{aligned}$$

From the inequality above we have

$$\omega_{T-1}^2 \varepsilon_T \leq S - \frac{\gamma^2}{8LR^2} \left(S - \delta_1 \frac{R}{\gamma} \sum_{k=0}^{T-1} \omega_k - 17 T^2 \delta_1 \right)^2 . \qquad (23)$$

Maximizing the right part of (23) on S we get

$$\omega_{T-1}^2 \varepsilon_T \leq \frac{2LR^2}{\gamma^2} + \delta_1 \frac{R}{\gamma} \sum_{k=0}^{T-1} \omega_k + 17 T^2 \delta_1 . \qquad (24)$$

Now from (4) we have

$$\begin{aligned}
\omega_{T-1}^2 \varepsilon_T &\leq \frac{2LR^2}{\gamma^2} + \delta_1 \frac{R}{\gamma} \sum_{k=0}^{T-1} \omega_k + 17 T^2 \delta_1 \\
&\leq \frac{2LR^2}{\gamma^2} + T^2 \left(\delta_1 \frac{R}{\gamma} + 17 \delta_1 \right) .
\end{aligned} \qquad (25)$$

Dividing both parts of this inequality by w_{T-1}^2 and using the lower estimate for it (4) we get

$$\varepsilon_T \leq \frac{8LR^2}{\gamma^2 T^2} + 4 \delta_1 \frac{R}{\gamma} + 68 \delta_1 = \frac{8LR^2}{\gamma^2 T^2} + 4 \left(\frac{R}{\gamma} + 17 \right) \delta_1 , \qquad (26)$$

Q.E.D.

3 Subspace Optimization Method with Inexact Solutions of Auxiliary Subproblems

The result of the previous section shows that the SESOP algorithm can work with additive noise in a gradient. It is essential that the method leads to the need to solve auxiliary low-dimensional optimization problems. So, there is an interesting case when the auxiliary problem (5) cannot be solved exactly. We consider this case in the following theorem.

Theorem 2. *Let the objective function f be L-smooth and γ-quasar-convex with respect to $\mathbf{x}^*$. Let τ_k be the step value obtained with the inexact solution of the auxiliary problem* (5) *on step 2 in Algorithm 1 on the k-th iteration. Namely, the following conditions for inexactness hold:*

(i) For the inexact gradient $g : \mathbb{R}^n \to \mathbb{R}^n$ there is some constant $\delta_1 \geq 0$ such that for all points $\mathbf{x} \in \mathbb{R}^n$ condition (14) *holds.*
(ii) The inexact solution τ_k meets the following condition:

$$|\langle \nabla f(\mathbf{x}_k), \mathbf{d}_{k-1}^2 \rangle| \leq k^2 \delta_2 \tag{27}$$

for some constant $\delta_2 \geq 0$ and each $k \in \mathbb{N}$. Note that $\mathbf{x}_k = \mathbf{x}_{k-1} + D_{k-1}\tau_{k-1}$.
(iii) The inexact solution τ_k meets the following condition for some constant $\delta_3 \geq 0$:

$$|\langle \nabla f(\mathbf{x}_k), \mathbf{x}_k - \mathbf{x}_0 \rangle| \leq \delta_3. \tag{28}$$

(iv) The problem from step 2 in Algorithm 1 is solved with accuracy $\delta_4 \geq 0$ on the function on each iteration, i.e. $f(\mathbf{x}_k) - \min_{\tau \in \mathbb{R}^n} f(\mathbf{x}_{k-1} + D_{k-1}\tau) \leq \delta_4$.

Then the sequence $\{\mathbf{x}_k\}$ generated by Algorithm 1 satisfies

$$f(\mathbf{x}_k) - f^* \leq \frac{8LR^2}{\gamma^2 k^2} + \left(\frac{R}{\gamma} + 10\right)\delta_1 + 4\sqrt{\delta_2} + \delta_3 + 5\sqrt{\frac{L\delta_4}{k}} \tag{29}$$

for each $k \geq 8$, where $R = \|\mathbf{x}^* - \mathbf{x}_0\|$.

Proof. The proof of this theorem is somewhat similar to the proof of Theorem 1 and was moved to the Appendix C in [12].

Remark 1. The obtained estimate of the rate of convergence for Algorithm 1 does not depend on the value $\max_k \|\mathbf{x}_k - \mathbf{x}^*\|$ and it depends only on R, L and $\gamma > 0$.

Remark 2. It is clear that when the auxiliary problem (5) in Algorithm 1 has an exact solution, then $\delta_2 = \delta_3 = \delta_4 = 0$. The constant before δ_1 was improved in comparison with the result from Theorem 1 because of more accurate work with constants in proofs (see Lemmas 1 and Appendix B in [12]).

According to Theorem 2, the SESOP method for a γ-convex function can find a solution with quality ε by function after $N = \sqrt{\frac{16LR^2}{\gamma^2\varepsilon}}$ iterations when the following condition holds:

$$\left(\frac{R}{\gamma} + 10\right)\delta_1 + 4\sqrt{\delta_2} + \delta_3 + 5\sqrt{\frac{L\delta_4}{k}} \leq \frac{\varepsilon}{2}.$$

In particular, the SESOP method finds solution with this quality after $N = \sqrt{\frac{16LR^2}{\varepsilon}}$ iterations for convex functions.

Now we want to discuss the relationship between conditions *(ii)*, *(iii)* and *(iv)* from Theorem 2. The condition on the accuracy of the subproblem *(iv)* is natural enough for such methods. Conditions *(ii)* and *(iii)* are caused by the form of the method and provide almost orthogonality of the gradient and vectors $\mathbf{d}_k^j, j = 1, 2$. We can prove the following simple result.

Theorem 3. *If condition (iv) from Theorem 2 holds, then we can choose* $\delta_2, \delta_3 \geq 0$ *according to the following estimates:*

$$\delta_3 \leq \sqrt{2L\delta_4}\left(\sqrt{\max_k(\|D_k\|\|\tau_k\|)} + \sqrt{\|\max_k \mathbf{d}_{k-1}^1\|}\right)$$

and

$$\delta_2 \leq \frac{1}{k^2}\sqrt{2L\max_k\|\mathbf{d}_k^3\|\delta_4}.$$

Proof. Now we want to express conditions (27) and (28) through the accuracy of the solution δ_4 of the subproblem solution (5). We need to introduce the following auxiliary function:

$$f_k(\tau) = f(\mathbf{x}_k + D_k\tau). \tag{30}$$

Note that $f_k : \mathbb{R} \to \mathbb{R}$ and its gradient is a one-dimension derivative. Let function f_k have a Lipschitz continuous gradient with constants L_k^j, $j = 1, 3$. We can derive these constants from L and the norms of directions d_k^j:

$$\left|\frac{d}{d\tau_j}f_k(\tau + \alpha e_j) - \frac{d}{d\tau_j}f_k(\tau)\right| = \left|\left\langle d_k^j, \nabla f(\mathbf{x}^1) - \nabla f(\mathbf{x}^2)\right\rangle\right| \leq L\|d_k^j\||\alpha|,$$

where $e_j \in \mathbb{R}^3$ is the j-th vector in the standard basis, $\alpha \in \mathbb{R}$ is some constant, $\mathbf{x}^1 = \mathbf{x}_k + D_k\tau$ and $\mathbf{x}^2 = \mathbf{x}_k + D_k(\tau + \alpha e_j)$. So we have the following expression for the Lipschitz constant of the gradient for f_k with respect to the j-th component:

$$L_k^j = L\|d_k^j\|. \tag{31}$$

It is easy to see that

$$\left|\frac{d}{d\tau_j}f_k(\tau)\right|^2 \leq 2L_k^j\left(f_k(\tau) - \min_{\tau_j} f_k(\tau)\right) \leq 2L_k^j\left(f_k(\tau) - \min_{\tau} f_k(\tau)\right) = 2L_k^j\delta_4$$

for all τ_j. From (31), the inequality above and the definition of f_k (30), we have the following expression:

$$|\langle \nabla f(\mathbf{x}_{k+1}), d_k^j \rangle| \leq \sqrt{2L\|\mathbf{d}_k^j\|\delta_4}. \tag{32}$$

It means if we choose $\delta_2 > 0$ in the following way:

$$\delta_2 \leq \frac{1}{k^2}\sqrt{2L \max_k \|\mathbf{d}_k^3\|\delta_4},$$

then condition *(ii)* in Theorem 2 is met.

In a similar way, we can obtain that f_k has a Lipschitz continuous gradient with constant L_k:

$$L_k = \|D_k\| L$$

and

$$|\nabla_\tau f_k(\tau)|^2 \leq 2L_k \left(f_k(\tau) - \min_\tau f_k(\tau) \right) = 2L_k \delta_4.$$

Note that

$$\mathbf{x}_k - \mathbf{x}_0 = D_{k-1}\tau_{k-1} + \mathbf{d}_{k-1}^1.$$

Finally, we can choose δ_3 in the following way:

$$\delta_3 \leq \sqrt{2L\delta_4}\left(\sqrt{\max_k (\|D_k\|\|\tau_k\|)} + \sqrt{\|\max_k \mathbf{d}_{k-1}^1\|} \right).$$

4 Numerical Experiments

In the current section we provide the results of numerical experiments. All experiments were carried out in the assumption that we can solve the subspace optimization problem at each iteration with some accuracy on function. For this, we used the quadratic test function

$$f(\mathbf{x}) = \mathbf{x}^\top A \mathbf{x} + 2\mathbf{b}^\top x$$

with $A \in \mathbb{S}_+^n$ (A is a symmetric positive semidefinite matrix), $\mathbf{b} \in \mathbf{R}^n$. Obviously, this function is convex and, consequently, 1-quasar-convex. The components of parameter $\mathbf{b}$ were generated randomly i.i.d. from uniform distribution $\mathcal{U}([-1, 1])$. The matrix $A = B^\top B$ where components $B \in \mathbb{R}^{n \times n}$ were generated in the same way as for vector $\mathbf{b}$.

The shift τ_k can be found as a solution of convex quadratic optimization problem

$$\min_{\tau \in \mathbf{R}^3} \tau^\top D_k^\top A D_k \tau - 2\left(\mathbf{b} + A\mathbf{x}_k\right)^\top D_k \tau$$

with any accuracy that we will vary in our experiments (see details below). The Lipschitz constant L of ∇f is also known and equals the maximal singular value of matrix A. For all experiments we take dimension $n = 500$.

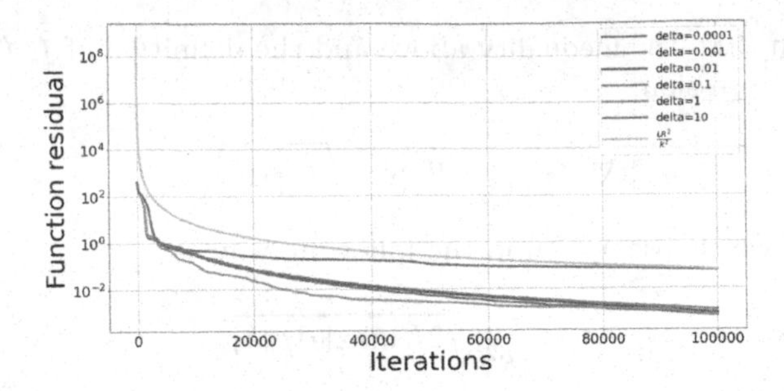

Fig. 1. The dependencies of convergence on gradient inexactness in the case of an exact solution of the subspace optimization problem. Convergence for different values of the constant δ_1.

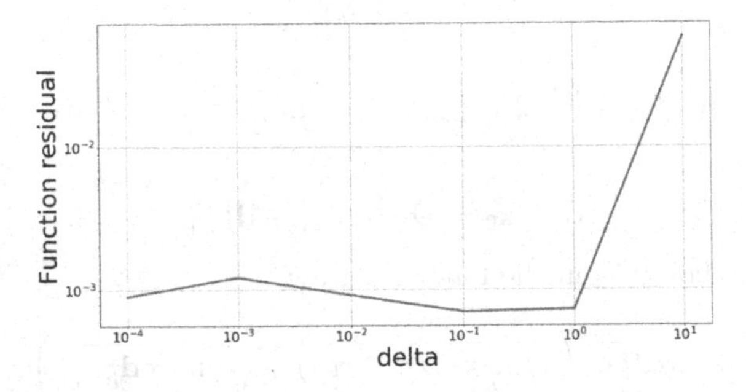

Fig. 2. Minimal values found for different values of the constant δ_1 in the case of an exact solution of the subspace optimization problem.

The first experiment compares the theoretical estimation from Theorem 1 and the real experiment in the case of inexactness in the gradient only. It means that we solve the quadratic optimization problem with machine accuracy that is significantly lower than the inexactness in the gradient. The inexact gradient will be given as an usual gradient with some noise $g(\mathbf{x}) = \nabla f(\mathbf{x}) + \delta\xi(\mathbf{x})$, where $\xi(\mathbf{x}) \sim \mathcal{U}(S_1(0))$ is a random vector from the unit sphere with uniform distribution. Obviously, such a vector meets the conditions of Theorem 1. The results of this experiment are presented in Figs. 1 and 2.

We can see in Fig. 1 that the convergence of the proposed variant of the SESOP method (Algorithm 1) at the first 100000 iterations is better than the theoretical convergence (the line $\frac{LR^2}{k^2}$ on the graph) without noise for any gradient inexactness for $\delta \in \left[10^{-4}, 10\right]$. Moreover, in Fig. 2 the dependence of the function residual on the gradient inexactness shows that there is no significant error accumulation for $\delta < 1$ at the first 100000 iterations. Such an optimistic result was obtained by Algorithm 1 due to the exact solution of the low-dimensional optimization subproblems (5).

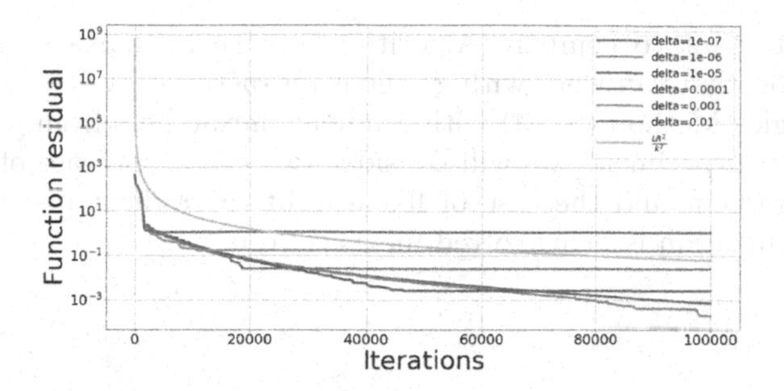

Fig. 3. The dependencies of convergence on the inexactness of the subspace optimization problem in Algorithm 1. Convergence for different for different values of the constant δ_4.

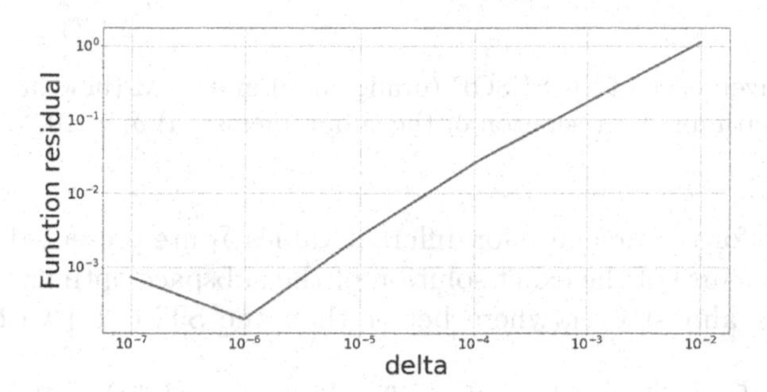

Fig. 4. Minimal values found for different for different values of the constant δ_4.

In the second experiment we studied the practical convergence rate for different inexactness $\delta_j, j = \overline{2,4}$ when δ_1 is fixed. In this experiments we take $\delta_1 = 10^{-3}$. Even in the ideal case, we cannot estimate the dependence of convergence on these parameters independently because when inexactness on the function of the subspace optimization problem (5) solution is small enough $\delta_4 \to 0$, then other inexactness also tends to zero. We varied the inexactness of subspace optimization solution δ_4. The results of the second experiment are shown in Figs. 3 and 4.

In this case in Fig. 3 we can see that the convergence is significantly better than the theoretical estimation only for accuracy values $\delta = 10^{-7}, 10^{-6}, 10^{-5}$. For values $10^{-2}, 10^{-3}$ there is no improvement after 20000 iterations and the theoretical estimation obtains better convergence. For value 10^{-4} the convergence stopped after 20000 iterations too but the theoretical convergence is not better due to a small number of iterations. In Fig. 4 we can see that the approached function value degrades with the linear rate depending on $\delta \geq 10^{-6}$, which corresponds to the results of Theorem 2. So, the proposed modification of the SESOP method is more sensitive to the accuracy of the solution of the subproblem (5) than to the inexactness of the gradient.

Finally, we want to compare Algorithm 1 with an inexact gradient with another method that can work with gradient inexactness. We choose the known Similar Triangles Method (STM) with gradient inexactness from [11]. Similarly to the previous experiment, we will consider two cases: the case of inexactness only in the gradient and the case of fixed additive gradient inexactness when subspace optimization is being solved inexactly too.

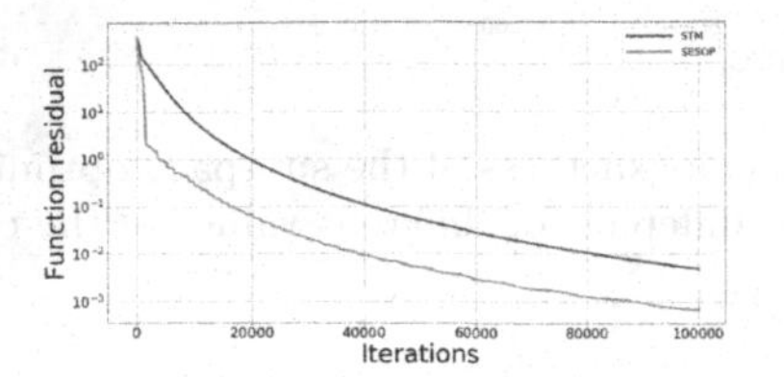

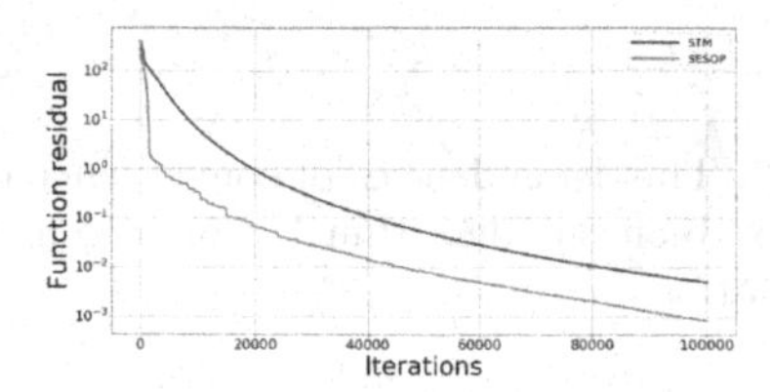

Fig. 5. The convergence of the SESOP (orange line) and STM (blue line) for different additive noise with an exact solution of the subproblems: (a) $\delta_1 = 10^{-3}$; (b) $\delta_1 = 0.1$.

The results for the first case for different values δ_1 are presented in Fig. 5. We can see that because of the exact solution of the subspace optimization problem Algorithm 1 is almost everywhere better than the STM [11] with an inexact gradient.

The results for the second case for different accuracy of the subspace problem solution are presented in Fig. 6. There is a natural result that for a solution exact enough at each iteration, Algorithm 1 stays better than the STM. Nevertheless, for the inexactness in the low-dimensional subproblems solution larger or equal to 10^{-4} the STM becomes better than the provided method (Algorithm 1).

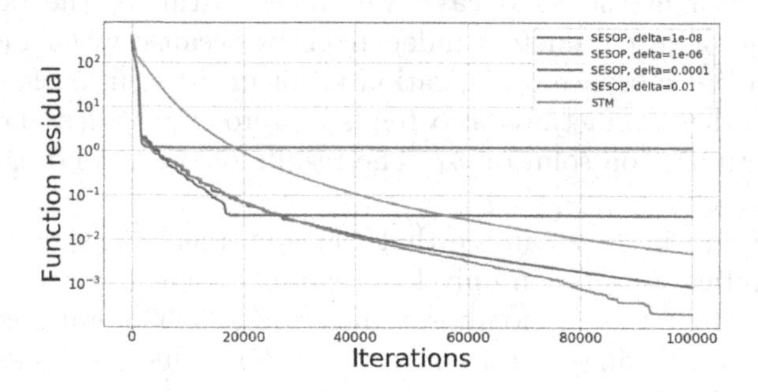

Fig. 6. The convergence of Algorithm 1 with an inexact solution of the subspace optimization problem and STM for additive noise $\delta_1 = 10^{-3}$.

5 Conclusion

The contributions of the paper can be summarized as follows:

- We propose one modification of the Sequential Subspace Optimization Method [8] with a δ-additive noise in the gradient (2). For the first time, the result was obtained describing the influence of this inexactness on the estimate of the convergence rate, whereby the quantity $O(\delta \max_k \|x_k - x^*\|)$ is replaced by the constant $O(\delta\|x_0 - x^*\|)$, $\|x_0 - x^*\| \leq \max_k \|x_k - x^*\|$.
- The influence of inexactness in solving auxiliary minimization problems (5) on the general theoretical estimate for Algorithm 1 is investigated.
- We provide numerical experiments which demonstrate the effectiveness of the approach proposed in this paper. Algorithm 1 is compared with another known Similar Triangles Method (STM) with an additive gradient noise.

Acknowledgments. The authors are grateful to Alexander Gasnikov and Mohammad Alkousa for very useful discussions.

References

1. Devolder, O.: Exactness, inexactness and stochasticity in first-order methods for large-scale convex optimization. Ph.D. thesis, ICTEAM and CORE, Universit'e Catholique de Louvain (2013)
2. Devolder, O., Glineur, F., Nesterov, Y.: First-order methods of smooth convex optimization with inexact oracle. Math. Program. **146**(1), 37–75 (2014)
3. d'Aspremont, A.: Smooth optimization with approximate gradient. SIAM J. Optim. **19**(3), 1171–1183 (2008)
4. Guminov, S., Gasnikov, A.: Accelerated methods for weakly-quasi-convex optimization problems. arxiv, https://arxiv.org/pdf/1710.00797.pdf (2020)
5. Hardt, M., Ma, T., Recht, B.: Gradient descent learns linear dynamical systems. J. Mach. Learn. Res. **19**(29), 1–44 (2018)
6. Hinder, O., Sidford, A., Sohoni N.S.: Near-optimal methods for minimizing star-convex functions and beyond. Proc. Mach. Learn. Res. **125**, 1–45 (2020). http://proceedings.mlr.press/v125/hinder20a/hinder20a.pdf
7. Kabanikhin, S.I.: Inverse and Ill-Posed Problems: Theory and Applications. Walter De Gruyter (2011)
8. Narkiss, G., Zibulevsky, M.: Sequential subspace optimization method for large-scale unconstrained problems. Technion-IIT, Department of Electrical Engineering (2005). http://spars05.irisa.fr/ACTES/PS2-4.pdf
9. Nemirovsky, A.S., Yudin, D.B.: Problem Complexity and Optimization Method Efficiency. Nauka, Moscow (1979)
10. Polyak, B.T.: Introduction to Optimization. Optimization Software, New York (1987)
11. Vasin, A., Gasnikov, A., Spokoiny, V.: Stopping rules for accelerated gradient methods with additive noise in gradient. arxiv, https://arxiv.org/pdf/2102.02921.pdf (2021)
12. Kuruzov, I., Stonyakin, F.: Sequential Subspace Optimization for Quasar-Convex Optimization Problems with Inexact Gradient. arxiv, https://arxiv.org/pdf/2108.06297.pdf (2021)

Global Optimization

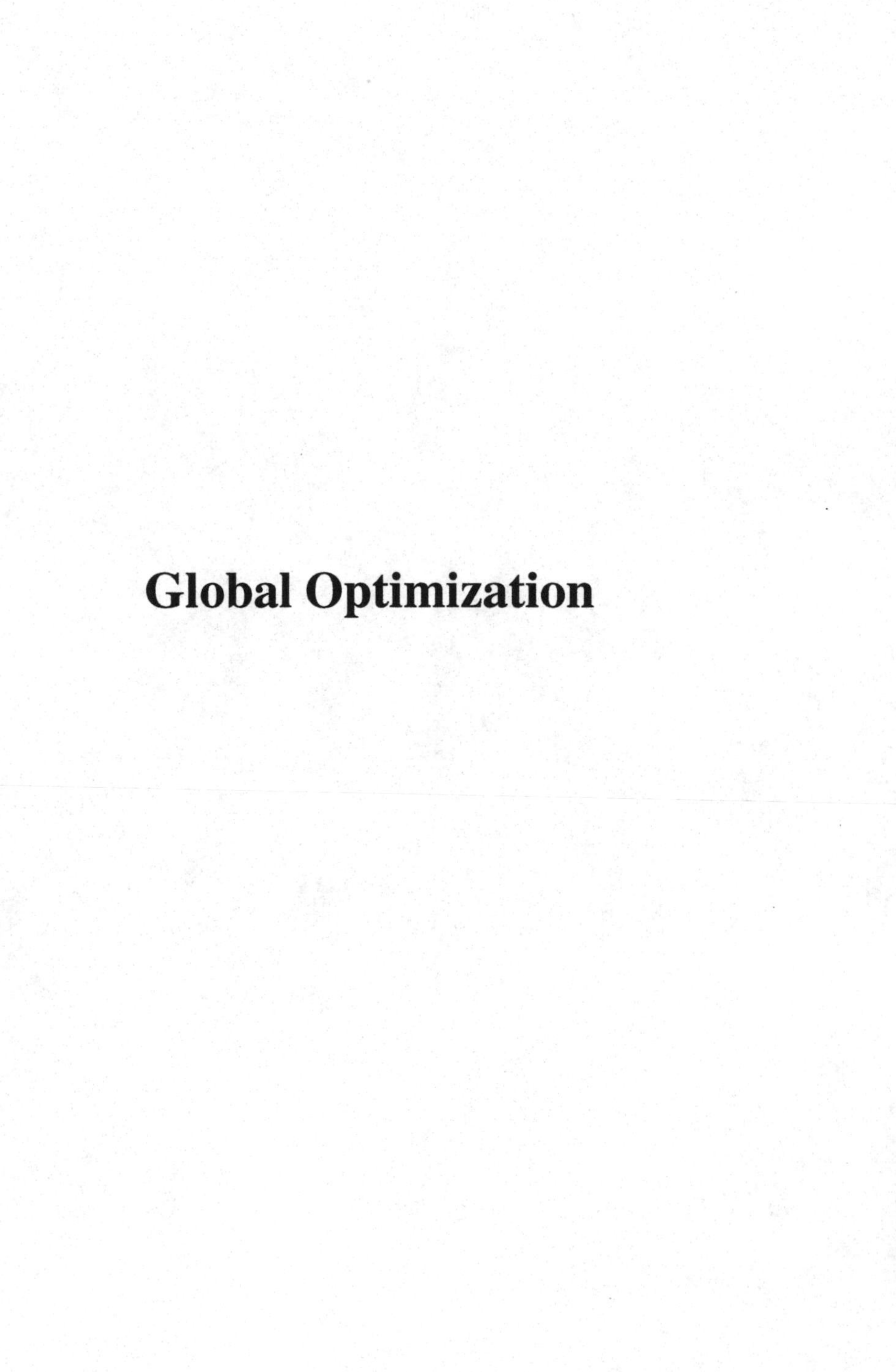

A Search Algorithm for the Global Extremum of a Discontinuous Function

Konstantin Barkalov(✉) and Marina Usova

Lobachevsky State University of Nizhny Novgorod, Nizhny Novgorod, Russia
konstantin.barkalov@itmm.unn.ru

Abstract. This article discusses the problem of finding the global minimum of a one-dimensional function that may have several finite jump discontinuity points. The discontinuities of the objective function may be due to the nature of the optimized object in the mathematical model (for example, shock effects, resonance phenomena, jumps in geometric dimensions or material properties, etc.). In some cases, the discontinuity points are known; but at the same time, there are problems with no a priori estimates of discontinuity points. It is known, however, that such points do exist. The paper gives a description of the global search algorithm for solving such class of problems. In addition, the authors conducted an experimental comparison of the proposed algorithm with the methods of MATLAB Global Optimization Toolbox.

Keywords: Global optimization · Multiextremal functions · Discontinuous functions · Algorithms' comparison

1 Introduction

This paper considers the problem of finding the global minimum x^* of a one-dimensional function $\varphi(x)$ of the form

$$\varphi(x^*) = \min\{\varphi(x) : x \in [a,b]\}. \tag{1}$$

Global optimization problems involving many local extrema arise in many applications (see, e.g. [6,17,18,20]). As a rule, the calculation of even a single value of the objective function in these problems takes considerable time, since the analytical form of the function is not known and the calculation of its values involves time-consuming numerical modelling. That is why these computationally expensive optimization problems are often called black-box problems. They are extremely complicated and are the subject of intensive research.

Despite their complex nature, these problems play an important role in many aspects of mathematical research. First, as has already been mentioned, they are applied for multiple tasks in a variety of fields (see, e.g., [10,26]). Secondly, many approaches to solving multidimensional optimization problems, in one way

This work was supported by the Russian Science Foundation, project No. 21-11-00204.

N. N. Olenev et al. (Eds.): OPTIMA 2021, CCIS 1514, pp. 37–49, 2021.
https://doi.org/10.1007/978-3-030-92711-0_3

or another, are based on reducing the solution of the original multidimensional problem to solving a series of related one-dimensional optimization problems (see, for example, popular approaches in [22,24,27,30]).

The literature describes various methods for solving the problem (1) depending on the assumptions about the properties of the objective function [13,14,24]. One of such assumptions, which is often made about applied optimization problems, is that the objective function $\varphi(x)$ satisfies the Lipschitz condition

$$|\varphi(x') - \varphi(x'')| \leq L\,|x' - x''|\,,\; x', x'' \in [a, b],$$

where the constant L is unknown a priori. This feature of the objective function is used for the development of algorithms for multiextremal optimization [7,8, 15,23].

However, in some applied problems, the Lipschitz condition may not be satisfied due to the presence of discontinuous jumps in the values of the function at certain points of the search domain. Such jumps can be caused by sharp changes in the characteristics of the optimized object which trigger minor changes in its parameters, for example:

- changes in the thickness of material (or other geometric parameters of the object);
- changes in physical properties (both when different materials are used in a single design project, and when the properties of one material change as a result of technological operations, for example, welding, etc.);
- changes in external sources of impact (temperature, force, etc.).

We will interpret such sharp changes as jump discontinuities. For these cases, the set of points, at which the characteristics of the object jump, is typically known in advance. However, there are problems with no a priori estimates of discontinuity points. It is known, however, that such points may exist.

The known methods for solving such problems, as a rule, either generalize the concept of a gradient for discontinuous functions [4,5,21], or belong to the class of bio-inspired algorithms [2,32]. Generally speaking, these methods only search for a local solution to the problem.

In this paper, we will consider a deterministic algorithm for solving global optimization problems with both given and unspecified discontinuity points. This algorithm was developed on the basis of the *global search algorithm* proposed by Strongin [31], and it can be used to find the global optimum. The scope of this article is limited to the one-dimensional case. Generalization of the new algorithm to the multidimensional case can be performed in accordance with the already existing standard approaches from [9,12], which use a dimensionality reduction based on the adaptive nested optimization scheme. The indicated dimensionality reduction scheme allows replacing the solution of the original multidimensional problem with the solution of a family of one-dimensional problems that are recursively related to each other.

The main part of the article is organized as follows. Section 2 describes the algorithm for functions with given discontinuities. Section 3 discusses how to identify unspecified discontinuities. Section 4 describes how the discontinuity detection method can be used in the original algorithm. Section 5 presents the results of a comparison of the developed algorithm with the global search methods implemented in MATLAB Global Optimization Toolbox [19]. Section 6 concludes the paper.

2 Algorithm for Functions with Given Discontinuity Points

Let us consider the problem (1) assuming that the objective function $\varphi(x), x \in [a,b]$, has jump discontinuities at the given points

$$a = \omega_0 < \omega_1 < ... < \omega_s < \omega_{s+1} = b \tag{2}$$

and satisfies the Lipschitz condition between these points, i.e.

$$|\varphi(x') - \varphi(x'')| \leq L\,|x' - x''|\,,\; x', x'' \in (\omega_{i-1}, \omega_i),\; 1 \leq i \leq s+1. \tag{3}$$

Points ω_0 and ω_{s+1} are included in the general list regardless of the presence of function discontinuities at them for the convenience of subsequent presentation. Let us denote the resulting set of points Ω, i.e. $\Omega = \{\omega_0, ..., \omega_{s+1}\}$, where $\omega_0 = a,\ \omega_{s+1} = b$.

In [31], the global search algorithm (GSA) was applied for solving problems without discontinuities. The authors also proposed a method for accounting for discontinuities based on a smoothing transformation of the objective function. Below we consider an algorithm based on GSA, in which the discontinuous function is minimized without smoothing of the discontinuities, which substantially simplifies the computational scheme of the algorithm.

In the preliminary step of the method, the first $s+1$ trials are carried out at arbitrary interior points $x^i \in (\omega_{i-1}, \omega_i),\ 1 \leq i \leq s+1$. A point $x^{k+1},\ k \geq s+1$, for the next trial is selected as follows.

1. Arrange prior trial points $x^1, ..., x^k$ and points $\omega_0, ..., \omega_{s+1}$ in ascending coordinate order. Designate the points of the resulting single series $x_i, 0 \leq i \leq m = k+s+1$,

$$a = x_0 < x_1 < ... < x_{m-1} < x_m = b, \tag{4}$$

and match them with values $z_i = \varphi(x_i)$ for $x_i \notin \Omega$. Note that, in accordance with the choice of trial points at the preliminary step of the algorithm, there will be no interval (x_{i-1}, x_i), the boundary points of which simultaneously belong to the set Ω.
2. For each interval $(x_{i-1}, x_i), 1 \leq i \leq m$, calculate the values

$$\mu_i = (1 - \gamma_i - \gamma_{i-1})\frac{|z_i - z_{i-1}|}{(x_i - x_{i-1})} \tag{5}$$

where

$$\gamma_i = \begin{cases} 1, & x_i \in \{\omega_0, ..., \omega_{s+1}\}, \\ 0, & x_i \notin \{\omega_0, ..., \omega_{s+1}\}. \end{cases} \tag{6}$$

Find the maximum value

$$\mu = \max\{\mu_i : 1 \le i \le m\}.$$

When $\mu = 0$, use the value $\mu = 1$. Note that rule (6) ensures that the inequality $0 \le \gamma_i + \gamma_{i-1} \le 1$ holds for every $1 \le i \le m$. Therefore, the value μ_i will be equal to 0 if one of the endpoints of the interval (x_{i-1}, x_i), $1 \le i \le m$, belongs to Ω.

3. For each interval (x_{i-1}, x_i), $1 \le i \le m$, calculate the characteristic

$$R(i) = (1 + \gamma_i + \gamma_{i-1})\Delta_i + (1 - \gamma_i - \gamma_{i-1})\frac{(z_i - z_{i-1})^2}{(r\mu)^2\Delta_i} - \frac{2[(1-\gamma_i)(1+\gamma_{i-1})z_i + (1+\gamma_i)(1-\gamma_{i-1})z_{i-1}]}{r\mu}, \quad (7)$$

where $r > 1$ is the method parameter.

4. Determine the interval to which the maximum characteristic corresponds

$$R(t) = \max\{R(i) : 1 \le i \le m\}.$$

5. Carry out next trial at the point

$$x^{k+1} = \frac{(x_t + x_{t-1})}{2} - (1 - \gamma_t - \gamma_{t-1})\frac{(z_t - z_{t-1})}{2r\mu}.$$

6. Check the stopping condition $x_t - x_{t-1} \le \epsilon$, where t is the number of the interval with the maximal characteristic and $\epsilon > 0$ is the predefined accuracy.

Remark 1. In accordance with (6), the value $\gamma_i + \gamma_{i-1}$ can be equal to either 0 (both boundary points of the interval belong to the domain of continuity of the function), or 1 (one of the boundary points is the jump point). In the first case, the formula (7) will have the form

$$R(i) = \Delta_i + \frac{(z_i - z_{i-1})^2}{(r\mu)^2\Delta_i} - \frac{2(z_i + z_{i-1})}{r\mu}.$$

And in the second case we will have

$$R(i) = 2\Delta_i - \frac{4z_i}{r\mu}, \quad R(i) = 2\Delta_i - \frac{4z_{i-1}}{r\mu},$$

in case of discontinuity at the point x_{i-1} or x_i respectively.

Remark 2. The above formulas for calculating the characteristics of intervals can be used (in a modified form) in the algorithm for solving problems with non-convex constraints [3,25,28,29]. Here, the solution to the problem with partially defined constraints is implicitly reduced to minimizing a discontinuous function, in which the discontinuity points are the boundaries of the admissible set of the constrained problem.

Remark 3. The erroneous assignment of some discontinuity points from the series (2) will not lead to the loss of convergence of the method. The algorithm will interpret such points as a recoverable discontinuity case since the left and right limits of the objective function at such points coincide.

It can be proved that the convergence conditions for this algorithm correspond to the convergence conditions for its prototype – the global search algorithm from [31]. The formulation and rigorous proof of the convergence conditions will be the subject of further publications.

Let us illustrate the operation of the algorithm during the search for the minimum of the function

$$\varphi(x) = 0.1\sum_{i=1}^{5} i\sin\left(10(i+1)x+i\right) + \begin{cases} -4, & 0 \le x < 3.1 \\ 10, & 3.1 \le x < 4.6 \\ 0, & 4.6 \le x < 7.0 \\ 30, & 7.0 \le x < 9.0 \\ 0, & 9.0 \le x \le 10.0 \end{cases} \tag{8}$$

at $x \in [0, 10]$. Here, all the discontinuity points were considered to be known, and the parameters of the method were $r = 2.0$, $\epsilon = 0.001$. Figure 1 shows a graph of this function; the discontinuities are marked with a dotted line. The dashes under the graph indicate the points of 70 trials required for the algorithm to solve problem with the specified accuracy.

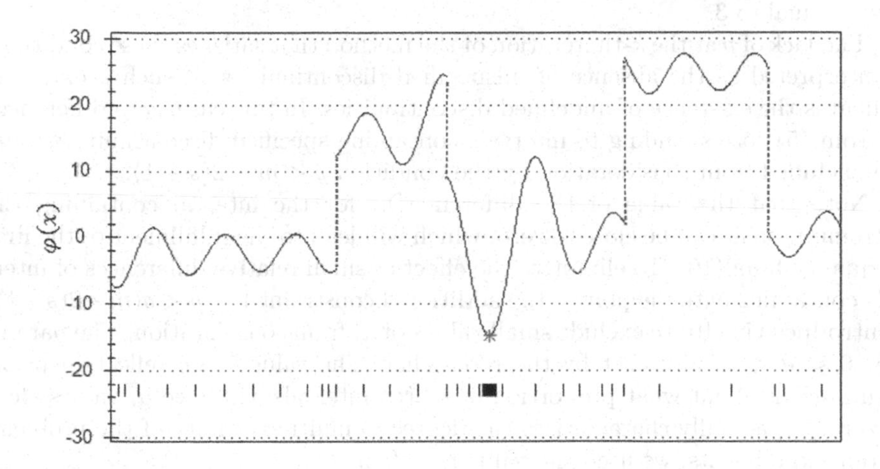

Fig. 1. Discontinuous objective function graph and trial points in case of given discontinuity points

3 Identifying Unspecified Discontinuities

Let us suppose that the location of the discontinuity points for the function $\varphi(x)$ is not known, but it is known that such points are possible. Let the function have a discontinuity at an a priori unknown point $\omega \in (x_{i-1}, x_i)$, then the relative difference μ_i from (5) will significantly exceed the relative differences for other intervals, since within these intervals the function satisfies the Lipschitz condition. Therefore, such instances can be identified based on a comparison of the relative differences μ_i from (5) corresponding to different search intervals (x_{i-1}, x_i).

Let us enumerate the differences μ_i from (5) in decreasing order (by the index in brackets)

$$\mu(1) \geq \mu(2) \geq ... \geq \mu(m) \tag{9}$$

and define the minimum number p satisfying the conditions

$$\frac{\mu(p)}{\mu(p+1)} \geq Q, \; 1 \leq p < q(m - 2(s+1)), \tag{10}$$

where $Q > 1$ and $0 < q < 1$ are parameters. The value of the parameter Q determines the a priori assumptions about how many times the relative difference μ_i, calculated on the interval (x_{i-1}, x_i) with the jump point, will exceed the maximum difference μ, found over all intervals belonging to the domain of continuity of the function. This value is set by the researcher. In our experiments it was equal to 3.

The lack of p at the k-th iteration of the method that satisfies these conditions is interpreted as the absence of unspecified discontinuities. If such p exists, it indicates the existence of undefined discontinuities. In this case, zero differences μ_i from (5) corresponding to intervals containing specified discontinuity points, are excluded from consideration by the condition $p < m - 2(s+1)$.

Note that the value of the difference μ_i for the interval containing the extremum point can be close to zero, which will lead to the fulfillment of the first inequality from (10). To eliminate the effect of small relative differences of intervals containing extreme points, the additional constraint $1 \leq p < q(m-2(s+1))$ is introduced in (10) to exclude small values of μ_i from consideration. The parameter $0 < q < 1$ is also set by the researcher. The value $1 - q$ reflects a priori assumptions about what proportion of search intervals will have μ_i values close to zero, i.e. actually characterizes the degree of multi-extremity of the problem. In our experiments, we used the value $q = 0.3$.

Let us select a subset of numbers

$$I = \{i : 1 \leq i \leq m, \; \mu_i = \mu(j), \; 1 \leq j \leq p\} \tag{11}$$

of the intervals (x_{i-1}, x_i), which correspond to large relative differences μ_i. Large difference values are interpreted as the presence of undefined discontinuities in the intervals $(x_{i-1}, x_i), i \in I$. Moreover, the set I can change in the process of solving the problem.

4 Algorithm for Unspecified Discontinuities

This algorithm is a modification of the algorithm described in Sect. 2 for given discontinuities and consists of the following.

The first $s+1$ trials are carried out at arbitrary interior points $x^i \in (\omega_{i-1}, \omega_i)$, $1 \leq i \leq s+1$, where the set $\Omega = \{\omega_0, ..., \omega_{s+1}\}$ contains the given discontinuity points, as well as boundary points $\omega_0 = a$ and $\omega_{s+1} = b$. A point for the next trial x^{k+1}, $k \geq s+1$, is selected as follows.

1. Sort points of a set $\{x^1, ..., x^k\} \cup \Omega$ in ascending order by their coordinates

$$a = x_0 < x_1 < ... < x_{m-1} < x_m = b, \tag{12}$$

   where $m = s+k+1$. Each point is matched with an attribute γ_i from (6) (the presence or absence of a given discontinuity) and with the value of the function $z_i = \varphi(x_i)$ calculated only for $x_i \notin \Omega$, i.e. at $\gamma_i = 0$.
2. Calculate the relative differences μ_i, $1 \leq i \leq m$, from (5) and order them in descending order, i.e. build a series (9).
3. Determine p which satisfies the inequalities (10), where $0 < q < 1 < Q$ are the parameters of the algorithm. Construct a set I from (11), the elements of which are interpreted as numbers of intervals that can contain a discontinuity. Generate interval attributes

$$\delta_i = \begin{cases} \text{sign}\,(z_i - z_{i-1}), & i \in I, \\ 0, & i \notin I. \end{cases} \tag{13}$$

   The value $\delta_i = -1$ corresponds to a sharp decrease in the function in the interval (x_{i-1}, x_i), $\delta_i = 1$ – to an increase, $\delta_i = 0$ – to the lack of jump. Note that in accordance with the rules (6) and (13) for calculating attributes γ_i and δ_i the inequalities $0 \leq \gamma_i + \gamma_{i-1} + |\delta_i| \leq 1$, $1 \leq i \leq m$, are satisfied.
4. Determine the maximum value

$$\mu = \max\left\{(1 - \gamma_i - \gamma_{i-1} - |\delta_i|)\frac{|z_i - z_{i-1}|}{(x_i - x_{i-1})} : 1 \leq i \leq m\right\}. \tag{14}$$

   When $\mu = 0$, use the value $\mu = 1$.
5. For each interval (x_{i-1}, x_i), $1 \leq i \leq m$, calculate the characteristic

$$R(i) = (1 + \gamma_i + \gamma_{i-1} + |\delta_i|)\Delta_i + (1 - \gamma_i - \gamma_{i-1} - |\delta_i|)\frac{(z_i - z_{i-1})^2}{(r\mu)^2 \Delta_i}$$
$$- \frac{2[((1 - \gamma_i)(1 + \gamma_{i-1})(1 - \delta_i))z_i + ((1 + \gamma_i)(1 - \gamma_{i-1})(1 + \delta_i))z_{i-1}]}{r\mu},$$

   where $r > 1$ is the method parameter, μ is from (14) and $\Delta_i = x_i - x_{i-1}$.
6. Determine the interval corresponding to the maximum characteristic

$$R(t) = \max\{R(i) : 1 \leq i \leq m\}.$$

7. Carry out next trial at the point

$$x^{k+1} = \frac{(x_t + x_{t-1})}{2} - (1 - \gamma_t - \gamma_{t-1} - |\delta_t|)\frac{(z_t - z_{t-1})}{2r\mu}. \quad (15)$$

8. Check the stopping condition $x_t - x_{t-1} \leq \epsilon$, where t is the number of the interval with the maximal characteristic and $\epsilon > 0$ is the predefined accuracy.

Remark. According to (6) and (13) the value $\gamma_i + \gamma_{i-1} + |\delta_i|$ can be 0 or 1, depending on the presence or absence of a jump in the interval (x_{i-1}, x_i). So, if the given interval does not contain jumps, i.e. $\gamma_i + \gamma_{i-1} + |\delta_i| = 0$, then its characteristic (7) takes the form

$$R(i) = \Delta_i + \frac{(z_i - z_{i-1})^2}{(r\mu)^2 \Delta_i} - \frac{2(z_i + z_{i-1})}{r\mu}.$$

If one of the boundary points of the interval is a given discontinuity, i.e. $\gamma_{i-1} = 1$ or $\gamma_i = 1$, then we get the characteristics

$$R(i) = 2\Delta_i - \frac{4z_i}{r\mu},\ R(i) = 2\Delta_i - \frac{4z_{i-1}}{r\mu}. \quad (16)$$

Similar formulas for the characteristics of intervals will be obtained if (x_{i-1}, x_i) is an interval that contains an unspecified discontinuity, i.e. when $|\delta_i| = 1$. In this case, we will also obtain formulas (16) for $\delta_i = -1$ and $\delta_i = 1$, which correspond to using the trial result at the right point of the interval to compute its characteristic in the case of decreasing function and choosing left point as the function increases.

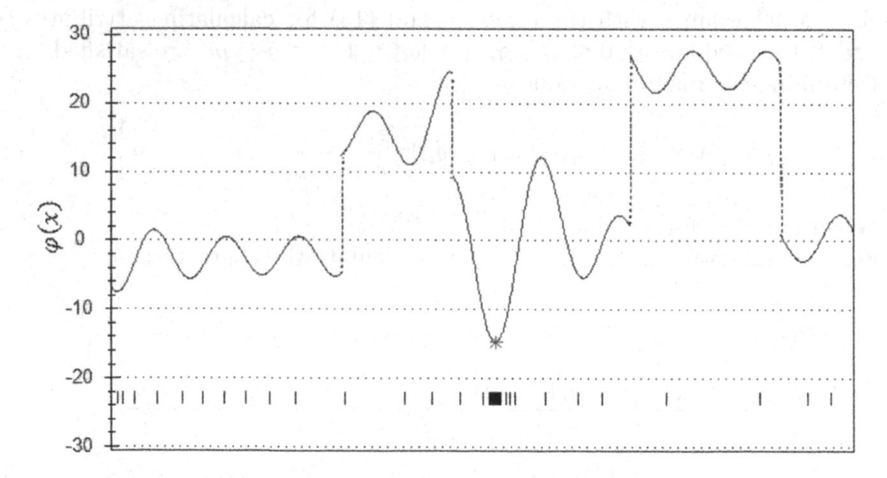

Fig. 2. Discontinuous objective function graph and trial points in case of unspecified discontinuity points

Figure 2 shows the results of solving the problem from the previous example (8), in which the discontinuity points were unspecified. Here we used the parameters of the method $r = 2.0, \epsilon = 0.001, Q = 3.0, q = 0.3$. The dashes under the graph indicate the points of 70 trials required for the algorithm to solve problems with the specified accuracy.

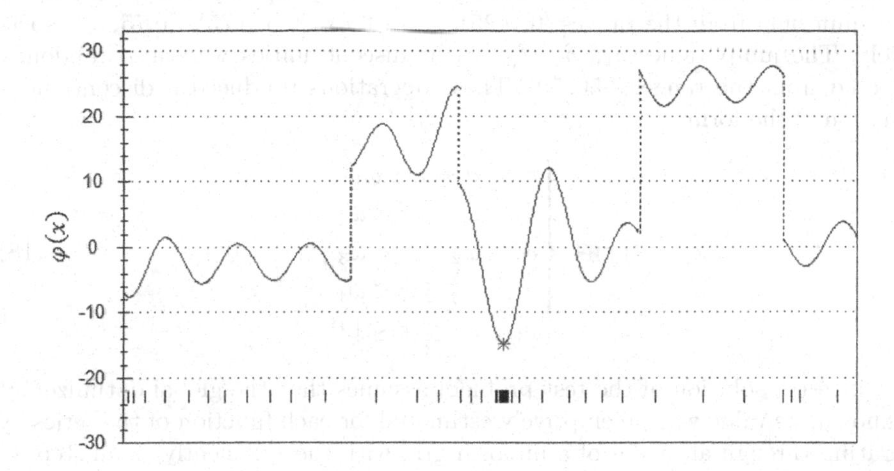

Fig. 3. Discontinuous objective function graph and trial points in case of partially given discontinuity points

The proposed algorithm also allows for partial specification of known discontinuities. Figure 3 shows the results of solving the same problem, for which the discontinuities were considered partially given (the discontinuities at the points 4.6 and 9.0 were considered to be known). Here we used the same parameters of the method $r = 2.0, \epsilon = 0.001, Q = 3.0, q = 0.3$. The dashes under the graph indicate the points of 67 trials required for the algorithm to solve problems with the specified accuracy.

5 Results of Numerical Experiments

To demonstrate the efficiency of the global search algorithm for discontinuous functions (GSA-D) described in the previous section, we compared it with Direct Search (DS) [1], Simulated Annealing (SA) [16], and Genetic Algorithm (GA) [11] implemented in MATLAB Global Optimization Toolbox [19].

We used the number of search trials K (i.e., the number of calculations of the objective function) performed by the method before it converges as the main comparison criterion.

The algorithms were compared on 1000 problems with a discontinuous objective function, which were constructed as follows. First, we generated continuous functions of the form

$$f(x)=\sum_{j=0}^{4}(j+1)A_j\sin(5\pi jx+j)+(j+1)B_j\cos(3\pi jx+j), x\in[0,1], \quad (17)$$

where the values of the coefficients A_j, B_j were chosen randomly and uniformly from the interval $[-6, 6]$. Then 4 discontinuity points were added to the continuous functions. The coordinates of these points ω_1, ω_2, ω_3, ω_4 were set randomly and uniformly from the ranges $[0, 0.25)$, $[0.25, 0.5)$, $[0.5, 0.75)$, $[0.75, 1]$ respectively. The jump values δ_1, δ_2, δ_3, δ_4 at discontinuities were also randomly selected from the range $[-50, 50]$. These operations produced a discontinuous function of the form

$$\varphi(x)=f(x)+\begin{cases}\delta_1, & 0\le x<\omega_1\\ \delta_2, & \omega_1\le x<\omega_2\\ \delta_3, & \omega_2\le x<\omega_3\\ \delta_4, & \omega_3\le x<\omega_4\\ \delta_1, & \omega_4\le x\le 1.0\end{cases}, \quad x\in[0,1]. \quad (18)$$

Since the solution of the test problem assumes that the global optimizer x^* is known, its value was preemptively estimated for each function of the series by iterating through all nodes of a uniform grid with the sufficiently small step.

The first experiment was performed on a series of 1000 problems with continuous functions of the form (17). Table 1 shows the number of problems solved, the average number of iterations performed by different methods, and the number of trials performed by the methods during the search. The problem was considered solved if the condition $|x^k - x^*| \le \delta$ was satisfied for the trial point x^k, where x^* is the known solution of the problem and $\delta = 10^{-2}$. At the same time, the accuracy used in the termination criteria for all methods was $\epsilon = 10^{-3}$.

In order to exclude the influence inherent randomness when running algorithms from MatLab, the GA, SA, and DS methods were run 5 times for each problem, and only the best result was recorded (in terms of the proximity of the found solution to the true one). In this case, the GA and SA methods were run with default parameters, while in the DS method the parameter x_0 (the starting point of the search) varied from 0 to 1 with a step of 0.2.

With multiple runs, the methods from MATLAB Global Optimization Toolbox solved almost all the problems in the series, while with a single run of these methods the number of correctly solved problems was almost 2 times lower. At the same time, the GSA-D method successfully solved all the problems in the series (the reliability parameter $r = 2.2$ was used for the method). At the same time, the number of trials spent was comparable to the DS method and significantly less than the number of trials of the GA and SA methods.

The next experiment was performed on a series of 1000 problems with discontinuous functions of the form (18). GSA-D used the parameters $r = 3.7$, $Q = 3.0$, $q = 0.3$, the other experimental conditions were similar to the previous run. The results of solving a series of problems with discontinuous functions are shown in Table 2.

Table 1. Results of solving a series of problems with continuous functions

Method	Problems solved	Average number of iterations	Average number of trials
GSA-D	1000	52	53
SA	1000	774	777
GA	999	24	1290
DS	970	38	71

Table 2. Results of solving a series of problems with discontinuous functions

Method	Problems solved	Average number of iterations	Average number of trials
GSA-D	1000	79	80
SA	998	765	770
GA	993	25	1310
DS	964	38	71

The results show that the Global Optimization Toolbox methods are somewhat worse at solving discontinuous problems than continuous ones, while the GSA-D algorithm proposed in the paper successfully solves the entire series of problems. The GA and SA algorithms show high reliability of finding the global optimizer after several runs, but these methods are significantly inferior to the GSA-D algorithm in terms of the number of objective function calculations. The DS method, while being able to solve the problem in a comparable number of trials, is inferior in the number of correctly solved problems.

Although the methods from the MATLAB Global Optimization Toolbox are claimed to work with discontinuous functions, they do not guarantee convergence to the global minimum. The convergence rate (in terms of the number of trials) for the GA and SA methods is significantly lower than that for the DS method. However, the DS algorithm, while surpassing the GA and SA methods by this indicator, has the lowest reliability among the considered methods. The GSA-D method proposed in the article demonstrated both the guaranteed convergence to the global minimizer for the problems with discontinuous functions and uses less trials to achieve convergence.

6 Conclusion

In this paper, we propose an algorithm for finding the minimum of a multiextremal function with jump discontinuities. The proposed GSA-D algorithm, unlike many known approaches, does not rely on the concept of a gradient and provides a search for a global solution to the problem. We carried out computational experiments confirming the convergence of the algorithm for a wide class of discontinuous multiextremal problems. The convergence conditions of the proposed algorithm, in general, correspond to the convergence conditions of its prototype – the global search algorithm for continuous problems, described in detail in [31]. Justification and research of the theoretical properties of the proposed GSA-D algorithm will be the subject of future publications.

References

1. Audet, C., Dennis, J.E.: Analysis of generalized pattern searches. SIAM J. Optim. **13**(3), 889–903 (2003)
2. Ban, N., Yamazaki, W.: Development of efficient global optimization method for discontinuous optimization problems with infeasible regions using classification method. J. Adv. Mech. Des. Syst. Manuf. **13**(1), JAMDSM0017 (2019)
3. Barkalov, K.A., Strongin, R.G.: A global optimization technique with an adaptive order of checking for constraints. Comput. Math. Math. Phys. **42**(9), 1289–1300 (2002)
4. Batukhtin, V.D.: On solving discontinuous extremal problems. J. Optim. Theory Appl. **77**, 575–589 (1993)
5. Batukhtin, V.D., Bigil'deev, S.I., Bigil'deeva, T.B.: Approximate gradient methods and the necessary conditions for the extremum of discontinuous functions. IFAC Proc. **31**, 25–34 (1998)
6. Cavoretto, R., De Rossi, A., Mukhametzhanov, M.S., Sergeyev, Y.D.: On the search of the shape parameter in radial basis functions using univariate global optimization methods. J. Global Optim. **79**(2), 305–327 (2021)
7. Evtushenko, Y.G., Malkova, V.U., Stanevichyus, A.A.: Parallel global optimization of functions of several variables. Comput. Math. Math. Phys. **49**(2), 246–260 (2009)
8. Evtushenko, Y.G., Posypkin, M.A.: A deterministic approach to global box-constrained optimization. Optim. Lett. **7**, 819–829 (2013)
9. Gergel, V.P., Grishagin, V., Gergel, A.: Adaptive nested optimization scheme for multidimensional global search. J. Glob. Optim. **66**(1), 35–51 (2016)
10. Gillard, J.W., Kvasov, D.E.: Lipschitz optimization methods for fitting a sum of damped sinusoids to a series of observations. Stat. Interface **10**(1), 59–70 (2017)
11. Goldberg, D.E.: Genetic Algorithms in Search. Optimization and Machine Learning, Addison-Wesley, Boston (1989)
12. Grishagin, V., Israfilov, R., Sergeyev, Y.D.: Convergence conditions and numerical comparison of global optimization methods based on dimensionality reduction schemes. Appl. Math. Comput. **318**, 270–280 (2018)
13. Horst, R., Pardalos, P.M.: Handbook of Global Optimization. Kluwer Academic Publishers, Dordrecht (1995)
14. Horst, R., Tuy, H.: Global Optimization - Deterministic Approaches. Springer, Heidelberg (1996). https://doi.org/10.1007/978-3-662-03199-5
15. Jones, D.R.: Direct global optimization algorithm. In: Floudas, C., Pardalos, P. (eds.) Encyclopedia of Optimization, pp. 725–735. Springer, Boston (2008). https://doi.org/10.1007/978-0-387-74759-0_128
16. Kirkpatrick, S., Gelatt, C.D., Vecchi, M.P.: Optimization by simulated annealing. Science **220**(4598), 671–680 (1983)
17. Kvasov, D.E., Sergeyev, Y.D.: Lipschitz global optimization methods in control problems. Autom. Remote Control **74**(9), 1435–1448 (2013)

18. Kvasov, D.E., Sergeyev, Y.D.: Deterministic approaches for solving practical black-box global optimization problems. Adv. Eng. Softw. **80**, 58–66 (2015)
19. Matlab global optimization toolbox. https://www.mathworks.com/help/gads/index.html
20. Modorskii, V.Y., Gaynutdinova, D.F., Gergel, V.P., Barkalov, K.A.: Optimization in design of scientific products for purposes of cavitation problems. In: AIP Conference Proceedings, vol. 1738 (2016)
21. Moreau, L., Aeyels, D.: Optimization of discontinuous functions: a generalized theory of differentiation. SIAM J. Optim. **11**(1), 53–69 (2000)
22. Paulavičius, R., Žilinskas, J.: Simplicial Global Optimization. Springer, New York (2014). https://doi.org/10.1007/978-1-4614-9093-7
23. Paulavičius, R., Žilinskas, J., Grothey, A.: Parallel branch and bound for global optimization with combination of Lipschitz bounds. Optim. Methods Softw. **26**(3), 487–498 (2011)
24. Pinter, J.D.: Global optimization in action (Continuous and Lipschitz Optimization: Algorithms, Implementations and Applications). Kluwer Academic Publishers, Dordrecht (1996)
25. Sergeyev, Y.D.: Univariate global optimization with multiextremal non-differentiable constraints without penalty functions. Comput. Optim. Appl. **34**(2), 229–248 (2006)
26. Sergeyev, Y.D., Candelieri, A., Kvasov, D.E., Perego, R.: Safe global optimization of expensive noisy black-box functions in the δ-Lipschitz framework. Soft. Comput. **24**(23), 17715–17735 (2020)
27. Sergeyev, Y.D., Kvasov, D.E.: Deterministic Global Optimization: An Introduction to the Diagonal Approach. Springer, New York (2017). https://doi.org/10.1007/978-1-4939-7199-2
28. Sergeyev, Y.D., Kvasov, D.E., Khalaf, F.: A one-dimensional local tuning algorithm for solving go problems with partially defined constraints. Optim. Lett. **1**(1), 85–99 (2007)
29. Sergeyev, Y.D., Markin, D.L.: An algorithm for solving global optimization problems with nonlinear constraints. J. Global Optim. **7**(4), 407–419 (1995)
30. Sergeyev, Y.D., Strongin, R.G., Lera, D.: Introduction to Global Optimization Exploiting Space-Filling Curves. Springer Briefs in Optimization, Springer, New York (2013). https://doi.org/10.1007/978-1-4614-8042-6
31. Strongin, R.G., Sergeyev, Y.D.: Global Optimization with Non-convex Constraints. Sequential and Parallel Algorithms, Kluwer Academic Publishers, Dordrecht (2000)
32. Zhang, J., Xu, J.: A new differential evolution for discontinuous optimization problems. In: Third International Conference on Natural Computation, ICNC 2007. IEEE Computer Society (2007)

A Novel Algorithm with Self-adaptive Technique for Solving Variational Inequalities in Banach Spaces

Yana Vedel, Vladimir Semenov(✉), and Sergey Denisov

Taras Shevchenko National University of Kyiv, Kyiv, Ukraine

Abstract. A novel algorithm for solving variational inequalities in a Banach space is proposed and studied. The proposed algorithm is an adaptive variant of the forward-reflected-backward algorithm, where the used rule for updating the step size does not require knowledge of the Lipschitz continuous constant of the operator. In addition, the Alber generalized projection is used instead of the metric projection onto the feasible set. For variational inequalities with monotone and Lipschitz continuous operators acting in a 2-uniformly convex and uniformly smooth Banach space, a theorem on the weak convergence of the method is proved.

Keywords: Variational inequality · Monotone operator · Algorithm · Convergence · 2-uniformly convex Banach space · Uniformly smooth Banach space

1 Introduction

Many problems of operations research and mathematical physics can be written in the form of variational inequalities [1–5]. The development and study of variational inequalities is an actively developing area of applied nonlinear analysis [4,6–26,28–37]. Note that often non-smooth optimization problems can be effectively solved if they are reformulated as saddle point problems and algorithms for solving variational inequalities are applied [7]. With the advent of generative adversarial networks, a steady interest in algorithms for solving variational inequalities arose among specialists in the field of machine learning [8–10].

The most famous method for solving variational inequalities is the Korpelevich extra-gradient algorithm [11]. A large number of publications are devoted to the study of the extra-gradient algorithm and its modifications [6,7,12–26]. An effective modern version of the extra-gradient method is the proximal mirror method of Nemirovski [7]. This method can be interpreted as a variant of the extra-gradient method with projection understood in the sense of Bregman divergence [27]. Also an interesting method of dual extrapolation for solving

This work was supported by the Ministry of Education and Science of Ukraine (project "Mathematical Modeling and Optimization of Dynamical Systems for Defense, Medicine and Ecology", 0119U100337).

N. N. Olenev et al. (Eds.): OPTIMA 2021, CCIS 1514, pp. 50–64, 2021.
https://doi.org/10.1007/978-3-030-92711-0_4

variational inequalities was proposed by Yu. Nesterov [28]. Adaptive variants of the Nemirovski mirror-prox method were studied in [19–26]. In the early 1980s, L.D. Popov proposed an interesting modification of the classical Arrow-Hurwitz algorithm for finding saddle points of convex-concave functions [29]. A modification of Popov's method for solving variational inequalities with monotone operators was studied in [30]. And in the article [31], a two-stage proximal algorithm for solving the equilibrium programming problem is proposed, which is an adaptation of the method [30] to the general Ky Fan inequalities. In [32–34], the two-stage proximal mirror method was studied, which is a modification of the two-stage proximal algorithm [31] using Bregman divergence instead of the Euclidean distance. Note that recently Popov's algorithm for variational inequalities has become well known among machine learning specialists under the name "Extrapolation from the Past" [9]. Further development of this circle of ideas led to the emergence of the so-called forward-reflected-backward algorithm [35] and related methods [36,37].

In this paper, we propose a new algorithm for solving variational inequalities in a Banach space. Variational inequalities in Banach spaces arise and are intensively studied in mathematical physics and the theory of inverse problems [1,2,4]. Recently, there has been progress in the study of algorithms for problems in Banach spaces [4,15–18]. This is due to the wide involvement of the results and constructions of the geometry of Banach spaces [38–40]. The proposed algorithm is an adaptive variant of the forward-reflected-backward algorithm [35], where the rule for updating the step size does not require knowledge of the Lipschitz constant of operator. Moreover, instead of the metric projection onto the feasible set, the Alber generalized projection is used [40]. An attractive feature of the algorithm is only one computation at the iterative step of the projection onto the feasible set. For variational inequalities with monotone Lipschitz operators acting in a 2-uniformly convex and uniformly smooth Banach space, a theorem on the weak convergence of the method is proved.

2 Preliminaries

We recall several concepts and facts of the geometry of Banach spaces that are necessary for the formulation and proof of the results [38–42].

Everywhere E denotes a real Banach space with the norm $\|\cdot\|$, E^* dual to E space, $\langle x^*, x\rangle$ is value of functional $x^* \in E^*$ on element $x \in E$. We denote norm in E^* as $\|\cdot\|_*$.

Let $S_E = \{x \in E : \ \|x\| = 1\}$. Banach space E is strictly convex if for all x, $y \in S_E$ and $x \neq y$ we have $\left\|\frac{x+y}{2}\right\| < 1$. The modulus of convexity of the space E is defined as follows

$$\delta_E(\varepsilon) = \inf\left\{1 - \left\|\frac{x+y}{2}\right\| : \ x, y \in S_E, \ \|x-y\| = \varepsilon\right\} \quad \forall \varepsilon \in (0,2].$$

Banach space E is uniformly convex if $\delta_E(\varepsilon) > 0$ for all $\varepsilon \in (0,2]$. Banach space E is called 2-uniformly convex if exists $c > 0$ that

$$\delta_E(\varepsilon) \geq c\varepsilon^2$$

for all $\varepsilon \in (0, 2]$. Obviously, a 2-uniformly convex space is uniformly convex. It is known that a uniformly convex Banach space is reflexive.

A Banach space E is called smooth if the limit

$$\lim_{t \to 0} \frac{\|x + ty\| - \|x\|}{t} \tag{1}$$

exists for all $x, y \in S_E$. A Banach space E is called uniformly smooth if the limit (1) exists uniformly in $x, y \in S_E$. There is a duality between the convexity and smoothness of the Banach space E and its dual E^* [38,39]:

- E^* is strictly convex space $\Rightarrow E$ is smooth space;
- E^* is smooth space $\Rightarrow E$ is strictly convex space;
- E is uniformly convex space $\Leftrightarrow E^*$ is uniformly smooth space;
- E is uniformly smooth space $\Leftrightarrow E^*$ is uniformly convex space.

Note that if the space E is reflexive, the first two implications can be reversed. It is known that Hilbert spaces and spaces L_p $(1 < p \leq 2)$ are 2-uniformly convex and uniformly smooth (spaces L_p are uniformly smooth for $p \in (1, \infty)$) [38,39].

Multivalued operator $J : E \to 2^{E^*}$, acting as follows

$$Jx = \left\{ x^* \in E^* : \ \langle x^*, x \rangle = \|x\|^2 = \|x^*\|_*^2 \right\},$$

is called the normalized duality mapping. It is known that [41]:

- if the space E is smooth, then the mapping J is single valued;
- if the space E is strictly convex, then the mapping J is injective and strictly monotone;
- if the space E is reflexive, then the mapping J is surjective;
- if the space E is uniformly smooth, then the mapping J is uniformly continuous on bounded subsets of E.

Let E be a smooth Banach space. Consider the functional introduced by Yakov Alber [40]

$$\phi(x, y) = \|x\|^2 - 2\langle Jy, x \rangle + \|y\|^2 \quad \forall x, y \in E.$$

A useful identity follows from the definition of ϕ:

$$\phi(x, y) - \phi(x, z) - \phi(z, y) = 2\langle Jz - Jy, x - z \rangle \quad \forall x, y, z \in E.$$

If the space E is strictly convex, then for $x, y \in E$ we have $\phi(x, y) = 0 \Leftrightarrow x = y$.

Lemma 1 ([40]). *Let E be a uniformly convex and uniformly smooth Banach space, (x_n), (y_n) are bounded sequences of E elements. Then*

$$\|x_n - y_n\| \to 0 \quad \Leftrightarrow \quad \|Jx_n - Jy_n\|_* \to 0 \quad \Leftrightarrow \quad \phi(x_n, y_n) \to 0.$$

Lemma 2 ([42]). *Let E be a 2-uniformly convex and smooth Banach space. Then, for some $\mu \geq 1$, the inequality holds*

$$\phi(x, y) \geq \frac{1}{\mu} \|x - y\|^2 \quad \forall x, y \in E.$$

Remark 1. For Banach spaces ℓ_p, L_p, and W_p^k $(1 < p \leq 2)$, μ is $\frac{1}{p-1}$.

Let K be a non-empty closed and convex subset of a reflexive, strictly convex and smooth space E. It is known [40] that for each $x \in E$ there is a unique point $z \in K$ such that

$$\phi(z, x) = \inf_{y \in K} \phi(y, x).$$

This point z is denoted by $\Pi_K x$, and the corresponding operator $\Pi_K : E \to K$ is called the generalized projection of E onto K (Alber generalized projection) [40]. Note that if E is a Hilbert space, then Π_K coincides with the metric projection onto the set K.

Lemma 3 ([40]). *Let K be a closed and convex subset of a reflexive, strictly convex and smooth space E, $x \in E$, $z \in K$. Then*

$$z = \Pi_K x \quad \Leftrightarrow \quad \langle Jz - Jx, y - z \rangle \geq 0 \quad \forall y \in K. \tag{2}$$

Remark 2. The inequality (2) is equivalent to the following [40]:

$$\phi(y, \Pi_K x) + \phi(\Pi_K x, x) \leq \phi(y, x) \quad \forall y \in K.$$

Basic information about monotone operators and variational inequalities in Banach spaces can be found in [1,2,4,40,41].

3 Algorithm

Let E be 2-uniformly convex and uniformly smooth Banach space, C be non-empty subset of space E, A be an operator from E to E^*. Consider variational inequality:

$$\text{find } x \in C : \ \langle Ax, y - x \rangle \geq 0 \quad \forall y \in C. \tag{3}$$

We denote set of solutions of (3) by S.

Assume that the following conditions are satisfied:

- set $C \subseteq E$ is convex and closed;
- operator $A : E \to E^*$ is monotone and Lipschitz-type with $L > 0$ on C;
- set S is non-empty.

Remark 3. We can formulate (3) as fixed point problem [40]:

$$x = \Pi_C J^{-1} (Jx - \lambda Ax), \tag{4}$$

where $\lambda > 0$. Formulation (4) is useful because is contains an obvious algorithmic idea.

Consider dual variational inequality:

$$\text{find } x \in C: \ \langle Ay, x-y\rangle \leq 0 \quad \forall y \in C. \tag{5}$$

We denote set of solutions of (5) by S^d. Note that set S^d is closed and convex [2]. Inequality (5) is sometimes called weak or dual formulation of (3) (or Minty inequality) and solutions (5) are weak solutions (3). For monotone operators A we always have $S \subseteq S^d$. In our conditions $S^d = S$ [2].

We assume that the following is satisfied:

- normalized duality mapping $J : E \to E^*$ sequentially weakly continuous, i.e. from $x_n \to x$ weak in E then $Jx_n \to Jx$ weak* in E^*.

Consider now a new algorithm for solving the variational inequality (3). We will use a simple rule for updating the parameters λ_n without information about the Lipschitz constant of the operator A. The proposed algorithm is a modification of the forward-reflected-backward algorithm recently proposed in [35] for solving operator inclusions with the sum of the maximal monotone and Lipschitz continuous monotone operators acting in a Hilbert space. Let us know the constant $\mu \geq 1$ from the Lemma 2. Recall that for the space L_p $(1 < p \leq 2)$, the value of μ is $\frac{1}{p-1}$.

Algorithm 1. Choose $x_0 \in E$, $x_1 \in E$, $\tau \in \left(0, \frac{1}{2\mu}\right)$ and λ_0, $\lambda_1 > 0$. Let $n = 1$.

1. Calculate

$$x_{n+1} = \Pi_C J^{-1}\left(Jx_n - \lambda_n A x_n - \lambda_{n-1}\left(Ax_n - Ax_{n-1}\right)\right).$$

2. If $x_{n-1} = x_n = x_{n+1}$, then STOP, else go to 3.
3. Calculate

$$\lambda_{n+1} = \begin{cases} \min\left\{\lambda_n, \tau\frac{\|x_{n+1}-x_n\|}{\|Ax_{n+1}-Ax_n\|_*}\right\}, & \text{if } Ax_{n+1} \neq Ax_n, \\ \lambda_n, & \text{otherwise.} \end{cases}$$

Let $n := n + 1$ and go to 1.

Sequence generated by rule of calculation (λ_n) is non-increasing and lower bounded by $\min\left\{\lambda_1, \tau L^{-1}\right\}$. Then exists $\lim_{n\to\infty} \lambda_n > 0$.

For sequence (x_n) generated by Algorithm 1 takes place the inequality

$$-2\left\langle \lambda_n A x_n + \lambda_{n-1}\left(Ax_n - Ax_{n-1}\right), y - x_{n+1}\right\rangle$$

$$\leq \phi\left(y, x_n\right) - \phi\left(x_{n+1}, x_n\right) - \phi\left(y, x_{n+1}\right) \quad \forall y \in C. \tag{6}$$

Inequality (6) shows a rule of finishing the algorithm. Indeed if

$$x_{n-1} = x_n = x_{n+1}$$

then from (6) it follows then $\langle Ax_n, y - x_n\rangle \geq 0$ for all $y \in C$, i.e. $x_n \in S$.

Now we go to the proof of convergence of Algorithm 1.

4 Basic Inequality

In this section, we state and prove the inequality on which the proof of Algorithm 1 weak convergence is based.

Lemma 4. *For sequence generated by Algorithm 1 (x_n) the following inequality holds:*

$$\phi(z, x_{n+1}) + 2\lambda_n \langle Ax_n - Ax_{n+1}, x_{n+1} - z\rangle + \tau\mu\frac{\lambda_n}{\lambda_{n+1}}\phi(x_{n+1}, x_n)$$

$$\leq \phi(z, x_n) + 2\lambda_{n-1}\langle Ax_{n-1} - Ax_n, x_n - z\rangle + \tau\mu\frac{\lambda_{n-1}}{\lambda_n}\phi(x_n, x_{n-1})$$

$$-\left(1 - \tau\mu\frac{\lambda_{n-1}}{\lambda_n} - \tau\mu\frac{\lambda_n}{\lambda_{n+1}}\right)\phi(x_{n+1}, x_n),$$

where $z \in S$.

Proof. Let $z \in S$. We have

$$\phi(z, x_{n+1}) \leq \phi(z, x_n) - \phi(x_{n+1}, x_n)$$
$$+ 2\langle \lambda_n Ax_n + \lambda_{n-1}(Ax_n - Ax_{n-1}), z - x_{n+1}\rangle. \tag{7}$$

From monotonicity of operator A we have

$$\langle \lambda_n Ax_n + \lambda_{n-1}(Ax_n - Ax_{n-1}), z - x_{n+1}\rangle = \lambda_n\langle Ax_n - Ax_{n+1}, z - x_{n+1}\rangle$$

$$+\lambda_{n-1}\langle Ax_n - Ax_{n-1}, z - x_{n+1}\rangle + \underbrace{\lambda_n\langle Ax_{n+1}, z - x_{n+1}\rangle}_{\leq 0}$$

$$\leq \lambda_n\langle Ax_n - Ax_{n+1}, z - x_{n+1}\rangle + \lambda_{n-1}\langle Ax_n - Ax_{n-1}, z - x_n\rangle$$
$$+ \lambda_{n-1}\langle Ax_n - Ax_{n-1}, x_n - x_{n+1}\rangle. \tag{8}$$

Applying (8) to (7) we obtain

$$\phi(z, x_{n+1}) \leq \phi(z, x_n) - \phi(x_{n+1}, x_n)$$

$$+2\lambda_n\langle Ax_n - Ax_{n+1}, z - x_{n+1}\rangle$$

$$+ 2\lambda_{n-1}\langle Ax_n - Ax_{n-1}, z - x_n\rangle + 2\lambda_{n-1}\langle Ax_n - Ax_{n-1}, x_n - x_{n+1}\rangle. \tag{9}$$

From rule of calculation λ_n we have upper estimation for

$$2\lambda_{n-1}\langle Ax_n - Ax_{n-1}, x_n - x_{n+1}\rangle$$

in (9). We have

$$2\lambda_{n-1}\langle Ax_n - Ax_{n-1}, x_n - x_{n+1}\rangle \leq 2\lambda_{n-1}\|Ax_n - Ax_{n-1}\|_*\|x_n - x_{n+1}\|$$

$$\leq 2\tau\frac{\lambda_{n-1}}{\lambda_n}\|x_n - x_{n-1}\|\|x_{n+1} - x_n\|$$

$$\leq \tau \frac{\lambda_{n-1}}{\lambda_n} \|x_n - x_{n-1}\|^2 + \tau \frac{\lambda_{n-1}}{\lambda_n} \|x_n - x_{n+1}\|^2$$

$$\leq \tau\mu \frac{\lambda_{n-1}}{\lambda_n} \phi(x_n, x_{n-1}) + \tau\mu \frac{\lambda_{n-1}}{\lambda_n} \phi(x_{n+1}, x_n).$$

We obtain

$$\phi(z, x_{n+1}) + 2\lambda_n \langle Ax_n - Ax_{n+1}, x_{n+1} - z\rangle + \tau\mu \frac{\lambda_n}{\lambda_{n+1}} \phi(x_{n+1}, x_n)$$

$$\leq \phi(z, x_n) + 2\lambda_{n-1} \langle Ax_{n-1} - Ax_n, x_n - z\rangle + \tau\mu \frac{\lambda_{n-1}}{\lambda_n} \phi(x_n, x_{n-1})$$

$$- \left(1 - \tau\mu \frac{\lambda_{n-1}}{\lambda_n} - \tau\mu \frac{\lambda_n}{\lambda_{n+1}}\right) \phi(x_{n+1}, x_n).$$

The proof is complete.

Remark 4. We can change rule of updating for step 3 of Algorithm 1 to the following:

$$\lambda_{n+1} = \begin{cases} \min\left\{\lambda_n, \tau \frac{\sqrt{\mu\, \phi(x_{n+1}, x_n)}}{\|Ax_{n+1} - Ax_n\|_*}\right\}, & \text{if } Ax_{n+1} \neq Ax_n, \\ \lambda_n, & \text{otherwise.} \end{cases} \quad (10)$$

Lemma 4 takes place also for variant of Algorithm 1 with the rule (10).

5 Convergence

Let us formulate the main result.

Theorem 1. *Let C be a non-empty convex and closed subset of 2-uniformly convex and uniformly smooth Banach space E, $A : E \to E^*$ is monotone Lipschitz continuous operator, $S \neq \emptyset$. Assume that normalized duality mapping J is sequentially weakly continuous. Then sequence generated by Algorithm 1 (x_n) converge weakly to $z \in S$.*

Proof. Let $z' \in S$. Assume

$$a_n = \phi(z', x_n) + 2\lambda_{n-1} \langle Ax_{n-1} - Ax_n, x_n - z'\rangle + \tau\mu \frac{\lambda_{n-1}}{\lambda_n} \phi(x_n, x_{n-1}),$$

$$b_n = \left(1 - \tau\mu \frac{\lambda_{n-1}}{\lambda_n} - \tau\mu \frac{\lambda_n}{\lambda_{n+1}}\right) \phi(x_{n+1}, x_n).$$

Inequality from Lemma 4 takes form

$$a_{n+1} \leq a_n - b_n.$$

Since there exists $\lim_{n\to\infty} \lambda_n > 0$, then

$$1 - \tau\mu\frac{\lambda_{n-1}}{\lambda_n} - \tau\mu\frac{\lambda_n}{\lambda_{n+1}} \to 1 - 2\tau\mu \in (0,1)\,, \quad n \to \infty.$$

Show that $a_n \geq 0$ for all large $n \in \mathbb{N}$. We have

$$a_n = \phi(z', x_n) + 2\lambda_{n-1}\langle Ax_{n-1} - Ax_n, x_n - z'\rangle + \tau\mu\frac{\lambda_{n-1}}{\lambda_n}\phi(x_n, x_{n-1})$$

$$\geq \frac{1}{\mu}\|x_n - z'\|^2 - 2\lambda_{n-1}\|Ax_{n-1} - Ax_n\|_*\|x_n - z'\| + \tau\frac{\lambda_{n-1}}{\lambda_n}\|x_{n-1} - x_n\|^2$$

$$\geq \frac{1}{\mu}\|x_n - z'\|^2 - 2\tau\frac{\lambda_{n-1}}{\lambda_n}\|x_n - x_{n-1}\|\|x_n - z'\| + \tau\frac{\lambda_{n-1}}{\lambda_n}\|x_{n-1} - x_n\|^2$$

$$\geq \left(\frac{1}{\mu} - \tau\frac{\lambda_{n-1}}{\lambda_n}\right)\|x_n - z'\|^2.$$

Since there exists such $n_0 \in \mathbb{N}$ that

$$\frac{1}{\mu} - \tau\frac{\lambda_{n-1}}{\lambda_n} > 0 \quad \text{for all } n \geq n_0,$$

then $a_n \geq 0$ starting from n_0. So we came into conclusion that there exists a limit

$$\lim_{n\to\infty}\left(\phi(z', x_n) + 2\lambda_{n-1}\langle Ax_{n-1} - Ax_n, x_n - z'\rangle + \tau\mu\frac{\lambda_{n-1}}{\lambda_n}\phi(x_n, x_{n-1})\right)$$

and

$$\sum_{n=1}^{\infty}\left(1 - \tau\mu\frac{\lambda_{n-1}}{\lambda_n} - \tau\mu\frac{\lambda_n}{\lambda_{n+1}}\right)\phi(x_{n+1}, x_n) < +\infty.$$

Hence we obtain that the sequence (x_n) is bounded and

$$\lim_{n\to\infty}\phi(x_{n+1}, x_n) = \lim_{n\to\infty}\|x_{n+1} - x_n\| = 0.$$

Since

$$\lim_{n\to\infty}\left(2\lambda_{n-1}\langle Ax_{n-1} - Ax_n, x_n - z'\rangle + \tau\mu\frac{\lambda_{n-1}}{\lambda_n}\phi(x_n, x_{n-1})\right) = 0,$$

then sequences $(\phi(z', x_n))$ converge for all $z' \in S$.

Show that all cluster points of sequence (x_n) are in the set S. Consider subsequence (x_{n_k}) which converges weakly to $z \in E$. Easy to see that $z \in C$. Show that $z \in S$. We have

$$\langle Jx_{n+1} - Jx_n + \lambda_n Ax_n + \lambda_{n-1}(Ax_n - Ax_{n-1}), y - x_{n+1}\rangle \geq 0 \quad \forall y \in C.$$

Hence using monotonicity of operator A we have an inequality

$$\langle Ay, y - x_n\rangle + \langle Ax_n, x_n - x_{n+1}\rangle \geq \langle Ax_n, y - x_{n+1}\rangle$$

$$\geq \frac{1}{\lambda_n} \langle Jx_n - Jx_{n+1}, y - x_{n+1} \rangle - \frac{\lambda_{n-1}}{\lambda_n} \langle Ax_n - Ax_{n-1}, y - x_{n+1} \rangle \quad \forall y \in C.$$

From $\lim_{n\to\infty} \|x_n - x_{n-1}\| = 0$ and Lipschitz property of operator A it follows

$$\lim_{n\to\infty} \|Ax_n - Ax_{n-1}\|_* = 0.$$

From uniform continuity of normalized duality mapping J on bounded sets we get

$$\lim_{n\to\infty} \|Jx_n - Jx_{n+1}\|_* = 0.$$

Hence,

$$\varliminf_{n\to\infty} \langle Ay, y - x_n \rangle \geq 0 \quad \forall y \in C.$$

From other side

$$\langle Ay, y - z \rangle = \lim_{k\to\infty} \langle Ay, y - x_{n_k} \rangle \geq \varliminf_{n\to\infty} \langle Ay, y - x_n \rangle \geq 0 \quad \forall y \in C.$$

Then it follows that $z \in S$.

Show that sequence (x_n) converges weakly to z. Arguing by contradiction. Let exists the subsequence (x_{m_k}) such that $x_{m_k} \to z'$ weakly and $z \neq z'$. Easy to see that $z' \in S$. We have

$$2 \langle Jx_n, z - z' \rangle = \phi(z', x_n) - \phi(z, x_n) + \|z\|^2.$$

From that we see the existence of $\lim_{n\to\infty} \langle Jx_n, z - z' \rangle$. From sequentially weak continuity of normalized duality mapping J we get

$$\langle Jz, z - z' \rangle = \lim_{k\to\infty} \langle Jx_{n_k}, z - z' \rangle = \lim_{k\to\infty} \langle Jx_{m_k}, z - z' \rangle = \langle Jz', z - z' \rangle,$$

i.e.,

$$\langle Jz - Jz', z - z' \rangle = 0.$$

Then it follows that $z = z'$.

The weak convergence of the variant of the algorithm with a constant parameter $\lambda > 0$ is similarly substantiated.

Algorithm 2. Choose $x_0 \in E$, $x_1 \in E$, $\lambda \in \left(0, \frac{1}{2\mu L}\right)$. Let $n = 1$.

1. Calculate
$$x_{n+1} = \Pi_C J^{-1} (Jx_n - 2\lambda Ax_n + \lambda Ax_{n-1}).$$
2. If $x_{n-1} = x_n = x_{n+1}$, then STOP, else let $n := n + 1$ and go to 1.

Remark 5. A special case of Algorithm 2 is the optimistic gradient descent ascent (OGDA) algorithm, popular among machine learning specialists [8,9].

Theorem 2. *Let C be a nonempty convex and closed subset of 2-uniformly convex and uniformly smooth Banach space E, operator $A : E \to E^*$ is monotone and Lipschitz continuous with constant $L > 0$. Let $S \neq \emptyset$ and normalized duality mapping J is sequentially weakly continuous. Then sequence generated by Algorithm 2 (x_n) converge weakly to $z \in S$.*

6 Numerical Experiments

In this section, we present three numerical examples to compare the performance of our algorithm with some other algorithms in the literature. Comparisons will be made on toy problems in Euclidean space. We compare Algorithms 1 and 2 with forward-backward-forward method proposed by Tseng in [14] (Alg. 3) and its self-adaptive version without relaxation [43] (Alg. 4). The numerical experiments are performed in Python 3.8.5 with numpy 1.19 on a 64-bit PC with an Intel Core i7-1065G7 1.3–3.9 GHz and 16 GB RAM.

Example 1 ([43]). Let $C = \{x \in [-5,5]^3 : x_1 + x_2 + x_3 = 0\}$ and $A : \mathbb{R}^3 \to \mathbb{R}^3$ be defined as $Ax = \left(e^{-\|x\|_2^2} + q\right) Mx$, where $q = 0.2$ and

$$M = \begin{pmatrix} 1 & 0 & -1 \\ 0 & 1.5 & 0 \\ -1 & 0 & 2 \end{pmatrix}.$$

The operator A is Lipschitz continuous with constant $L \approx 5.0679$ and pseudo-monotone on $\mathbb{R}^3$ but not monotone. The variational inequality with A and C has a unique solution $x^* = (0,0,0)$.

We use $D_n = \|x_n - x^*\|$, $n = 0, 1, 2, \ldots$, to show the computational performance of the algorithms. The parameters are chosen as follows: $\lambda = 0.9/L$ for Alg. 3 and $0.9/2L$ for Alg. 2. $\tau = 0.9$ for Alg. 4 and $\tau = 0.45$ for Alg. 1. Starting λ is set to 1.0 for the both adaptive algorithms, and $\varepsilon = 10^{-8}$ for residual stopping criterion. We take the starting point $(-4,3,5) \in \mathbb{R}^3$. The results are shown in Fig. 1.

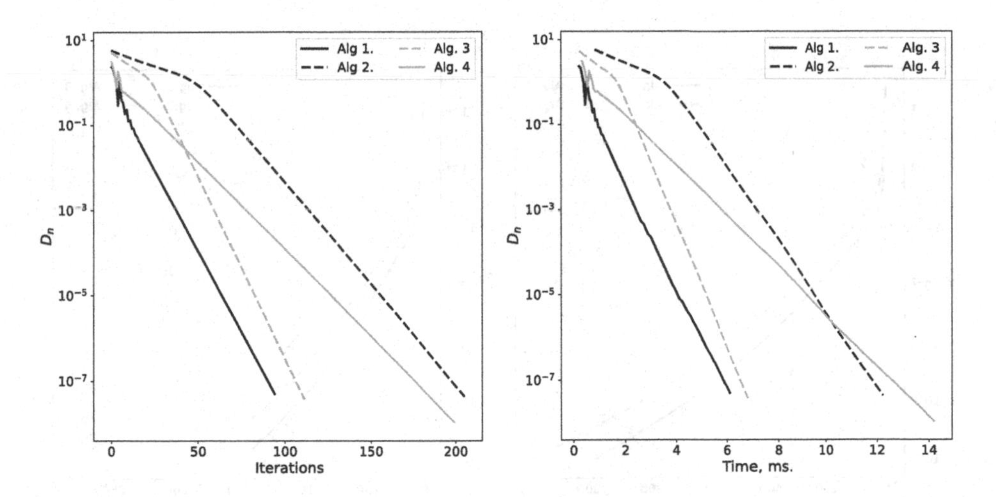

Fig. 1. Numerical results for Example 1.

Obviously, exact computation time differs from run to run – but the relation between algorithms was always close to the result shown on the right side of Fig 1. Also, it should be noted, that we have shown results with τ chosen near its maximal value for both of the adaptive algorithms, so they are in the same conditions. But in our tests the best convergence of Alg. 4 was reached with τ between 0.6 and 0.7 (for the problem above).

Example 2 ([17]). Let $C = \mathbb{R}_+^m = \{x \in \mathbb{R}^m : x_i \geq 0\}$ $(m = 1000)$, $A : \mathbb{R}^m \to \mathbb{R}^m$ be defined as $Ax = Mx + q$ with $q \in \mathbb{R}^m$ and

$$M = NN^* + S + D,$$

where N is an $m \times m$ matrix, S is an $m \times m$ skew-symmetric matrix and D is an $m \times m$ positive definite diagonal matrix. The operator A is Lipschitz continuous with constant $L = \|M\|$ and monotone. All entries N and S are generated randomly in $(-5, 5)$, of D are in $(0, 0.3)$, of q uniformly generated from $(-5, 0)$. We use $D_n = \|x_{n+1} - x_n\|_2$ (Alg. 1, 2) and $D_n = \|y_n - x_n\|_2$ (Alg. 3, 4) to measure the error of the n-th iteration since we don't know the exact solution, and the maximum iteration of 1000 as the stopping criterion. We take the starting point $(1, 1, ..., 1) \in \mathbb{R}^m$. The parameters are chosen as follows: λ set to $0.9/L$ for Alg. 3 and $0.9/2L$ for Alg. 2, τ set to 0.9 for Alg. 4 and 0.45 for Alg. 1. Starting λ is set to $2/L$ for both adaptive algorithms, $\varepsilon = 10^{-5}$ for residual stopping criterion. Numerical results are reported in Fig. 2. In our experiments with the problem above, Alg. 4 (and Alg. 3 with nearly maximal theoretically feasible λ, as above) converge much faster in terms of iterations count. For smaller m, such as 100 or 500, operator computation time is quite small, so they outperform Alg. 1 and Alg. 4 by time also. But for $m = 1000$ and greater, Alg. 1 converges stably faster in time measure.

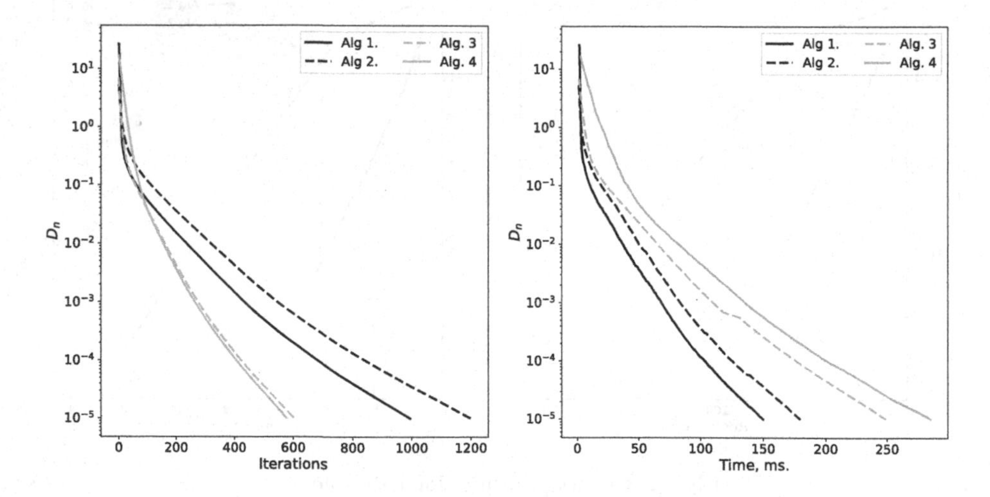

Fig. 2. Numerical results for Example 2.

Example 3. Consider a bilinear saddle point problem

$$\min_{x \in X} \max_{y \in Y} (Px, y),$$

where P is an $m \times n$ matrix, $X = \Delta^n = \{x \in \mathbb{R}^n : \sum_{i=1}^{n} x_i = 1, x_i \geq 0\}$ and $Y = \Delta^m$. We can formulate this problem as a variational inequality by setting

$$Az = A(x, y) = \begin{pmatrix} P^* y \\ Px \end{pmatrix}, \quad C = X \times Y.$$

For performance measure, we use duality gap $G(z) = \max_{v \in C}(Av, z - v)$, which can be simply computed as $\max_i (Px)_i - \min_j (P^* y)_j$ due to simplex constraints. All entries P are generated randomly in $[-3, 3]$. The parameters are chosen as follows: λ set to $0.9/L$ for Alg. 3 and $0.9/2L$ for Alg. 2., τ set to 0.95 for Alg. 4 and 0.45 for Alg. 1. Starting λ is set to 1.0 for both adaptive algorithms, $\varepsilon = 10^{-5}$ for residual stopping criterion. We take the starting points $(1/n, ..., 1/n) \in \Delta^n$, $(1/m, ..., 1/m) \in \Delta^m$. The results for dimension $n = m = 1000$ are shown in Fig. 3.

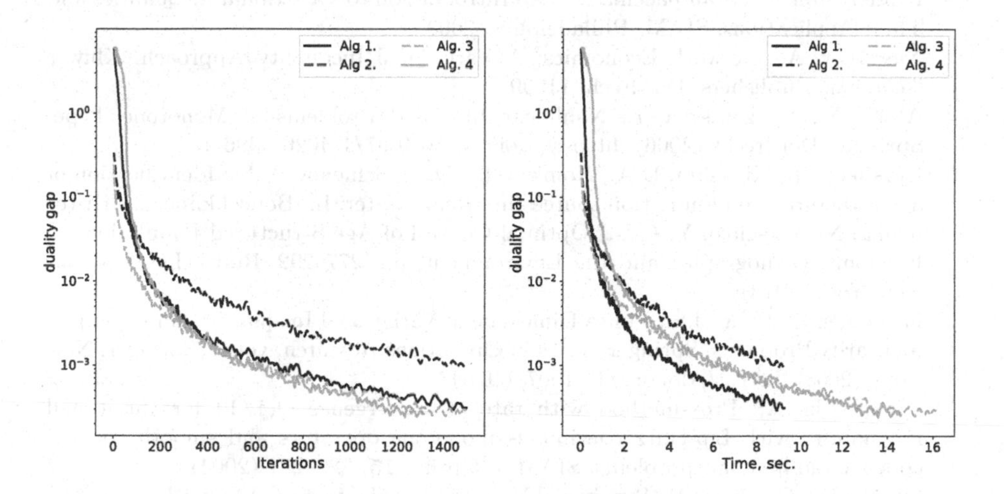

Fig. 3. Numerical results for Example 3.

As expected, Alg. 1 gains advantage with increased operator computation time (as seen from Fig. 3, it does more iterations, but needs less time to reach the same error rate). Behavior, shown on the right image of Fig. 3, was almost stable between runs. Also, with the test problem in Example 3, decreasing τ makes both adaptive algorithms slower, but relative behavior remains the same.

7 Conclusions

In this paper, we have proposed and studied a new algorithm for solving variational inequalities in a Banach space. The proposed algorithm is an adaptive variant of the forward-reflected-backward algorithm [35], where the rule for updating the step size does not require knowledge of the Lipschitz continuous operator constant. Moreover, instead of the metric projection onto the admissible set, the Alber generalized projection is used [40]. An attractive feature of the algorithm is only one computation at the iterative step of the generalized projection onto the feasible set. For variational inequalities with monotone Lipschitz continuous operators acting in a 2-uniformly convex and uniformly smooth Banach space, a theorem on the weak convergence of the method is proved.

Based on the technique [43], similar results can most likely be obtained for problems with pseudo-monotone, Lipschitz continuous, and sequentially weakly continuous operators acting in a uniformly convex and uniformly smooth Banach space.

References

1. Lions, J.-L.: Some Methods of Solving Non-Linear Boundary Value Problems. Dunod-Gauthier-Villars, Paris (1969)
2. Kinderlehrer, D., Stampacchia, G.: An Introduction to Variational Inequalities and Their Applications. SIAM, Philadelphia (2000)
3. Nagurney, A.: Network Economics: A Variational Inequality Approach. Kluwer Academic Publishers, Dordrecht (1999)
4. Alber, Y., Ryazantseva, I.: Nonlinear Ill Posed Problems of Monotone Type. Springer, Dordrecht (2006). https://doi.org/10.1007/1-4020-4396-1
5. Lyashko, S.I., Klyushin, D.A., Nomirovsky, D.A., Semenov, V.V.: Identification of age-structured contamination sources in ground water. In: Boucekkline, R., Hritonenko, N., Yatsenko, Y. (eds.) Optimal Control of Age-Structured Populations in Economy, Demography, and the Environment, pp. 277–292. Routledge, London-New York (2013)
6. Facchinei, F., Pang, J.S.: Finite-Dimensional Variational Inequalities and Complementarity Problems. Springer Series in Operations Research, vol. 1. Springer, New York (2003). https://doi.org/10.1007/b97543
7. Nemirovski, A.: Prox-method with rate of convergence O(1/T) for variational inequalities with Lipschitz continuous monotone operators and smooth convex-concave saddle point problems. SIAM J. Optim. **15**, 229–251 (2004)
8. Daskalakis, C., Ilyas, A., Syrgkanis, V., Zeng, H.: Training GANs with optimism. In: International Conference on Learning Representations (ICLR) (2018)
9. Gidel, G., Berard, H., Vincent, P., Lacoste-Julien, S.: A variational inequality perspective on generative adversarial nets. In: International Conference on Learning Representations (ICLR) (2019)
10. Liu, M., et al.: A decentralized parallel algorithm for training generative adversarial nets. In: Advances in Neural Information Processing Systems (NeurIPS) (2020)
11. Korpelevich, G.M.: An extragradient method for finding saddle points and for other problems. Matecon **12**(4), 747–756 (1976)
12. Censor, Y., Gibali, A., Reich, S.: The subgradient extragradient method for solving variational inequalities in Hilbert space. J. Optim. Theory Appl. **148**, 318–335 (2011). https://doi.org/10.1007/s10957-010-9757-3
13. Semenov, V.V.: Modified extragradient method with Bregman divergence for variational inequalities. J. Autom. Inf. Sci. **50**(8), 26–37 (2018). https://doi.org/10.1615/JAutomatInfScien.v50.i8.30

14. Tseng, P.: A modified forward-backward splitting method for maximal monotone mappings. SIAM J. Control. Optim. **38**, 431–446 (2000)
15. Shehu, Y.: Convergence results of forward-backward algorithms for sum of monotone operators in Banach spaces. Results Math. **74**, 138 (2019). https://doi.org/10.1007/s00025-019-1061-4
16. Shehu, Y.: Single projection algorithm for variational inequalities in Banach spaces with application to contact problem. Acta Math. Sci. **40**, 1045–1063 (2020)
17. Yang, J., Cholamjiak, P., Sunthrayuth, P.: Modified Tseng's splitting algorithms for the sum of two monotone operators in Banach spaces. AIMS Math. **6**(5), 4873 4900 (2021). https://doi.org/10.3934/math.2021286
18. Cholamjiak, P., Shehu, Y.: Inertial forward-backward splitting method in Banach spaces with application to compressed sensing. Appl. Math. **64**, 409–435 (2019)
19. Verlan, D.A., Semenov, V.V., Chabak, L.M.: A strongly convergent modified extragradient method for variational inequalities with non-lipschitz operators. J. Autom. Inf. Sci. **47**(7), 31–46 (2015). https://doi.org/10.1615/JAutomatInfScien.v47.i7.40
20. Bach, F., Levy, K.Y.: A universal algorithm for variational inequalities adaptive to smoothness and noise. arXiv preprint arXiv:1902.01637 (2019)
21. Antonakopoulos, K., Belmega V., Mertikopoulos, P.: An adaptive mirror-prox method for variational inequalities with singular operators. In: Advances in Neural Information Processing Systems (NeurIPS), vol. 32, pp. 8455–8465. Curran Associates, Inc. (2019)
22. Stonyakin, F., Gasnikov, A., Dvurechensky, P., Alkousa, M., Titov, A.: Generalized mirror prox for monotone variational inequalities: universality and inexact Oracle. arXiv preprint arXiv:1806.05140 (2019)
23. Stonyakin, F.S., Vorontsova, E.A., Alkousa, M.S.: New version of mirror prox for variational inequalities with adaptation to inexactness. In: Jaćimović, M., Khachay, M., Malkova, V., Posypkin, M. (eds.) OPTIMA 2019. CCIS, vol. 1145, pp. 427–442. Springer, Cham (2020). https://doi.org/10.1007/978-3-030-38603-0_31
24. Denisov, S.V., Semenov, V.V., Stetsyuk, P.I.: Bregman extragradient method with monotone rule of step adjustment. Cybern. Syst. Anal. **55**(3), 377–383 (2019). https://doi.org/10.1007/s10559-019-00144-5
25. Denisov, S.V., Nomirovskii, D.A., Rublyov, B.V., Semenov, V.V.: Convergence of extragradient algorithm with monotone step size strategy for variational inequalities and operator equations. J. Autom. Inf. Sci. **51**(6), 12–24 (2019). https://doi.org/10.1615/JAutomatInfScien.v51.i6.20
26. Vedel, Y.I., Golubeva, E.N., Semenov, V.V., Chabak, L.M.: Adaptive Extraproximal algorithm for the equilibrium problem in the Hadamard spaces. J. Autom. Inf. Sci. **52**(8), 46–58 (2020). https://doi.org/10.1615/JAutomatInfScien.v52.i8.40
27. Bregman, L.M.: The relaxation method of finding the common point of convex sets and its application to the solution of problems in convex programming. USSR Comput. Math. Math. Phys. **7**(3), 200–217 (1967). https://doi.org/10.1016/0041-5553(67)90040-7
28. Nesterov, Yu.: Dual extrapolation and its applications to solving variational inequalities and related problems. Math. Program. **109**(2–3), 319–344 (2007). https://doi.org/10.1007/s10107-006-0034-z
29. Popov, L.D.: A modification of the Arrow-Hurwicz method for search of saddle points. Math. Notes Acad. Sci. USSR **28**(5), 845–848 (1980). https://doi.org/10.1007/BF01141092
30. Malitsky, Y.V., Semenov, V.V.: An extragradient algorithm for monotone variational inequalities. Cybern. Syst. Anal. **50**(2), 271–277 (2014). https://doi.org/10.1007/s10559-014-9614-8

31. Lyashko, S.I., Semenov, V.V.: A new two-step proximal algorithm of solving the problem of equilibrium programming. In: Goldengorin, B. (ed.) Optimization and Its Applications in Control and Data Sciences. SOIA, vol. 115, pp. 315–325. Springer, Cham (2016). https://doi.org/10.1007/978-3-319-42056-1_10
32. Chabak, L., Semenov, V., Vedel, Y.: A new non-euclidean proximal method for equilibrium problems. In: Chertov, O., Mylovanov, T., Kondratenko, Y., Kacprzyk, J., Kreinovich, V., Stefanuk, V. (eds.) ICDSIAI 2018. AISC, vol. 836, pp. 50–58. Springer, Cham (2019). https://doi.org/10.1007/978-3-319-97885-7_6
33. Nomirovskii, D.A., Rublyov, B.V., Semenov, V.V.: Convergence of two-stage method with Bregman divergence for solving variational inequalities. Cybern. Syst. Anal. **55**(3), 359–368 (2019). https://doi.org/10.1007/s10559-019-00142-7
34. Gibali, A., Thong, D.V.: A new low-cost double projection method for solving variational inequalities. Optim. Eng. **21**, 1613–1634 (2020). https://doi.org/10.1007/s11081-020-09490-2
35. Malitsky, Y., Tam, M.K.: A forward-backward splitting method for monotone inclusions without Cocoercivity. SIAM J. Optim. **30**(2), 1451–1472 (2020). https://doi.org/10.1137/18M1207260
36. Csetnek, E.R., Malitsky, Y., Tam, M.K.: Shadow Douglas-Rachford splitting for monotone inclusions. Appl. Math. Optim. **80**, 665–678 (2019)
37. Cevher, V., Vu, B.C.: A reflected forward-backward splitting method for monotone inclusions involving Lipschitzian operators. Set-Valued Variational Anal. **29**, 163–174 (2021). https://doi.org/10.1007/s11228-020-00542-4
38. Diestel, J.: Geometry of Banach Spaces. Springer, Heidelberg (1975). https://doi.org/10.1007/BFb0082079
39. Beauzamy, B.: Introduction to Banach Spaces and Their Geometry. North-Holland, Amsterdam (1985)
40. Alber, Y.I.: Metric and generalized projection operators in Banach spaces: properties and applications. In: Theory and Applications of Nonlinear Operators of Accretive and Monotone Type, vol. 178, pp. 15–50. Dekker, New York (1996)
41. Vainberg, M.M.: Variational Method and Method of Monotone Operators in the Theory of Nonlinear Equations. Wiley, New York (1974)
42. Aoyama, K., Kohsaka, F.: Strongly relatively nonexpansive sequences generated by firmly nonexpansive-like mappings. Fixed Point Theory Appl. **2014**, 95 (2014). https://doi.org/10.1186/1687-1812-2014-95
43. Bot, R.I., Csetnek, E.R., Vuong, P.T.: The forward-backward-forward method from continuous and discrete perspective for pseudo-monotone variational inequalities in Hilbert spaces. Eur. J. Oper. Res. **287**(1), 49–60 (2020). https://doi.org/10.1016/j.ejor.2020.04.035

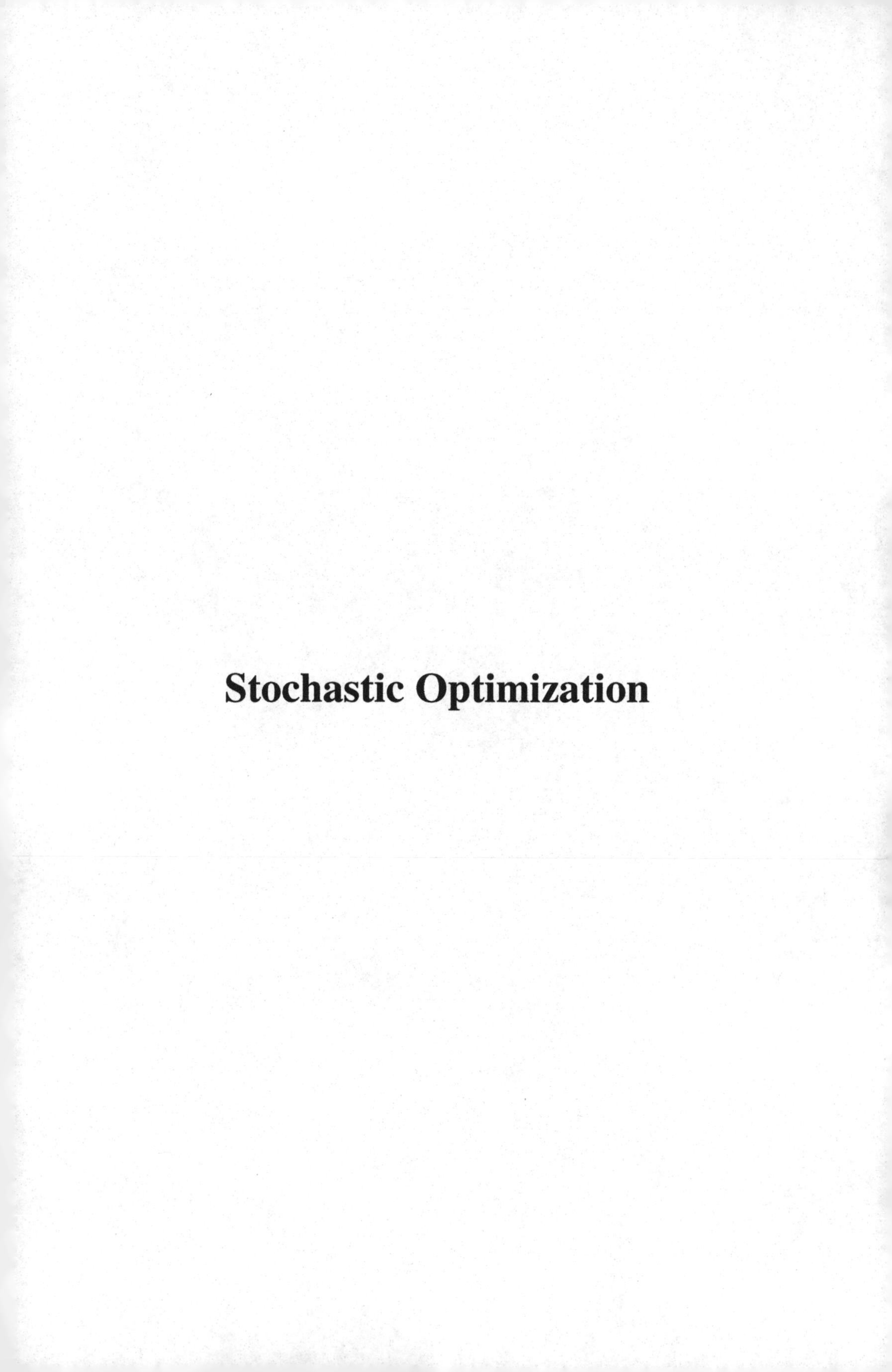

Stochastic Optimization

On One Method of Optimization of Quantum Systems Based on the Search for Fixed Points

Alexander Buldaev(✉) and Ivan Kazmin

Buryat State University, Smolin Street, 24a, 670000 Ulan-Ude, Russia

Abstract. A class of optimal control problems for quantum systems is considered, which is described by bilinear differential equations with a quadratic optimality criterion. A method for non-local improvement of control is proposed, which, in contrast to gradient and other local methods, does not require the operation of calculating the target functional to achieve the improvement goal. The method is based on the construction of a control improvement condition in the form of a fixed point problem in the control space. The obtained condition allows us to apply and modify the well-known theory and methods of fixed points for constructing an iterative algorithm for solving the optimal control problems under consideration. Based on the proposed fixed-point approach, the well-known maximum principle and a strengthened necessary condition for optimality of control are described. Conditions for the convergence of iterative control sequences are given. A comparative analysis of the proposed optimization method with known methods is carried out using model examples.

Keywords: Control of quantum systems · Fixed point problem · Optimization method

1 Introduction

In recent decades, various problems of control of quantum systems have become increasingly important. Reviews of research in this direction can be found in the works of Butkovsky A. G. [1], A. N. Pechen [2] et al. In the works of V. F. Krotov, V. I. Gurman, and their disciples and followers [3–5], classes of controlled quantum systems described by linear in state and control ordinary differential controls with quadratic optimality criteria are considered. The characteristic features of the class under consideration are the high dimension of the system state vector ($n \approx 10^4 - 10^6$); no restrictions on the state, including terminal restrictions; the scalar control function. In the class under consideration, the application of the necessary optimality conditions in the form of the maximum

This work was supported by the Russian Foundation for Basic Research, project 18-41-030005, and Buryat State University, the project of the 2021 year.

N. N. Olenev et al. (Eds.): OPTIMA 2021, CCIS 1514, pp. 67–81, 2021.
https://doi.org/10.1007/978-3-030-92711-0_5

principle for the search for extremal controls due to the large dimension causes significant difficulties. Therefore, methods of improving control are considered as a tool for finding solutions to problems. In particular, the gradient method and the global Krotov method.

In [6], based on the construction of non-standard formulas for the increment of the objective functional that do not contain the remainder of the expansions, effective methods for non-local improvement of control in bilinear control systems with quadratic control quality functionals are developed. Improving control is achieved by solving special Cauchy problems for phase and vector-matrix standard conjugate systems in the state space.

In the present paper, in the class of problems under consideration, a method of non-local control improvement is proposed, based on the construction of a formula for incrementing the functional without residual terms of expansions using a modification of the standard conjugate system. This formula allows us to construct conditions for improving control in the form of operator equations that have the form of a problem about a fixed point in the control space. This makes it possible to apply and modify the well-known theory and methods of fixed points for constructing iterative optimization algorithms for quantum-controlled systems of the class under consideration.

2 Conditions for Improvement and Optimality of Control

Following [3–5], a model class of optimal control problems for quantum systems is considered, written in the general form:

$$\dot{x}(t) = (A + u(t)B)x(t), x(t_0) = x^0, u(t) \in U \subset R, t \in T = [t_0, t_1], \quad (1)$$

$$\Phi(u) = \langle x(t_1), Lx(t_1)\rangle \to \inf_{u \in V}, \quad (2)$$

where $x(t) = (x_1(t), ..., x_n(t))$ is the state vector of the system, L is a real symmetric matrix, A and B are real matrices. As admissible controls $u(t)$, $t \in T$, consider a set V of piecewise continuous scalar functions on an interval T with values in a compact and convex set $U \subset R$. The initial state x^0 and the interval T are fixed.

The Pontryagin function with a conjugate variable ψ and the standard conjugate system has the form:

$$H(\psi, x, u, t) = \langle \psi, (A + uB)x\rangle, \psi \in R^n,$$

$$\dot{\psi}(t) = -(A^T + u(t)B^T)\psi(t), t \in T, \psi(t_1) = -2Lx(t_1). \quad (3)$$

For an admissible control $v \in V$ denote by $x(t, v)$, $t \in T$, the solution of the system (1) for $u(t) = v(t)$; by $\psi(t, v)$, $t \in T$, the solution of the standard conjugate system (3) for $x(t) = x(t, v)$, $u(t) = v(t)$. Denote by P_Y the projection operator on a set $Y \subset R^k$ in the Euclidean norm:

$$P_Y(z) = \arg\min_{y \in Y}(\|y - z\|), z \in R^k.$$

An important property of the projecting operator is the fulfillment of the inequality:

$$\langle y - P_Y(z), z - P_Y(z) \rangle \leq 0, y \in Y.$$

The well-known [6,7] necessary optimality condition (maximum principle) for control $u \in V$ in the class of problems (1), (2) can be represented in the form:

$$u(t) = \arg\max_{w \in U} \langle \psi(t,u), Bx(t,u) \rangle w, t \in T. \quad (4)$$

Using a mapping u^*, defined by the ratio:

$$u^*(\psi, x) = \arg\max_{w \in U} \langle \psi, Bx \rangle w, \psi \in R^n, x \in R^n, t \in T,$$

condition (4) can be written as follows:

$$u(t) = u^*(\psi(t,u), x(t,u)), t \in T.$$

Condition (4) using the projecting operation can be written in an equivalent form with the parameter $\alpha > 0$:

$$u(t) = P_U(u(t) + \alpha \langle \psi(t,u), Bx(t,u) \rangle), t \in T. \quad (5)$$

Note that to fulfill the maximum principle (4), it is sufficient to check the condition (5) for at least one $\alpha > 0$. Conversely, condition (4) implies the fulfillment of condition (5) for all $\alpha > 0$.

Let's define mapping u^α with the parameter $\alpha > 0$ using the relation:

$$u^\alpha(\psi, x, w) = P_U(w + \alpha \langle \psi, Bx \rangle), x \in R^n, \psi \in R^n, w \in U, t \in T.$$

Using the mapping u^α, the maximum principle in the projection form (5) can be written as:

$$u(t) = u^\alpha(\psi(t,u), x(t,u), u(t)), t \in T. \quad (6)$$

Consider the problem of improving control $u \in V$: find the control $v \in V$ with the condition $\Phi(v) \leq \Phi(u)$.

Introduce a differential-algebraic conjugate system, which in the problem (1), (2), following the work [8], can be represented in the form:

$$\dot{p}(t) = -(A^T + w(t)B^T)p(t), \quad (7)$$

$$p(t_1) = -2Lx(t_1) - q, \quad (8)$$

$$\langle 2Lx(t_1) + q, y(t_1) - x(t_1) \rangle = \langle y(t_1), Ly(t_1) \rangle - \langle x(t_1), Lx(t_1) \rangle. \quad (9)$$

The algebraic Eq. (9) can always be resolved analytically for the value q in the form of explicit or conditional formulas (not uniquely for $n > 1$). Thus, the differential-algebraic conjugate system (7)–(9) can always be reduced (not in the only way for $n > 1$) to a differential conjugate system with a uniquely determined value q. In particular, for the standard solution of Eq. (9), which has the form $q = L(y(t_1) - x(t_1))$, obtain a differential conjugate system:

$$\dot{p}(t) = -(A^T + w(t)B^T)p(t), \tag{10}$$

$$p(t_1) = -L(x(t_1) + y(t_1)). \tag{11}$$

For controls $u \in V$, $v \in V$ denote by $p(t,u,v)$, $t \in T$, the solution of the modified conjugate system (7)–(9) for $x(t) = x(t,u)$, $y(t) = x(t,v)$, $w(t) = u(t)$. The definition implies obvious equality $p(t,u,u) = \psi(t,u)$, $t \in T$.

Following [8], based on the modified conjugate system (7)–(9) in the class of problems under consideration, obtain a special formula for the increment of the functional, which does not contain the remainder of the expansions:

$$\Phi(v) - \Phi(u) = -\int_T \langle p(t,u,v), Bx(t,u)\rangle (v(t) - u(t))dt. \tag{12}$$

Based on the formula (12), following [8], construct the conditions for nonlocal improvement of control.

Consider the equation:

$$v(t) = u^*(p(t,u,v), x(t,v)), t \in T. \tag{13}$$

Suppose that Eq. (13) has a solution $v \in V$. Then, by the definition of the mapping u^*, based on the formula (12), get an improvement of the control: $\Phi(v) - \Phi(u) \leq 0$.

Thus, improving the control of $u \in V$ it is sufficient to solve the fixed point problem (13).

For a given $\alpha > 0$ consider the equation based on the projecting operation:

$$v(t) = u^\alpha(p(t,u,v), x(t,v), u(t)), t \in T. \tag{14}$$

Suppose that Eq. (14) has a solution $v^\alpha \in V$. Then, by the definition of the mapping and the above-known property of the projecting operation, based on formula (12), obtain an improvement in the functional with an estimate:

$$\Phi(v) - \Phi(u) \leq -\frac{1}{\alpha}\int_T (v(t) - u(t))^2 dt. \tag{15}$$

Equation (14) can also be interpreted as a problem about a fixed point in the control space.

Thus, to improve the control of $u \in V$ it is possible to solve the fixed point problem (14).

The constructed control improvement conditions have the following connection with the known conditions of the maximum principle (4) and (6).

Denote $V(u) \subseteq V$ the set of admissible fixed points of the problem (13). Let $u \in V(u)$. Then the equality is fulfilled:

$$u(t) = u^*(\psi(t,u), x(t,u)), t \in T,$$

that is, the control u satisfies the condition of the maximum principle (4).

Conversely, if the control u satisfies the condition of the maximum principle (4), then it is obvious $u \in V(u)$.

Therefore, the following statement is true.

Lemma 1. *In the problem (1), (2) the control $u \in V$ satisfies the condition of the maximum principle (4) if and only if $u \in V(u)$.*

Thus, the maximum principle (4) can be formulated in terms of the fixed point problem (13) in the following form.

Theorem 1. *Let the control $u \in V$ be optimal in problem (1), (2). Then $u \in V(u)$.*

Corollary 1. *1) The fixed point problem (13) is always solvable for control $u \in V$, in satisfying the maximum principle (4); 2) the absence of fixed points in the problem (13) or the non-fulfillment of the condition $u \in V(u)$ indicates that the control $u \in V$ is not optimal; 3) in the case of non-uniqueness of the solution of the fixed point problem (13), there is a fundamental possibility of a strict improvement of the control $u \in V$, that satisfies the maximum principle (4).*

Based on the projection problem about a fixed point (14) in problem (1), (2) obtain similar statements. Denote $V^{\alpha}(u) \subseteq V$ the set of admissible fixed points of the problem (14).

Lemma 2. *$u \in V^{\alpha}(u)$ if and only if $u \in V$ satisfies the condition of the maximum principle in the projection form (6).*

Thus, the necessary optimality condition (6) can be formulated in the form of the following statement.

Theorem 2. *Let the control $u \in V$ be optimal in the problem (1), (2). Then $u \in V^{\alpha}(u)$ for some $\alpha > 0$.*

Corollary 2. *1) The fixed point problem (14) is always solvable for control $u \in V$, which satisfies the necessary optimality condition (6); 2) the absence of fixed points in the problem (14) or the failure to fulfill the condition $u \in V^{\alpha}(u)$ for all $\alpha > 0$ indicates that the control $u \in V$ is not optimal; 3) in the case of non-uniqueness of the solution of the fixed point problem (14) the control $u \in V$, satisfying the necessary condition (14) is strictly improved on the solution $v \in V^{\alpha}(u)$, $v \neq u$ according to the estimate (15).*

Thus, the projection problem of a fixed point (14), in contrast to the problem of a fixed point based on the maximization operation (13), in the case of the existence of a solution $v \in V^{\alpha}(u)$ allows concluding based on the estimate (15) that the control $u \in V$ is strictly improved without calculating the objective function.

Estimation (15) allows formulating a strengthened necessary optimality condition for the problem (1), (2) in terms of the fixed point problem (14).

Theorem 3. *(enhanced necessary optimality condition). Let the control $u \in V$ be optimal in the problem (1), (2). Then for all $\alpha > 0$, the control $u \in V$ is the only solution to the fixed point problem (14), i.e. $V^\alpha(u) = \{u\}$, $\alpha > 0$.*

Indeed, in the case of the existence for some $\alpha > 0$ of a fixed point $v \in V^\alpha(u)$, $v \neq u$, by the estimate (15) obtain a strict improvement $\Phi(v) < \Phi(u)$, which contradicts the optimality of the control $u \in V$.

The strengthened necessary optimality condition can be used to test for the optimality of extreme controls (i.e. satisfying the maximum principle). In the case of the existence of fixed points $v \in V^\alpha(u)$, $v \neq u$ for some $\alpha > 0$, it follows that extreme control $u \in V$ is not optimal.

3 Iterative Algorithms

The proposed approach to optimizing the models of controlled quantum systems under consideration consists in the sequential construction and calculation of control improvement problems in the form of constructed fixed point problems with uniquely determined control operators. To calculate the problems of a fixed point in the control space, one can apply and modify the methods of successive approximations known in computational mathematics. In particular, simple iteration methods and their modifications.

For the numerical solution of the problem of a fixed point of an operator $G : V_E \to V_E$, acting on a set V_E in a complete normalized space E with a norm $\|\cdot\|_E$, in a form $v = G(v), v \in V_E$, can use a simple iteration method with an index $k \geq 0$, that has the form $v^{k+1} = G(v^k), v^0 \in V_E$.

The convergence of the iterative process can be analyzed using the well-known principle of compressive mappings.

Let represent the condition for improving control (13) in the form of the problem of a fixed point of some control operator:

$$v = G_1^*(v), v \in V.$$

Let's define the first mapping X by the relation $X(v) = x, v \in V, x(t) = x(t, v), t \in T$. Construct the second mapping P in the same way $P(v) = p, v \in V, p(t) = p(t, u, v), t \in T$. The third mapping V^* is constructed in the form $V^*(p, x) = v^*, p \in C(T), x \in C(T), v^*(t) = u^*(p(t), x(t)), t \in T$, where $C(T)$ is the space of continuous functions on T.

As a result, problem (13) can be represented as an operator equation in the control space in the form of a fixed point problem:

$$v = V^*(P(v), X(v)) = G_1^*(v), v \in V. \quad (16)$$

Construct a new operator problem about a fixed point, equivalent to the condition (13).

Introduce the mapping X^* as follows $X^*(p) = x, p \in C(T), x \in C(T)$, where $x(t)$, $t \in T$, is the solution of the special Cauchy problem:

$$\dot{x}(t) = (A + u^*(p(t), x(t))B)x(t), x(t_0) = x^0.$$

Consider the operator equation:

$$v = V^*(P(v), X^*(P(v))) = G_2^*(v), v \in V. \tag{17}$$

It can be shown, similarly to [9], that Eq. (17) is equivalent to Eq. (16).

The methods of simple iteration under $k \geq 0$ for solving operator problems about a fixed point (16) and (17) have the following form, respectively:

$$v^{k+1} = V^*(P(v^k), X(v^k)), v^0 \in V,$$

$$v^{k+1} = V^*(P(v^k), X^*(P(v^k))), v^0 \in V.$$

By the definition of mappings, the following relation holds:

$$X(V^*(P(v), X^*(P(v)))) = X^*(P(v)), v \in V. \tag{18}$$

From (18) get:

$$X^*(P(v^k)) = X(V^*(P(v^k), X^*(P(v^k)))) = X(v^{k+1}).$$

Therefore, the second method of simple iteration is represented in the following implicit form:

$$v^{k+1} = V^*(P(v^k), X(v^{k+1})), v^0 \in V.$$

In pointwise form, simple iteration methods can be respectively written as:

$$v^{k+1}(t) = u^*(p(t, u, v^k), x(t, v^k)), v^0 \in V, t \in T, \tag{19}$$

$$v^{k+1}(t) = u^*(p(t, u, v^k), x(t, v^{k+1})), v^0 \in V, t \in T. \tag{20}$$

To assess the computational efficiency of iterative algorithms, it is important to note that the complexity of implementing one iteration of the implicit method (20) is similar to the complexity of implementing the explicit method (19) and amounts to two Cauchy problems for phase and conjugate variables.

Indeed, at the k-th iteration for $k \geq 0$ of the process (20), after calculating the solution of the Cauchy problem $p(t, u, v^k)$, $t \in T$, the solution is found $x(t)$, $t \in T$, of the phase system:

$$\dot{x}(t) = (A + u^*(p(t, u, v^k), x(t))B)x(t), x(t_0) = x^0.$$

Then the output control is built according to the rule:

$$v^{k+1}(t) = u^*(p(t, u, v^k), x(t)), t \in T.$$

In this case, according to the construction, the ratio is fulfilled:

$$x(t) = x(t, v^{k+1}), t \in T.$$

The condition for improving control in projection form (14) can also be represented in the form of equivalent fixed point problems on the set of admissible controls.

Introduce an additional operator V^α, $\alpha > 0$, by the relation:

$$V^\alpha(p, x, w) = v^\alpha, p \in C(T), x \in C(T), w \in V,$$

$$v^\alpha(t) = u^\alpha(p(t), x(t), w), t \in T.$$

Define the operator X^α, $\alpha > 0$, $X^\alpha(p, w) = x^\alpha, p \in C(T), w \in V, x^\alpha(t) = x^\alpha(t, p, w), t \in T$, where $x^\alpha(t, p, w), t \in T$, is the solution of the Cauchy problem:

$$\dot{x}(t) = (A + u^\alpha(p(t), x(t), w(t))B)x(t), x(t_0) = x^0.$$

Let's consider the problems of a fixed point:

$$v = V^\alpha(P(v), X(v), u) = G_1^\alpha(v), v \in V, \alpha > 0, \tag{21}$$

$$v = V^\alpha(P(v), X^\alpha(P(v), u), u) = G_2^\alpha(v), v \in V, \alpha > 0. \tag{22}$$

The equivalence of problems (21) and (22) to the control improvement condition (14) is proved similarly.

Emphasize the following important features of the constructed projection operator equations. The constructed projection control operators, due to the properties of the projecting operation, are continuous and satisfy the Lipschitz condition, in contrast to discontinuous and multivalued control operators based on the maximum operation in the general case.

Extreme controls interpreted as fixed points of the operator projection problems (21) and (22) can be determined for any values of the projecting parameter $\alpha > 0$, including for sufficiently small values.

These features are essential factors for improving the efficiency of numerical search for extreme controls based on projection operator equations.

The methods of simple iteration under $k \geq 0$ for solving operator problems about a fixed point (21) and (22), respectively, have the form:

$$v^{k+1} = V^\alpha(P(v^k), X(v^k), u), v^0 \in V, \alpha > 0,$$

$$v^{k+1} = V^\alpha(P(v^k), X^\alpha(P(v^k), u), u), v^0 \in V, \alpha > 0.$$

The second method of simple iteration for finding fixed points of the problem (22) can be written implicitly:

$$v^{k+1} = V^\alpha(P(v^k), X(v^{k+1}), u), v^0 \in V, \alpha > 0.$$

In the pointwise form, iterative methods for finding fixed points take the form:

$$v^{k+1}(t) = u^\alpha(p(t, u, v^k), x(t, v^k), u(t)), v^0 \in V, \alpha > 0, t \in T, \tag{23}$$

$$v^{k+1}(t) = u^\alpha(p(t, u, v^k), x(t, v^{k+1}), u(t)), v^0 \in V, \alpha > 0, t \in T. \tag{24}$$

The complexity of the computational implementation of one iteration of the explicit and implicit projection methods is made up of two Cauchy problems for phase and conjugate variables.

The convergence of the constructed iterative processes (19), (20), (23), and (24) can be analyzed using the well-known principle of compressive mappings in the full space of measurable functions $V \subset V_L = \{v \in L_\infty(T) : v(t) \in U, t \in T\}$, with the norm $\|v\|_\infty = ess\, \sup_{t \in T} \|v(t)\|$, $v \in V_L$.

In particular, it can be shown, that with sufficiently small projecting parameters $\alpha > 0$ the processes (23) and (24) can converge in the norm $\| \cdot \|_\infty$ to the solutions of the corresponding fixed point problems (21) and (22).

Under the assumption that the iterative processes converge to find fixed points, consider the following methods of successive control approximations for solving the problem (1), (2).

In the proposed methods of fixed points, index $k \geq 0$ iterations are carried out before the first strict improvement of control $u \in V$ over the target functionality: $\Phi(v^k) < \Phi(u)$. Next, a new fixed point problem is constructed to improve the obtained computational control and the iterative process is repeated. As an initial approximation v^0 at $k = 0$ the resulting calculation control is also selected. As a result, relaxation sequences of controls u^l, $l \geq 0$, arise with the property $\Phi(v^{l+1}) < \Phi(u^l)$, formed as a result of the sequential calculation of control improvement tasks.

In problem (1), (2), the family of phase trajectories of system (2) is collectively bounded, i.e. $x(t,u) \in X, t \in T, u \in V$, where $X \in R^n$ is a convex compact set. Due to the limitation of the family of phase trajectories of the sequence $\Phi(u^l)$, $l \geq 1$, are limited from below. Therefore, taking into account the relaxation, these sequences are convergent, i.e. $\Phi(u^{l+1}) - \Phi(u^l) \to 0, l \to \infty$.

Hence, the following criteria arise for completing the calculation of the problem (1), (2) by the proposed methods.

If the first strict improvement control $u^l \in V$ by index $k \geq 0 : \Phi(v^k) < \Phi(u^l)$, then $u^{l+1} = v^k$, and the condition for stopping the calculation by the functional is checked:

$$|\Phi(u^{l+1}) - \Phi(u^l)| \leq \varepsilon_1 |\Phi(u^l)|,$$

where $\varepsilon_1 > 0$ is the specified relative accuracy of calculating the target functional.

If the specified stopping criterion is met, then the calculation ends with the proposed modifications of the methods. Otherwise, a new fixed point problem is constructed to improve the obtained computational control u^{l+1} and the iterative process is repeated.

If strict improvement control $u^l \in V$ by index $k \geq 0$ does not occur, i.e. $\Phi(v^k) \geq \Phi(u^l)$, then the iterative process is carried out until the condition is met:

$$\|v^{k+1} - v^k\|_\infty \leq \varepsilon_2 \|v^k\|_\infty,$$

where $\varepsilon_2 > 0$ is the specified relative accuracy of calculating the fixed point problem. This is the end of the calculation of fixed points by the proposed methods.

Note the following comparative features of the proposed methods of fixed points.

In contrast to the well-known gradient methods, the proposed methods do not guarantee relaxation over the target functional at each iteration of successive control approximations. The relaxation property compensates for the non-locality of successive control approximations, the absence of a rather laborious operation of convex or needle-like control variation in the vicinity of the current control at each iteration, as well as the possibility of strict improvement of non-optimal extreme controls.

In contrast to the global method [2–4] and non-local methods [6], at each iteration of the proposed explicit methods and implicit fixed methods, the Cauchy problem with a continuous and uniquely defined right-hand side is solved based on the projecting operation.

4 Examples

Examples are considered that illustrate the characteristic features and computational efficiency of the proposed optimization methods in comparison with the known methods.

The numerical solution of phase and conjugate Cauchy problems was performed by the Runge-Kutta-Werner method of variable (5–6) order of accuracy using the DIVPRK program of the IMSL Fortran PowerStation 4.0 library [10]. The values of the controlled, phase and conjugate variables were stored in the nodes of a fixed uniform grid T_h with a sampling step $h = 10^{-5}$ on the interval T. In the intervals between neighboring grid nodes T_h the control value was assumed to be constant and equal to the value in the left node. The criteria for stopping the calculation were determined by the values $\varepsilon_1 = 10^{-6}$ and $\varepsilon_2 = 10^{-3}$.

Example 1. (method based on an algorithm (19)).

Consider the problem chosen in [11] as an illustration of the problem of optimal spin control in a well-known class of optimization models for quantum systems [3–5], which can be presented in the following form:

$$\Phi(u) = 1 - \langle x(t_1), Lx(t_1)\rangle \rightarrow min,$$

$$L = \begin{pmatrix} a_1^2+b_1^2 & a_1a_2+b_1b_2 & 0 & a_1b_2-b_1a_2 \\ a_1a_2+b_1b_2 & a_2^2+b_2^2 & a_2b_1-b_2a_1 & 0 \\ 0 & a_2b_1-b_2a_1 & b_1^2+a_1^2 & b_1b_2+a_1a_2 \\ a_1b_2-b_1a_2 & 0 & b_1b_2+a_1a_2 & b_2^2+a_2^2 \end{pmatrix},$$

$$\dot{x}_1(t) = u(t)x_3(t) + x_4(t), \dot{x}_2(t) = x_3(t) - u(t)x_4(t),$$

$$\dot{x}_3(t) = -u(t)x_1(t) - x_2(t), \dot{x}_4(t) = -x_1(t) + u(t)x_2(t),$$

$$x_1(0) = \frac{1}{\sqrt{2}}, x_2(0) = \frac{1}{\sqrt{2}}, x_3(0) = 0, x_4(0) = 0, t \in T = [0, t_1], t_1 = 1.5,$$

$$a_1 = 0.6, b_1 = -0.3, a_2 = 0.1, b_2 = \sqrt{0.54}.$$

The vector $x(t)$ describes the state of the quantum system, the function $u(t) \in U = [-30, 30]$ characterizes the effect of an external field.

In [11], to calculate the optimal control problem under consideration, the global Krotov method [3–5] was used, the efficiency of which was compared with the well-known gradient method. The control determined from physical considerations was chosen as the initial control approximation for the specified iterative methods:

$$u^0(t) = tg(2\gamma(2t - 1.5)), t \in T, \gamma = -\frac{1}{3}arctg(-30).$$

The Pontryagin function in the problem has the form:

$$H(p, x, u, t) = p_1(ux_3 + x4) + p_2(x_3 - ux_4) + p_3(-ux_1 - x_2) + p_4(-x_1 + ux_2).$$

We consider the modified conjugate system in an unambiguously defined form (10), (11).

The fixed point problem (13) for improving control $u(t)$, $t \in T$, has the form:

$$v(t) = u^*(p(t, u, v), x(t, v)), t \in T, \quad u^*(p, x) = \begin{cases} +30, g(p, x) > 0, \\ -30, g(p, x) < 0, \\ w \in U, g(p, x) = 0, \end{cases}$$

where $g(p, x) = p_1x_3 - p_2x_4 - p_3x_1 + p_4x_2$. The explicit iterative method of simple iteration for solving this fixed point problem at $k \geq 0$ respectively takes the form (19).

In the case of the existence of a time interval $[\Theta_1, \Theta_2] \subset T$ of a non-zero measure, where $g(p(t), x(t)) = 0$, $t \in [\Theta_1, \Theta_2]$, the control u^* is called degenerate on this interval. Degenerate controls are determined by the sequential differentiation by an argument $t \in T$ of the identity $g(p(t), x(t)) = 0$ taking into account the phase and conjugate systems. In practical calculations, similarly to the work [11], the equality of the switching function $g(p, x)$, to zero, which determines a degenerate mode, is understood in the sense of belonging to some small ε - neighborhood of zero. Thus, obtain the following practical calculation formula for the simple iteration method:

$$v^{k+1}(t) = \begin{cases} +30, g(p(t, u, v^k), x(t, v^k)) > \varepsilon, \\ -30, g(p(t, u, v^k), x(t, v^k)) < -\varepsilon, \\ w \in U, |g(p(t, u, v^k), x(t, v^k))| \leq \varepsilon, \end{cases}$$

where the value $w \in U$ is determined by the following rule.

The value is calculated:

$$a_k = -p_1(t, u, v^k)x_4(t, v^k) - p_2(t, u, v^k)x_3(t, v^k) + p_3(t, u, v^k)x_2(t, v^k) + p_4(t, u, v^k)x_1(t, v^k).$$

If $|a_k| > \varepsilon$, then the value is calculated $c_k = \frac{b_k}{a_k}$, where:

$$b_k = -p_1(t, u^0, v^k)x_3(t, v^k) + p_2(t, u^0, v^k)x_4(t, v^k) + p_3(t, u^0, v^k)x_1(t, v^k) - p_4(t, u^0, v^k)x_2(t, v^k).$$

The value $w \in U$ is calculated using the formula:

$$w = \begin{cases} +30, c_k > 30, \\ -30, c_k < -30, \\ c_k, |c_k| \leq 30. \end{cases}$$

If $|a_k| \leq \varepsilon$ then the value is calculated:

$$d_k = p_1(t, u, v^k)x_2(t, v^k) - p_2(t, u, v^k)x_1(t, v^k) + p_3(t, u, v^k)x_4(t, v^k) - p_4(t, u, v^k)x_3(t, v^k).$$

If $|d_k| > \varepsilon$, then the value $w = 0$. If $|d_k| \leq \varepsilon$, then the value $w \in U$ cannot be determined by the differentiation rule and is selected arbitrarily from the interval U.

The calculation of the optimal control problem was carried out by the method of fixed points (MMNT) based on the explicit method of simple iteration (19) from various initial starting controls. For an adequate comparison of the methods, the ε - neighborhood of zero was determined by the value $\varepsilon = 0.001$, specified in [11].

In [11], the final calculated value of the functional $\Phi^* \approx 0.000952$, obtained by the global method at the 9th iteration of control improvement is indicated. In this case, a degenerate section of the final control, determined according to the rules [11], is the interval $[0.0667, t_1]$.

The method of fixed points based on the maximization operation with the specified calculation parameters reaches an approximate value of the functional $\Phi^* \approx 0.001009$ with the number of computational improvement problems equal to 13. Figure 1 shows the starting control and the calculated final control obtained by the method of fixed points. In this case, the degenerate interval of the final control is approximately equal to $[0.0682, 1.4712]$.

Note that for the starting control $u(t) = 0$, $t \in T$, have the value $\Phi(u) \approx 0.6605$, and the method of fixed points from this starting control reaches the calculated value $\Phi^* \approx 0.001234$ with a qualitatively similar final control with a degenerate interval $[0.0691, 1.4738]$.

Thus, within the framework of the example under consideration, the proposed method of fixed points based on the explicit method of simple iteration (19) allows us to achieve similar quantitative and qualitative results with the global method [11]. The methods differ in the number of calculated iterations of control improvement. But at each iteration of the global method [11], it is necessary to solve a complex Cauchy problem for a phase state variable of the system state with a special right-hand side based on a multi-valued and discontinuous

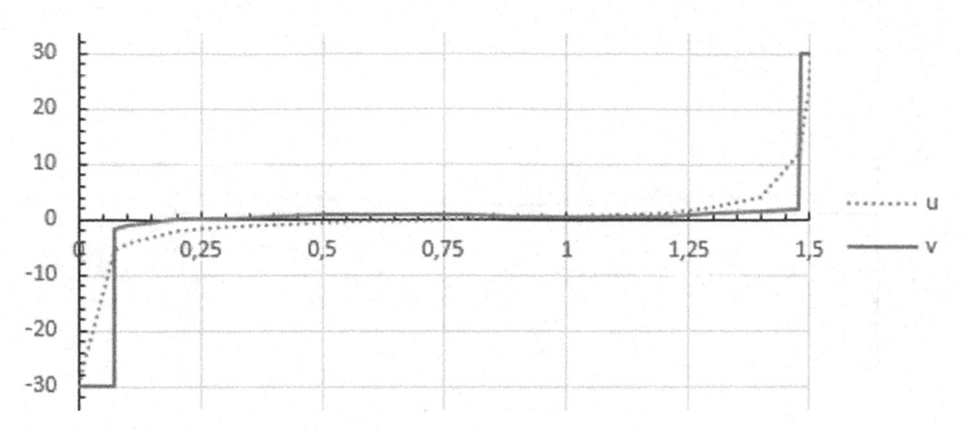

Fig. 1. u – starting control, v – calculated final control by MMNT.

operation at the maximum for calculating control values. In the proposed fixed point method, significantly simpler Cauchy problems with a pre-defined control function are solved at each iteration, which makes the considered fixed-point method significantly easier to implement than the global method.

Example 2. (method based on the algorithm (24)).

The problem from the previous example is considered to illustrate the operation of the fixed point method based on the implicit method of simple iteration (24).

The fixed point problem (14) for improving control $u(t)$, $t \in T$ takes the following form:

$$v(t) = P_U(u(t) + \alpha(p_1(t,u,v)x_3(t,v) - p_2(t,u,v)x_4(t,v) - p_3(t,u,v)x_1(t,v) + p_4(t,u,v)x_2(t,v))).$$

The implicit iterative method of simple iteration (24) for solving this fixed point problem at $k \geq 0$, respectively, takes the form:

$$v^{k+1}(t) = P_U(u(t) + \alpha(p_1(t,u,v^k)x_3(t,v^{k+1}) - p_2(t,u,v^k)x_4(t,v^{k+1}) - p_3(t,u,v^k)x_1(t,v^{k+1}) + p_4(t,u,v^k)x_2(t,v^{k+1}))).$$

The calculation of the optimal control problem was carried out from various initial starting controls.

Figure 2 shows the final computational control $v1(t)$, $t \in T$, obtained by the considered projection method of fixed points (PMNT1) with $\alpha = 10^{-2}$, and the number of iterations of improvement equal to 16, and the value of the functional $\Phi^* \approx 0.000641$.

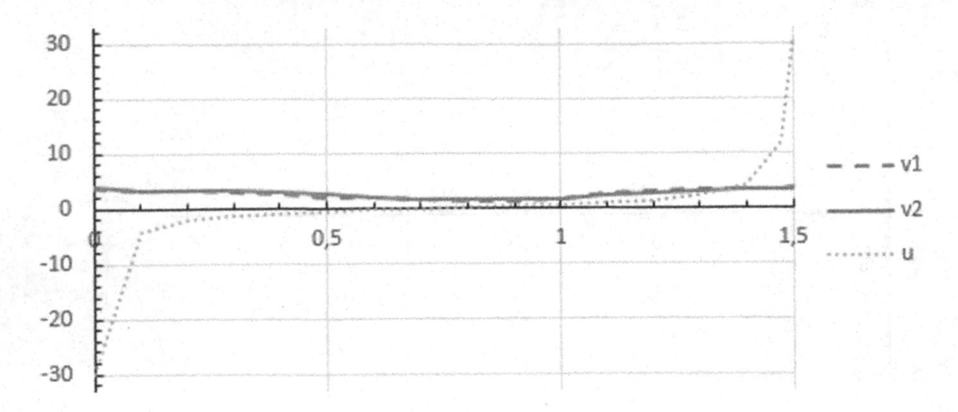

Fig. 2. u – starting control, $v1$ – control by PMNT1, $v2$ – control by PMNT2.

Figure 2 also shows the final computational control $v2(t)$, $t \in T$, obtained by the PMNT2 modification, which takes into account degenerate control intervals,

$$v^{k+1}(t) = \begin{cases} v^{k+1}(t) = P_U(u(t) + \alpha(p_1(t,u,v^k)x_3(t,v^{k+1}) \\ -p_2(t,u,v^k)x_4(t,v^{k+1}) - p_3(t,u,v^k)x_1(t,v^{k+1}) \\ +p_4(t,u,v^k)x_2(t,v^{k+1}))), |g(p(t,u,v^k),x(t,v^k))| > \varepsilon, \\ w, |g(p(t,u,v^k),x(t,v^k))| \leq \varepsilon, \end{cases}$$

where the value $w \in U$ is determined by the rule specified in the previous example, with the number of improvement iterations equal to 14, the functional value $\Phi^* \approx 0.000537$ and a degenerate control interval $[0.0868, 1.4430]$.

The calculations performed within the framework of the model problem show the high efficiency of the proposed projection methods of fixed points based on the implicit method of simple iteration (24) and their modifications to take into account special control sections characteristic of the class of quantum controlled systems under consideration. The main feature of the proposed method is the solution at each iteration of the Cauchy problem with a special right-hand side based on a uniquely defined and continuous projecting operation for calculating control values, in contrast to the global method [2–4]. In quantum systems with multidimensional control, the structure of the proposed fixed-point methods and the global method remains the same, but the computational advantage of implementing special Cauchy problems in the proposed fixed-point methods compared to the global method increases significantly.

5 Conclusion

In the considered model class of controlled quantum systems:

1. new constructive conditions for non-local improvement of controls are constructed in the form of fixed point problems in the control space;

2. new and strengthened forms of the maximum principle are obtained based on the constructed fixed point problems;
3. new iterative optimization methods based on the search for fixed points have been developed.

Emphasize the main computational features of the proposed methods of fixed points in the considered models of optimization of quantum systems.

1. Nonlocality of improvement of control and absence of time-consuming operation of convex or needle variation of control at each iteration, which is typical for gradient methods.
2. The possibility of strict improvement of extreme controls in contrast to gradient methods.
3. Numerical solution of Cauchy problems with a continuous and uniquely defined right-hand side at each iteration of the constructed explicit methods and implicit projection methods in contrast to the well-known global Krotov method.

These properties of the developed methods of fixed points are important factors for improving the efficiency of optimization of quantum-controlled systems.

References

1. Butkovsky, A., Samoylenko, Yu.: Control of Quantum Mechanical Processes. Nauka, Moscow (1984)
2. Morzhin, O., Pechen, A.: Krotov's method in problems of optimal control of closed quantum systems. Russ. Math. Surv. **5**(74), 83–144 (2019)
3. Krotov, V.: Control of quantum systems and some ideas of optimal control theory. Autom. Remote Control. **3**, 15–23 (2009)
4. Krotov, V., Bulatov, A., Baturina, O.: Optimization of linear systems with controlled coefficients. Autom. Remote Control. **6**, 64–78 (2011)
5. Gurman, V.: Turnpike solutions in optimal control problems for quantum mechanical systems. Autom. Remote Control. **6**, 115–126 (2011)
6. Srochko, V.: Iterative Methods for Solving Optimal Control Problems. Fizmatlit, Moscow (2000)
7. Vasiliev, O.: Optimization Methods. World Federation Publishers Company INC., Atlanta (1996)
8. Buldaev, A., Khishektueva, I.-K.: The fixed point method for the problems of nonlinear systems optimization on the managing functions and parameters. Bull. Irkutsk State Univ. Series Math. **19**, 89–104 (2017)
9. Buldaev, A.: Operator forms of the maximum principle and iterative algorithms in optimal control problems. In: Olenev, N., Evtushenko, Y., Khachay, M., Malkova, V. (eds.) OPTIMA 2020. CCIS, vol. 1340, pp. 101–112. Springer, Cham (2020). https://doi.org/10.1007/978-3-030-65739-0_8
10. Bartenev, O.: Fortran for Professionals. IMSL Mathematical Library. Dialog-MIFI, Moscow (2001)
11. Baturina, O., Morzhin, O.: Optimal control of the spin system based on the global improvement method. Autom. Remote Control. **6**, 79–86 (2011)

Stochastic Optimization for Dynamic Pricing

Dmitry Pasechnyuk[1(✉)], Pavel Dvurechensky[2], Sergey Omelchenko[1], and Alexander Gasnikov[1]

[1] Moscow Institute of Physics and Technology, Dolgoprudny, Russia
[2] Weierstrass Institute for Applied Analysis and Stochastics, Berlin, Germany

Abstract. We consider the problem of supply and demand balancing that is stated as a minimization problem for the total expected revenue function describing the behavior of both consumers and suppliers. In the considered market model we assume that consumers follow the discrete choice demand model, while suppliers are equipped with some quantity adjustment costs. The resulting optimization problem is smooth and convex making it amenable for application of efficient optimization algorithms with the aim of automatically setting prices for online marketplaces. We propose to use stochastic gradient methods to solve the above problem. We interpret the stochastic oracle as a response to the behavior of a random market participant, consumer or supplier. This allows us to interpret the considered algorithms and describe a suitable behavior of consumers and suppliers that leads to fast convergence to the equilibrium in a close to the real marketplace environment.

Keywords: Automatic pricing · Expected revenue function · Stochastic convex optimization · Supply and demand balancing

1 Introduction

With the development of platforms for online trading and services provision, the problem of dynamic pricing becomes more and more urgent. The environment of online marketplaces and financial services raises the question of establishing the most relevant prices for the items presented, and in such cases, relevance is understood as the possibility of market clearing, i.e. balancing supply and demand.

There are various approaches to finding equilibrium prices [1,3,8,9,18]. In this paper, we consider an approach based on the characterization of the market state by a potential function of total expected revenue, similar to the function of total excessive revenue from [17]. In [13], the authors, using this approach, analyze the case in which consumers follow a discrete choice model with imperfect behavior introduced by random noise in the assessment of utility, and suppliers seek to maximize their profit, taking into account the quantity adjustment costs.

The research is supported by the Ministry of Science and Higher Education of the Russian Federation (Goszadaniye) №075-00337-20-03, project No. 0714-2020-0005.

N. N. Olenev et al. (Eds.): OPTIMA 2021, CCIS 1514, pp. 82–94, 2021.
https://doi.org/10.1007/978-3-030-92711-0_6

This paper substantiates the convexity of the used potential as a function of prices. Hence, it follows that there are prices that minimize the potential, and it turns out that this minimum satisfies the condition of market clearing.

Thus, if there is an intermediary responsible for the formation of prices, it only needs to build a sequence of price values leading to a minimum. At the same time, it seeks to use the smallest possible amount of observations of market participants. Note that the introduced formulation is very convenient for applying the results of optimization theory to find the optimal prices. Indeed, the function under consideration is to be minimized and it simultaneously includes the characteristics of both consumers and suppliers, although in the classical game-theoretic approach the opposition of the interests of these parties leads through the Nash equilibrium scheme to minimax problems. Nevertheless, it turns out to be possible to propose exactly the potential function that describes the whole system and reduces its dynamic to only tendency to an extreme point, similar to physical systems. On the other hand, the considered function has the form of a sum including together the terms that characterize the behavior of both parties and each agent. This means that we can consider both consumers and suppliers as the same market participants and uniformly take into account their interests, without distinguishing between their types. Together, the described advantages lead to the ability to use the developments of convex optimization to find methods for constructing a sequence of prices, and analyze them in terms of the oracle complexity, which in this setting means the number of observations of the agents' behavior.

In this paper, continuing the ideas of [13], the problem of finding equilibrium prices is posed in a stochastic setting, and the number of observations of the intermediary for single market participants is determined as a measure of the effectiveness of the method for solving this problem. Section 2 describes the problem statement and introduces the stochastic oracle for considered function to optimize. Section 3 discusses various algorithms for stochastic convex optimization, proposes theoretical estimates of the efficiency of the algorithms, and describes the practical advantages of each of them. Section 4 generalizes the stochastic setting for the case of an infinite number of consumers, which makes it possible to take into account the previously unobserved consumers, and describes the algorithms for this setting. Section 5 considers a special case of the problem with zero quantity adjustment costs, in which the potential loses its smoothness property, and proposes several approaches to get around this difficulty. Section 6 shows the results of modeling of the behavior of the proposed algorithms for a synthetic problem and demonstrates the improvement in the efficiency of the methods, which can be achieved by considering the stochastic formulation of the problem, in comparison with the dynamics from [13].

2 Problem Statement

Let us imagine the marketplace environment. There are a number of suppliers (for example, shops) and a number of consumers (buyers). And every supplier offers some products (goods) in the assortment. Products are grouped by their type, every group contains some alternatives. In turn, every consumer can choose to buy one of the alternatives, guided by its subjective utility and price (with some element of randomness). For reasons of increasing profit, suppliers can change the assortment, taking into account the costs of these changes themselves. We use the corresponding mathematical models for the consumer and the supplier, and characterize them by the values of expected surplus and maximal revenue, respectively. Summing these terms up we obtain the function of total expected revenue. It describes the current imbalance in the market system. Its minimum corresponds to the prices at which the market is cleared. This means that we can formulate an optimization problem for finding equilibrium prices.

Let us consider such an optimization problem (in a simplified form) for the case of n product alternatives divided into m disjoint groups $G_i \subset \{1, ..., n\}$, S suppliers with convex costs functions $c_s : \mathbb{R}^n_+ \to \mathbb{R}_+$, closed and convex sets of capacity constraints $Y_s \subset \mathbb{R}^n_+$, typical supplies $\hat{y}_s \in Y_s$ and quantity adjustment costs equal to $-\Gamma_s \cdot \|y - \hat{y}_s\|_2^2$ for some parameter $\Gamma_s > 0$, D consumers with matrix $A = \{a_{id} > 0\}_{1,1}^{n,D}$ of alternatives subjective utility for each of them, and prices vector p:

$$\min_{p \in \mathbb{R}^n_+} \left\{ f(p) := \sum_{s=1}^{S} \pi_s(p) + \sum_{d=1}^{D} E_d(p) \right\}, \tag{1}$$

where

$$\pi_s(p) = \max_{y \in Y_s} \left\{ \langle y, p \rangle - c_s(y) - \Gamma_s \cdot \|y - \hat{y}_s\|_2^2 \right\} \tag{2}$$

is the maximal revenue of supplier with taking into account costs $c_s(\cdot)$ and quantity adjustment costs parametrized by Γ_s for given prices, and

$$E_d(p) = \ln \left(\sum_{j=1}^{m} \left(\sum_{i \in G_j} e^{(a_{id} - p_i)/\mu_j} \right)^{\mu_j} \right)$$

is the expected surplus $E_d(p) = \mathbb{E}_\epsilon \left[\max_i \{a_{id} - p_i + \epsilon_i\} \right]$ for the discrete choice demand model with noise and for corresponding nested logit distribution [11], with some correlation parameters $0 < \mu_j \leq 1$. We can also provide an explicit expression for the gradient of the introduced objective function:

$$\nabla f(p) = \sum_{s=1}^{S} y_s(p) - \sum_{d=1}^{D} x_d(p), \tag{3}$$

where $y_s(p)$ is the optimal solution of the optimization problem (2) (by Demyanov-Danskin theorem [4]), and

$$[x_d(p)]_i = \frac{e^{(a_{id}-p_i)/\mu_j}\left(\sum_{k\in G_j} e^{(a_{kd}-p_k)/\mu_j}\right)^{\mu_j-1}}{\sum_{h=1}^{m}\left(\sum_{k\in G_h} e^{(a_{kd}-p_k)/\mu_h}\right)^{\mu_h}}, \quad i \in \{1,...,n\}$$

are the probabilities to choose certain alternative by the consumer[1].

To use optimization methods for dynamic pricing we can utilize that f is of the form of sum. If the number of suppliers S and consumers D is big, evaluation of all the term in both sums is too expensive in term of working time and algorithmic complexity. But if we randomly pick up only one of $(S+D)$ terms at iteration, computing resources are used much more sparingly, which leads to faster operation of the algorithm. So, considering only one term of sum from (3), we introduce the stochastic gradient oracle:

$$\nabla f_i(p) = \begin{cases} y_s(p) & i = s \leq S \\ -x_d(p) & i = S+d \end{cases}, \quad i \in \{1,...,S+D\} \tag{4}$$

Note, that the use of stochastic oracle is very natural. Indeed, using the gradient dynamics of prices from [13], we evaluate all of the terms in (4) at iteration, i.e. it is necessary to consider the behaviour of all the consumers and suppliers to make one step of dynamic. But in practice we cannot guarantee that we collect all this information in short time (consumers may be impermanent, so the waiting time may be arbitrarily long). At the same time, the decisions of consumers and suppliers are not rigidly connected: we can observe some number of consumer's sales daily (represented by $x_d(\cdot)$), and much less often and independently the periodic store assortment changes (represented by $y_d(\cdot)$). Therefore, using the dynamics with stochastic oracle in the form of (4) we can immediately take into account newly observed behaviour and make iteration, that is now cheap both in computation and in the required downtime. However, in real-life environment we also cannot estimate the probability of customer's choice, represented by the vector $x_d(p)$ included in second case of (4). We can obtain only some single sales, those are in fact the random samples of the form $\mathcal{X}_d(p) = (0 \cdots 0\ 1\ 0 \cdots 0)$, where $[\mathcal{X}_d(p)]_i = 1$ w.p. $[x_d(p)]_i$, and $[\mathcal{X}_d(p)]_j = 0$ for $i \neq j$. Therefore, $\mathbb{E}[\mathcal{X}_d(p)] = x_d(p)$, and we can introduce another, more practical, stochastic gradient oracle for constructing our dynamics:

$$\widetilde{\nabla} f_i(p) = \begin{cases} y_s(p) & i = s \leq S \\ -\mathcal{X}_d(p) & i = S+d \end{cases}, \quad i \in \{1,...,S+D\}. \tag{5}$$

[1] Since $E_d(p) = \mathbb{E}_\epsilon\left[\max_i \{a_{id} - p_i + \epsilon_i\}\right] = \sum_i \mathbb{P}\left[a_{id} - p_i + \epsilon_i = \max_i \{a_{id} - p_i + \epsilon_i\}\right] \cdot \mathbb{E}_\epsilon\left[a_{id} - p_i + \epsilon_i\right]$, and therefore $\frac{\partial E_d(p)}{\partial p_i} = -\mathbb{P}\left[a_{id} - p_i + \epsilon_i = \max_i \{a_{id} - p_i + \epsilon_i\}\right]$.

Now, let us clarify properties of the objective function and introduced oracles. In our simplified setting the following result holds:

Lemma 1 *(Theorem 3.7 [13]). f has L-Lipschitz continuous gradient w.r.t.* $\|\cdot\|_2$ *with*

$$L = \sum_{s=1}^{S} \frac{1}{\Gamma_s} + \sum_{d=1}^{D} \frac{1}{\min_j \mu_j},$$

and each $\nabla f_i(p)$ for all $i \in \{1, ..., S+D\}$ is L_i-Lipschitz continuous w.r.t. $\|\cdot\|_2$ *with*

$$L_i = \begin{cases} \frac{1}{\Gamma_i} & i \leq S \\ \frac{1}{\min_j \mu_j} & i > S \end{cases}, \quad i \in \{1, ..., S+D\}$$

3 Algorithms and Theoretical Guarantees

3.1 Stochastic Gradient Descent

Let us consider the simplest dynamic of prices, based on the classical stochastic gradient descent method. A very natural interpretation for this dynamics is that in every iteration we can observe only one of the market participants, supplier or consumer. As soon as participant makes an economical decision (consumer chooses the product or supplier modifies the supply plan), we evaluate the stochastic oracle $\widetilde{\nabla} f_i(p_t)$, and make a step to the equilibrium prices. This dynamic is listed as Algorithm 1. We use the $[\cdot]_+$ notation for the positive part function, i.e. for $a = [b]_+$ we have $a_i = \max\{0, b_i\}$, and denote by $i \sim \mathcal{U}\{1, ..., S+D\}$ the i.i.d. random variables from discrete uniform distribution.

We analyse this dynamic as the projected stochastic gradient method with Polyak–Ruppert averaging with tunable parameter $C > 0$ to control the step size of the method. One practical advantage of this method is the robustness to the choice of C: it may be chosen regardless of theoretical value of L. However, the analysis additionally requires the condition of stochastic gradient's boundedness, but due to the smoothness of f we can bound the norm of (4), while the additional randomization in the second case of (5) acts on the standard simplex and therefore is also bounded.

Algorithm 1. SGD dynamic

Require: p_0 — starting prices values, N — number of iterations, C — parameter to control the step size

1: **for** $t = 0, 1, \ldots, N-1$ **do**
2: $\quad i \sim \mathcal{U}\{1, ..., S+D\}$
3: $\quad p_{t+1} = \left[p_t - \frac{C}{\sqrt{t+1}} \widetilde{\nabla} f_i(p_t)\right]_+$
4: **end for**
5: $\widetilde{p}_N = \frac{1}{N} \sum_{t=1}^{N} p_t$
6: **return** $\widetilde{p}_N$

Theorem 1 *(Theorem 7 [12]). Let us assume that used stochastic oracle is uniformly bounded: $\|\widetilde{\nabla} f_i(p_t)\|_2 \leq B$ for all $i \in \{1, ..., S+D\}$ and $t \in \{1, ..., N\}$. The suboptimality of prices $\widetilde{p}_N$ given by SGD dynamic (Algorithm 1) is decreasing as follows*

$$\mathbb{E}[f(\widetilde{p}_N)] - f(p_*) \leq \frac{\|p_0 - p_*\|_2^2 + CB^2(1 + C \ln N)}{2C\sqrt{N}}.$$

Moreover, to obtain the prices satisfying the suboptimality bound

$$\mathbb{E}[f(\widetilde{p}_N)] - f(p_*) \leq \varepsilon,$$

it is sufficient to call $\widetilde{\nabla} f_i$ oracle $\mathcal{O}\left(\frac{1}{\varepsilon^2}\right)$ times.

Therefore, considered SGD dynamic with Polyak–Ruppert averaging obtains convergence rate of $\mathcal{O}\left(N^{-1/2}\right)$, up to a logarithmic term. This also matches the result from [17].

3.2 Adaptive Stochastic Gradient Method

In this section we describe a slightly different AdaGrad [5] dynamics that has a different step size policy. More precisely, the stepsize is chosen based on the stochastic subgradients on the trajectory of the method. This allow the algorithm to adapt to the local information and possibly make longer steps.

Algorithm 2. AdaGrad dynamic

Require: p_0 — starting prices values, N — number of iterations, η — step size parameter, ϵ — small term to prevent zero division

1: $H_0 = 0$
2: **for** $t = 0, 1, \dots, N-1$ **do**
3: $\quad i \sim \mathcal{U}\{1, ..., S+D\}$
4: $\quad g_t = \widetilde{\nabla} f_i(p_t)$
5: $\quad H_{t+1} = H_t + \langle g_t, g_t \rangle$
6: $\quad p_{t+1} = \left[p_t - \frac{\eta}{\sqrt{H_{t+1} + \epsilon}} g_t\right]_+$
7: **end for**
8: $\widetilde{p}_N = \frac{1}{N} \sum_{t=1}^{N} p_t$
9: **return** $\widetilde{p}_N$

Theorem 2 *(Corollary 4.3.8 [6]). Let* $\|p_t - p_0\|_\infty \le R$ *for all* $t \in \{1, ..., N\}$, η *is proportional to* R. *The suboptimality of the prices* $\widetilde{p}_N$ *given by AdaGrad dynamic (Algorithm 2) is decreasing as follows*

$$\mathbb{E}[f(\widetilde{p}_N)] - f(p_*) \le \frac{3R}{2N} \cdot \sum_{i=1}^{n} \mathbb{E}\left[\left(\sum_{t=1}^{N} [g_t]_i^2\right)^{1/2}\right].$$

To obtain the prices satisfying the suboptimality bound

$$\mathbb{E}[f(\widetilde{p}_N)] - f(p_*) \le \varepsilon,$$

it is sufficient to call $\widetilde{\nabla} f_i$ *oracle* $\mathcal{O}\left(\frac{1}{\varepsilon^2}\right)$ *times.*

Hence the proposed AdaGrad dynamic demonstrates convergence rate of the order $\mathcal{O}\left(N^{-1/2}\right)$. This asymptotic is similar to that for SGD dynamic, but in practice such a simple modification of the step size allows to discernibly improve the convergence rate. At the same time, the step size hyperparameter η is still free and it allows to manually tune the algorithm for the best practical efficiency.

4 The Case of Infinite Number of Consumers

In general, the total expected revenue framework considered in [13] allows one to describe not only the individual consumers, but also the groups of consumers with close behavior. At the same time, their model covers only the setting of finite number of agents, which may be not completely practical. Indeed, if the number of agents is huge or new agents can enter the marketplace as time goes, it makes sense to consider the limit when the number of agents tends to infinity. This leads to a general, non-finite-sum objective given as an expectation.

To be more specific, we consider that consumers are represented by a random variable with some unknown distribution, i.e. $d \sim \mathcal{D}$. Then the characteristic vector $(a_{id})_{i=1}^{n}$ is also random. To maintain the property of market clearing at optimal prices we also assume that the number of suppliers is infinite and that they are represented by another random variable, i.e. $s \sim \mathcal{S}$. The next step is to take the limit in (1) and make transition to expectation w.r.t. distributions $\mathcal{D}, \mathcal{S}$ instead of sums. For the sake of normalization we introduce the parameter $0 < \beta < 1$ that is equal to the fraction of suppliers among all market participants. Informally, $\beta = \lim_{(S+D)\to\infty} S/(S+D)$.

In this way, considering function $\frac{1}{S+D}f(p)$ instead of $f(p)$ and taking the limit as $(S+D) \to \infty$, we have the new optimization problem in the form of expectation:

$$\min_{p\in\mathbb{R}_+^n} \left\{ \widetilde{f}(p) := \beta \cdot \mathbb{E}_{s\sim\mathcal{S}}[\pi_s(p)] + (1-\beta) \cdot \mathbb{E}_{d\sim\mathcal{D}}[E_d(p)] \right\}. \tag{6}$$

Note, that to preserve the convergence properties of the SGD dynamic considered in Sect. 3.1 it is sufficient just to generalize the used stochastic oracle (5) to the proposed setting by defining

$$\widetilde{\nabla}\widetilde{f}(p) := \begin{cases} y_s(p) & \text{for } s \sim \mathcal{S} \text{ w.p. } \beta \\ -\mathcal{X}_d(p) & \text{for } d \sim \mathcal{D} \text{ w.p. } 1-\beta \end{cases}. \tag{7}$$

Thus, to generate the stochastic gradient, we first with probability β choose to choose among suppliers or with probability $1-\beta$ we choose to choose among consumers. Then, in the former case we sample supplier from the distribution $\mathcal{S}$ and in the latter case we sample consumer from the distribution $\mathcal{D}$. Finally, for the chosen agent the stochastic gradient is defined as in (5).

Since problem (6) is a general stochastic optimization problem, we can apply Algorithm 1 and obtain the dynamic listed as Algorithm 3. Its interpretation is quite similar to that of Algorithm 1: at the every iteration we observe the behaviour of one market participant, consumer or supplier, and change the prices in a proper way. As in the SGD dynamic, samplings $d \sim \mathcal{D}$, $s \sim \mathcal{S}$ and switching w.p. β are provided by natural flow of participants, we assume that information about participants decisions arrives uniformly. Since Lemma 1 and the conditions of Theorem 1 still hold, the convergence rate of dynamic below is similar to the one given in Theorem 1.

Algorithm 3. SGD dynamic (online setting)

Require: β — fraction of suppliers, p_0 — starting prices values, N — number of iterations, C — parameter to control learning rate

1: **for** $t = 0, 1, \ldots, N-1$ **do**

2: $\quad p_{t+1} = \left[p_t - \frac{C}{\sqrt{t+1}}\widetilde{\nabla}\widetilde{f}(p_t)\right]_+$

3: **end for**

4: $\widetilde{p}_N = \frac{1}{N}\sum_{t=1}^{N} p_t$

5: **return** $\widetilde{p}_N$

5 The Case of Zero Quantity Adjustment Costs

In this section we return to the setting of Sect. 2 and consider the limiting case when in (2) $\Gamma_s = 0$. In this case there is no typical supply $\hat{y}_s$ and all the necessary information about the costs incurred by the supplier is set by the function $c_s(\cdot)$. In this case there is no guarantee that the function f in (1) is smooth (cf. Lemma 1) and the algorithms have to be properly modified to guarantee the convergence. This section is devoted to such modifications.

5.1 Mirror Descent Dynamic

Stochastic mirror descent (SMD) [7,14,15] is widely used and theoretically optimal algorithm for stochastic convex non-smooth optimization problems. We consider here a particular case of the SMD dynamic with Euclidean proximal setup, in which it looks quite similar to SGD dynamic, but uses a different step size policy. SMD dynamic is listed below as Algorithm 4.

Algorithm 4. SMD dynamic

Require: p_0 — starting prices values, N — number of iterations, C — parameter to control step size

1: **for** $t = 0, 1, \ldots, N-1$ **do**

2: $\quad i \sim \mathcal{U}\{1, ..., S+D\}$

3: $\quad p_{t+1} = \left[p_t - \dfrac{CR}{M\sqrt{t+1}}\widetilde{\nabla} f_i(p_t)\right]_+$

4: **end for**

5: $\widetilde{p}_N = \dfrac{1}{N}\sum_{t=1}^{N} p_t$

6: **return** $\widetilde{p}_N$

Theoretical guarantee below utilizes the Lipschitz continuity of function f, more precisely the boundedness of the stochastic gradient norm. Considering the expression (3), we can bound the first term by applying some economic reasoning, namely that the supply is always limited or scarcity principle [2]. The second term is also bounded since all x_d are probability vectors and belong to the standard simplex. Stochastic gradient given by (5) is bounded by the same reason.

Theorem 3 *(Proposition 1 [10]). Let* $\|p_t - p_0\|_2 \leq R$ *and* $\mathbb{E}[\|\widetilde{\nabla} f(p_t)\|^2] \leq M^2$ *for all* $t \in \{1, ..., N\}$. *The suboptimality of prices* $\widetilde{p}_N$ *given by SMD dynamic (Algorithm 4) is decreasing as follows*

$$\mathbb{E}[f(\widetilde{p}_N)] - f(p_*) \leq \frac{\max\{C, C^{-1}\}RM}{\sqrt{N}}.$$

Moreover, to obtain the prices satisfying the suboptimality bound

$$\mathbb{E}[f(\widetilde{p}_N)] - f(p_*) \le \varepsilon,$$

it is sufficient to call $\widetilde{\nabla} f_i$ *oracle* $\mathcal{O}\left(\frac{1}{\varepsilon^2}\right)$ *times.*

So, the described SMD dynamic has convergence rate of $\mathcal{O}\left(N^{-1/2}\right)$ in non-smooth case that takes place if $\Gamma_s = 0$ for some s. This bound matches that of the SGD dynamic from Theorem 1.

5.2 Dual Smoothing

The special structure of the function $\pi_s(p)$ in (2) allows us to use the Nesterov's smoothing technique [16]. The idea is, in the case when $\Gamma_s = 0$, to replace the $-\Gamma_s \cdot \|y - \hat{y}_s\|_2^2$ term in (2) with the synthetic penalty $-\eta \cdot \|y - y_0\|_2^2$, where $\eta = \varepsilon/(2R^2)$ for some fixed target suboptimality ε and R such that $\|y_* - y_0\|_2 \le R$. With this substitution for all s, following the argumentation of Lemma 1, we have that the modified function f_η has Lipschitz-continuous gradient with constant

$$L = \frac{2\,SR^2}{\varepsilon} + \sum_{d=1}^{D} \frac{1}{\min_j \mu_j}.$$

At the same time, if we minimize f_η up to accuracy $\frac{\varepsilon}{2}$, i.e. we have for some $\widetilde{p}$ that

$$f_\eta(\widetilde{p}) - \min_{p \in \mathbb{R}^n_+} f_\eta(p) \le \frac{\varepsilon}{2},$$

then it holds that

$$f(\widetilde{p}) - f(p_*) \le \varepsilon.$$

It means that we can obtain the solution of the problem (1) satisfying the target suboptimality bound by optimizing the modified function f_η that is smooth and therefore allows us to apply some of the methods described in previous sections.

Note that the transition from non-smooth setting to the smooth one with the described approach is not free in terms of convergence rate due to the dependence of the constant L on the target suboptimality ε.

6 Numerical Experiments

In this section, we focus on the problem (1), which is motivated by important applications to offering smart pricing options by online marketplaces and management of demand and supply by such financial intermediaries like brokers [13].

We provide the results of our numerical experiments, which are performed on a PC with processor Intel Core i7-8650U 1.9 GHz using pure Python 3.7.3 (without C code) under managing OS Windows 10 (64-bits). Numpy.float128 data type with precision $1e-18$ and with max element $\approx 1.19e+4932$ is used. Random seed is set to 17.

We compare Algorithm 1 and Algorithm 2 with pricing dynamics 4.2 and 4.4 from [13] on the problem (1) with the following settings: number of suppliers $S=5$, number of consumers $D=10$, number of products $n=20$, number of groups $m=5$. Γ_s is chosen like 10^{-4} (some small value that affects to convergence of pricing schemes from [13]). We generated also S vectors $\hat{y}_s$ from uniform distribution $\mathcal{U}[0.01,2]$ of size n. Initialization point p is chosen from uniform distribution $\mathcal{U}[0.01,5]$ of size n, but it is scaled by the maximum element. Matrix A consists of columns with each one is drawn from uniform distribution And μ values are drawn i.i.d. from uniform distribution $\mathcal{U}[0.1,1]$ of size m.

We consider $c_s(y)$ equal to $||y||_2^2$. Hence, the closed form solution of (2) is

$$y_s(p)=\frac{p+2\Gamma_s\hat{y}_s}{2(1+\Gamma_s)}.$$

To estimate the suboptimality we perform pricing dynamic 4.2 [13] (Gradient descent method) with stopping criterion $||p_{t+1}-p_t||_2\leq 10^{-10}$ and obtain lower bound for $f_*=f(p_*)$ from Theorem 4.3 [13].

During the experiments we store objective suboptimality and number of oracle calls for each algorithm at every iteration. The results are presented on the Fig. 1.

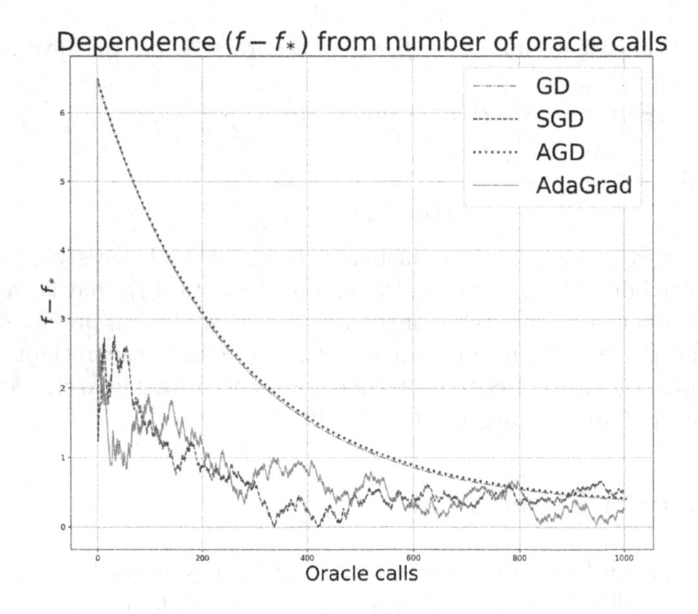

Fig. 1. Dependence of $f-f_*$ from the number of oracle calls.

As we can see from the Fig. 1, there is almost no difference between accelerated and non-accelerated pricing schemes from [13] and that the stochastic algorithms converge to the optimal value much faster than accelerated gradient method with respect to chosen parameters.

7 Conclusion

We propose a stochastic version of the formulation of the problem of finding equilibrium prices by minimizing the potential function of total expected revenue, proposed in a deterministic form in [13]. Thanks to the introduction of the stochastic oracle, we analyze and interpret in terms of observing the real-life marketplace environment several dynamics based on the efficient stochastic gradient optimization methods. We also analyze the case of a non-smooth potential function in the case of zero quantity adjustment costs. In addition, we propose a generalized setting allowing an infinite number of market participants and equally suitable application of optimization methods. Numerical experiments show that stochastic methods turn out to be discernibly more efficient in comparison with full-gradient dynamics.

References

1. Besbes, O., Zeevi, A.: Blind network revenue management. Oper. Res. **60**(6), 1537–1550 (2012)
2. Burke, E.: Thoughts and details on scarcity: originally presented to the Right Hon. William Pitt, in the month of November, 1795. No. 17937, T. Gillet (1800)
3. Chen, Y., Farias, V.F.: Robust dynamic pricing with strategic customers. Math. Oper. Res. **43**(4), 1119–1142 (2018)
4. Danskin, J.M.: The Theory of Max-Min and Its Application to Weapons Allocation Problems, vol. 5. Springer, Heidelberg (2012). https://doi.org/10.1007/978-3-642-46092-0
5. Duchi, J., Hazan, E., Singer, Y.: Adaptive subgradient methods for online learning and stochastic optimization. J. Mach. Learn. Res. **12**(7) (2011)
6. Duchi, J.C.: Introductory lectures on stochastic optimization. Math. Data **25**, 99 (2018)
7. Duchi, J.C., Agarwal, A., Johansson, M., Jordan, M.I.: Ergodic mirror descent. SIAM J. Optim. **22**(4), 1549–1578 (2012)
8. Ferreira, K.J., Simchi-Levi, D., Wang, H.: Online network revenue management using Thompson sampling. Oper. Res. **66**(6), 1586–1602 (2018)
9. Könönen, V.: Dynamic pricing based on asymmetric multiagent reinforcement learning. Int. J. Intell. Syst. **21**(1), 73–98 (2006)
10. Lan, G., Nemirovski, A., Shapiro, A.: Validation analysis of mirror descent stochastic approximation method. Math. Program. **134**(2), 425–458 (2012)
11. McFadden, D.: Modeling the choice of residential location. spatial interaction theory and planning models. In: Karlquist, A., Lundquist, L., Snickbars, F., Weibull, J.W. (eds.) North-Holland, Amsterdam 1(9), 7 (1978)
12. Moulines, E., Bach, F.: Non-asymptotic analysis of stochastic approximation algorithms for machine learning. Adv. Neural. Inf. Process. Syst. **24**, 451–459 (2011)
13. Müller, D., Nesterov, Y., Shikhman, V.: Dynamic pricing under nested logit demand. arXiv preprint arXiv:2101.04486 (2021)

14. Nemirovski, A., Juditsky, A., Lan, G., Shapiro, A.: Robust stochastic approximation approach to stochastic programming. SIAM J. Optim. **19**(4), 1574–1609 (2009). https://doi.org/10.1137/070704277
15. Nemirovskij, A.S., Yudin, D.B.: Problem Complexity and Method Efficiency in Optimization. Wiley-Interscience (1983)
16. Nesterov, Y.: Smooth minimization of non-smooth functions. Math. Program. **103**(1), 127–152 (2005)
17. Nesterov, Y., Shikhman, V.: Distributed price adjustment based on convex analysis. J. Optim. Theory Appl. **172**(2), 594–622 (2017)
18. Papanastasiou, Y., Savva, N.: Dynamic pricing in the presence of social learning and strategic consumers. Manage. Sci. **63**(4), 919–939 (2017)

Adaptive Gradient-Free Method for Stochastic Optimization

Kamil Safin[1(✉)], Pavel Dvurechensky[3,4], and Alexander Gasnikov[1,2,3]

[1] Moscow Institute of Physics and Technology, Dolgoprudny, Russia
kamil.safin.95@phystech.edu
[2] Caucasus Mathematical Center Adyghe State University, Maikop, Russia
[3] Institute for Information Transmission Problems RAS, Moscow, Russia
[4] Weierstrass Institute for Applied Analysis and Stochastics, Berlin, Germany

Abstract. In this paper we propose adaptive gradient-free coordinate-wise method for stochastic optimization. Adaptivity is based on line-search to simultaneously adjust to unknown Lipschitz constant and variance. In contrast to previous works, our method allows the guess for the Lipschitz constant to decrease to perform larger steps and benefit convergence. The method is practically applicable to adversarial attacks and reinforcement learning.

1 Introduction

Our problem is stochastic empirical risk optimization [30]. Specifically, we consider machine learning tasks where Empirical Risk Minimization (ERM) is used:

$$\min_{x \in \mathbb{R}^n} f(x) \overset{\text{def}}{=} \min_{x \in \mathbb{R}^n} \mathbb{E}_\xi f(x, \xi) = \min_{x \in \mathbb{R}^n} \frac{1}{m} \sum_{i=1}^{m} f(x, \xi_i), \tag{1}$$

where ξ is a random variable with a probability distribution $P(\xi)$. This is a common problem statement for a machine learning task, where $f(x, \xi)$ is supposed to be a loss value on a certain data point, which is represented by ξ and $f(x)$ is a total loss over a dataset which we want to minimize w.r.t. to model parameters x. In fact, we don't need the structure of finite sum in our objective but this setting allows easily interpret the random values ξ_i (which can be seen as i-th object in training dataset). All conclusions obtained in this paper can be easily applied to general setting of stochastic minimization.

We consider special case of so-called black-box tasks (zero-order minimization) where gradient is unavailable even for one term – we only have access to function values [2,4,9,27]. This type of optimization setting is important, for example, in reinforcement learning, where we only have access states of some

The work of P. Dvurechensky and A. Gasnikov was supported by the Russian Science Foundation (project 21-71-30005).

N. N. Olenev et al. (Eds.): OPTIMA 2021, CCIS 1514, pp. 95–108, 2021.
https://doi.org/10.1007/978-3-030-92711-0_7

environment [7,8,16,26], or in evolutionary algorithms in training neural networks [11]. Another popular field in recent applications is adversarial attacks, which test neural networks on generalizability [10,29].

There are plenty of methods for adaptive optimization [13,19,20,24]. But most of them are applicable to problems with at least first-order oracles (what means that we can get value of gradient of objective function). Only few adaptive methods are known for zero-order optimization [3,12,26].

Existing adaptive methods [15] adjust per-coordinate learning rate, but they can only increase Lipschitz constant during minimization, which lead to smaller steps of algorithm. In some cases, increase of step size during the minimization can lead to faster convergence [24,28].

Another way to deal with zero-order oracle is to somehow approximate true gradient. For example finite difference method [1] is used to build zero-order oracle version of Adaptive Momentum Method [5] or more general Exponential moving average methods [22].

In this paper we focus our attention on finite sum problem statement since it typical setting for machine learning tasks, where total loss is a sum of individual losses on each train sample. [21] combines variance reduction method with black-box optimization setting by two-point gradient estimation which leads to lower variance. Another way to estimate gradient is to use Gaussian smoothing [12] to rebuild the gradient surrogate.

In the following table we collected summarized information about this paper and related works.

PAPER	G-FREE[a]	F-SUM	PRM-ADAPT	BTCH-ADAPT	COORD	PROOF	IDEA
[21]	√	√	×	×	√	√	√
[6]	√	√	×	×	√	√	√
[14]	√	×	×	×	√	√	√
[24]	×	×	√	√	×	√	√
[3]	√	×	√	√	×	×	√
[18]	√	√	×	×	√	√	√
[5]	√	×	√	×	×	√	√
[22]	√	×	√	×	×	√	√
[17]	×	√	×	×	×	√	√
Our work	√	√	√	√	√	×	√

[a] G-Free stands for Gradient-Free method. F-Sum for Finite Sum problem. Prm-Adapt for Parameters Adaptive method. Btch-Adapt for Batch Adaptive selection method, Coord for Coordinate-wise method. Proof is for rigorous proof and Idea is for sketch of proof

In this paper we propose an algorithm similar to [24] but for gradient-free setting. We propose an adaptive gradient-free coordinate-wise descent for stochastic convex optimization. Our methods are flexible enough to use optimal choice of mini-batch size without additional information on the problem. Moreover, our procedure allows an increase of the step-size, which practically leads to faster convergence. The numerical experiments show that in gradient-free setting it is more profitable to use coordinate descent rather than restore full gradient by finite differences method.

1.1 Notation

In this article we consider unconstrained optimization problem over n-dimensional space E:

$$E = \mathbb{R}^n \tag{2}$$

We consider Euclidean space setup and by $\langle \cdot, \cdot \rangle$ and $\|\cdot\|$ we denote corresponding inner product and l_2-norm:

$$\begin{aligned} \langle x, y \rangle &\stackrel{\text{def}}{=} \sum_{i=1}^{n} x_i y_i \\ \|x\|^2 &\stackrel{\text{def}}{=} \langle x, x \rangle = \sum_{i=1}^{n} x_i^2 \end{aligned} \tag{3}$$

For a differentiable function $f(x)$ by $\nabla f(x)$ we denote its gradient, and by $\nabla_i f(x)$ – its directional gradient along a direction specified by vector e_i:

$$\nabla_i f(x) = \langle \nabla f(x), e_i \rangle e_i, \tag{4}$$

where e_i is the ith unit coordinate vector, i.e. is has the following form:

$$e_i = \overbrace{\underbrace{(0, \ldots, 0, 1}_{\text{i}}, 0, \ldots, 0)}^{\text{n}} \tag{5}$$

For our analysis we need to define weighted variants of norm and dot product, where ith vector component multiplied by corresponding Lipschitz constant:

$$\begin{aligned} \|x\|_L^2 &= \sum_{i=1}^{n} L_i x_i^2 \\ \langle x, y \rangle_L &= \sum_{i=1}^{n} L_i x_i y_i \end{aligned} \tag{6}$$

1.2 Assumptions and Definitions

We assume that $f(x, \xi)$ has coordinate-wise L_i-Lipschitz gradient $\nabla f(x, \xi)$, i.e. for a direction e_i:

$$\|\nabla_i f(x, \xi) - \nabla_i f(y, \xi)\| \leq L_i \|x - y\| \tag{7}$$

Since we don't have information about objective gradients we only can use substitutes. For a chosen direction specified by vector e_i (5) we can approximate directional derivative as:

$$\frac{f(x + te_i) - f(x)}{t} \simeq \langle \nabla f(x), e_i \rangle \tag{8}$$

Based on function observations over the selected batch of values $\{\xi_l\}_{l=1}^r$ we can form the approximation of its directional gradient as:

$$\widetilde{\nabla}_i^r f\left(x, \{\xi_l\}_{l=1}^r\right) \stackrel{\text{def}}{=} \frac{1}{r}\sum_{l=1}^{r} \frac{f(x+te_i,\xi_l) - f(x,\xi_l)}{t} e_i, \tag{9}$$

where r is the size of mini-batch. It is easy to check that (9) is an unbiased estimator for objective directional gradient:

$$\mathbb{E}_\xi \left[\widetilde{\nabla}_i^r f\left(x, \{\xi_l\}_{l=1}^r\right)\right] = \nabla_i f(x,\xi) \tag{10}$$

We assume that derivative-free approximation of directional gradient (9) well approximates the directional gradient and bounded from the above:

$$\left\|\nabla_i f(x,\xi) - \widetilde{\nabla}_i f(x,\xi)\right\|^2 \le \sigma^2 \tag{11}$$

By $f\left(x, \{\xi_l\}_{l=1}^r\right)$ we denote the value of $f(x)$ calculated over the batch with r randomly selected terms:

$$f\left(x, \{\xi_l\}_{l=1}^r\right) = \frac{1}{r}\sum_{l=1}^{r} f(x,\xi_l) \tag{12}$$

2 Adaptive Algorithm

In this section we solve the problem (1) for convex objective. We assume that all directional Lipschitz constants L_i are known. Based on this assumption we derive optimal batch size for estimating the directional derivative. In the following part of the article we will abandon this assumption and propose adaptive algorithm for unknown Lipschitz constant estimation.

2.1 Optimal Batch Size Selection

We start with coordinate descent with known Lipschitz constants:

$$x^{k+1} = x^k - \frac{1}{2L_i}\widetilde{\nabla}_{i_{k+1}}^r f\left(x^k, \{\xi_l\}_{l=1}^r\right), \tag{13}$$

where direction of current search i_k is selected over coordinate axes with probability p_i proportionally to directional Lipschitz constant:

$$p_i = \frac{L_i^\beta}{\sum\limits_{i=1}^{n} L_i^\beta} \tag{14}$$

Algorithm 1. Coordinate Gradient-Free Descent

Require: Number of iterations N, accuracy ε, σ^2, starting point x^0.
1: **for** $k = 0, \dots, N-1$ **do**
2: generate direction i_k
3: calculate $\widetilde{\nabla}^r_{i_{k+1}} f\left(x_k, \{\xi_l\}_{l=1}^r\right) = \frac{1}{r}\sum\limits_{i=1}^{r} \frac{c_{i_k}}{t}(f(x+te_{i_k},\xi_i) - f(x,\xi_i))$
4: $x^{k+1} = x^k - \frac{1}{2L_{i_{k+1}}}\widetilde{\nabla}^r_{i_{k+1}} f\left(x_k, \{\xi_l\}_{l=1}^r\right)$
5: **end for**

Since we want to use approximation $\widetilde{\nabla}^r_{i_{k+1}} f\left(x, \{\xi_l\}_{l=1}^r\right)$ of directional gradient $\nabla_{i_{k+1}} f\left(x\right)$ we need to bound the difference between them. For an arbitrary direction i and for $(x^{k+1} - x^k)$ having the same direction as e_i we can get the following:

$$\begin{aligned}
&\left\langle \nabla_{i_{k+1}} f\left(x^k\right) - \widetilde{\nabla}^r_{i_{k+1}} f\left(x^k, \{\xi_l\}_{l=1}^r\right),\ x^{k+1} - x^k \right\rangle \\
&\quad \le \frac{1}{2L_i}\left\| \nabla_{i_{k+1}} f\left(x^k\right) - \widetilde{\nabla}^r_{i_{k+1}} f\left(x^k, \{\xi_l\}_{l=1}^r\right)\right\|^2 + \frac{L_{i_{k+1}}}{2}\left\|x^{k+1} - x^k\right\|^2 \\
&\quad = \frac{1}{2L_{i_{k+1}}}\delta^{k+1} + \frac{L_{i_{k+1}}}{2}\left\|x^{k+1} - x^k\right\|^2, \qquad (15)
\end{aligned}$$

Where we denoted $\delta^{k+1} := \left\| \nabla_{i_{k+1}} f\left(x^k\right) - \widetilde{\nabla}^r_{i_{k+1}} f\left(x^k, \{\xi_l\}_{l=1}^r\right)\right\|^2$.

By standard reasoning we can show that for directional Lipschitz constant L_i from (7) it holds true:

$$\begin{aligned}
f\left(x^{k+1}, \{\xi_l\}_{l=1}^r\right) \le f\left(x^k, \{\xi_l\}_{l=1}^r\right) &+ \left\langle \nabla_{i_{k+1}} f\left(x^k, \{\xi_l\}_{l=1}^r\right),\ x^{k+1} - x^k \right\rangle \\
&+ \frac{L_{i_{k+1}}}{2}\left\|x^{k+1} - x^k\right\|^2
\end{aligned} \qquad (16)$$

Replacing in (16) directional gradient with its approximation and using bound (15) we get the following:

$$\begin{aligned}
f\left(x^{k+1}, \{\xi_l\}_{l=1}^r\right) \le f\left(x^k, \{\xi_l\}_{l=1}^r\right) &+ \left\langle \widetilde{\nabla}^r_{i_{k+1}} f\left(x^k, \{\xi_l\}_{l=1}^r\right),\ x^{k+1} - x^k \right\rangle \\
&+ L_{i_{k+1}}\left\|x^{k+1} - x^k\right\|^2 + \frac{1}{2L_i}\delta^{k+1}
\end{aligned} \qquad (17)$$

Using the fact from update rule: $x^{k+1} - x^k = -\frac{1}{2L_{i_{k+1}}}\widetilde{\nabla}^r_{i_{k+1}} f\left(x^k, \{\xi_l\}_{l=1}^r\right)$ from (17) we obtain:

$$\frac{1}{4L_{i_{k+1}}}\left\|\widetilde{\nabla}^r_{i_{k+1}} f\left(x^k, \{\xi_l\}_{l=1}^r\right)\right\|^2 \le f\left(x^k, \{\xi_l\}_{l=1}^r\right) - f\left(x^{k+1}, \{\xi_l\}_{l=1}^r\right) + \frac{1}{2L_{i_{k+1}}}\delta^{k+1} \qquad (18)$$

For the sequence $x^1, x^2, \ldots$, generated by algorithm (1) the following holds:

$$\left\|x^{k+1} - x\right\|_L^2 = \left\|x^k - x - \frac{1}{2L_{i_{k+1}}}\widetilde{\nabla}^r_{i_{k+1}} f\left(x^k, \{\xi_l\}_{l=1}^r\right)\right\|_L^2$$
$$= \left\|x^k - x\right\|_L^2 - \langle\widetilde{\nabla}^r_{i_{k+1}} f\left(x^k, \{\xi_l\}_{l=1}^r\right), x^k - x\rangle + \frac{1}{4L_{i_{k+1}}}\left\|\widetilde{\nabla}^r_{i_{k+1}} f\left(x^k, \{\xi_l\}_{l=1}^r\right)\right\|^2, \quad (19)$$

Where we used the definition of weighted norm and inner product (6).
From (18) and (19) we can derive inequality for inner product:

$$\begin{aligned}\langle\widetilde{\nabla}^r_{i_{k+1}} f\left(x^k, \{\xi_l\}_{l=1}^r\right), x^k - x\rangle &\leq \left[f\left(x^k, \{\xi_l\}_{l=1}^r\right) - f\left(x^{k+1}, \{\xi_l\}_{l=1}^r\right)\right] \\ &+ \frac{1}{2L_{i_{k+1}}}\delta^{k+1} + \left(\left\|x^k - x\right\|_L^2 - \left\|x^{k+1} - x\right\|_L^2\right)\end{aligned} \quad (20)$$

Taking the conditional expectation over the batch from left and right sides of (20):

$$\begin{aligned}\langle\nabla_{i_{k+1}} f\left(x^k\right), x^k - x\rangle \leq f(x^k) - f(x^{k+1}) &+ \frac{1}{2L_{i_{k+1}}}\delta^{k+1} \\ &+ \left\|x^k - x\right\|_L^2 - \left\|x^{k+1} - x\right\|_L^2\end{aligned} \quad (21)$$

Using convexity of $f(x)$ and definition of directional derivative (4):

$$f(x^k) - f(x) \leq \langle\nabla f(x^k), x^k - x\rangle \leq \sum_{i=1}^{n}\langle\nabla_{i_{k+1}} f\left(x^k\right), x^k - x\rangle \quad (22)$$

Taking expectation over i in (21) and substituting it in (22) we obtain:

$$\begin{aligned}f(x^k) - f(x) \leq nf(x^k) - n\mathbb{E}_i[f(x^{k+1})] &+ n\mathbb{E}_i\left[\frac{1}{2L_{i_{k+1}}}\delta^{k+1}\right] \\ + n\mathbb{E}_i[\left\|x^k - x\right\|_L^2] &- n\mathbb{E}_i[\left\|x^{k+1} - x\right\|_L^2]\end{aligned} \quad (23)$$

Summing above inequality from 0 to N we get:

$$f(\overline{x}^N) - Nf(x^*) \leq \frac{3n}{2}R^2 + \frac{n}{2\widetilde{L}}\mathbb{E}[\delta] \quad (24)$$

Here we introduced δ – an upper bound for a δ^{k+1}: $\delta^{k+1} \leq \delta$, R^2 is an upper bound for norm: $\left\|x^0 - x^*\right\|_L^2 \leq R^2$ and $\overline{x}^N = \frac{1}{N}\sum\limits_{k=1}^{N} x^k$. By $\widetilde{L}$ we denoted expectation of $\frac{1}{L_i}$: $\frac{1}{\widetilde{L}} = \frac{\sum L_i^{\beta-1}}{\sum L_i^{\beta}}$

We want to choose size of mini-batch r in respect with $\mathbb{E}[\delta] = \widetilde{L}\varepsilon$. Since $\mathbb{E}\left[\left\|\widetilde{\nabla}^r_{i_{k+1}} f\left(x, \{\xi_l\}_{l=1}^r\right) - \nabla f(x)\right\|^2\right] \leq \frac{\sigma^2}{r}$, we get the rule for batch size selection:

$$r = \max\left\{\frac{\sigma^2}{\varepsilon \widetilde{L}}, 1\right\} \tag{25}$$

2.2 Idea of Adaptivity for Lipschitz Constant

On each step of batch selection we need to know L_i and σ^2. In practice, true value of L_i is unknown. We provide an algorithm for adaptive choosing L_i on each step.

The idea of choosing L_i is to iteratively select the value of L_i until the following condition become true:

$$f(x^{k+1}) \leq f(x^k) + \langle \widetilde{\nabla}^r_{i_{k+1}} f\left(f, \{\xi_l\}_{l=1}^r\right) x^k,\ x^{k+1} - x^k \rangle + L_i^{k+1} \left\|x^{k+1} - x^k\right\|^2 + \frac{\varepsilon}{2} \tag{26}$$

So the adaptive procedure of Lipschitz constant estimation can be written as follows:

Algorithm 2. Adaptive L value selection

$L_i^{k+1} := \frac{L_i^k}{4}$
repeat
  $L_i^{k+1} := 2L_i^{k+1}$.
until

$$f(x^{k+1}) \leq f(x^k) + \langle \widetilde{\nabla}^r_{i_{k+1}} f\left(f, \{\xi_l\}_{l=1}^r\right) x_k,\ x^{k+1} - x^k \rangle + L_i^{k+1} \left\|x^{k+1} - x^k\right\|^2 + \frac{\varepsilon}{2}.$$

It is worth to mention that we select the value of L_i for chosen coordinate direction e_i. In this case we can better adjust L for each direction since function behavior may be different on different directions.

3 Algorithm

Now we can present the final version of algorithm for stochastic gradient-free optimization as follows:

Algorithm 3. Adaptive Stochastic Gradient-Free Descent

Require: Number of iterations N, accuracy ε, σ^2, starting point x^0.
1: **for** $k = 0, \dots, N-1$ **do**
2: generate $e_{k+1} \in E(1)$ independently
3: $L_i^{k+1} := \frac{L_i^k}{4}$
4: **repeat**
5: $L_i^{k+1} := 2L_i^{k+1}$.
6: $r_{k+1} = \max\left\{\frac{\sigma^2}{L_i^{k+1}\varepsilon}, 1\right\}$.
7: calculate $\widetilde{\nabla}^r_{i_{k+1}} f\left(x^k, \{\xi_l\}_{l=1}^r\right) = \frac{1}{r}\sum\limits_{l=1}^{r} \frac{e_{i_k}}{t}(f(x+te_{i_k}, \xi_i) - f(x, \xi_i))$
8: $x^{k+1} = x^k - \frac{1}{2L_i^{k+1}} \widetilde{\nabla}^r_{i_{k+1}} f\left(x_k, \{\xi_l\}_{l=1}^r\right)$.
9: **until**

$$f(x^{k+1}) \leq f(x^k) + \langle \widetilde{\nabla}^r_{i_{k+1}} f\left(x^k, \{\xi_l\}_{l=1}^r\right), x^{k+1} - x^k \rangle + L_i^{k+1}\|x^{k+1} - x^k\|^2 + \varepsilon/2. \tag{27}$$

10: **end for**

3.1 Accelerated Adaptive Algorithm

Using technique proposed in [25] we can propose accelerated version of our algorithm (4).

3.2 Coordinate-Wise Selection of L

We can illustrate the importance of coordinate-wise selection of constant L by example from [23]. Suppose that we optimize the following function:

$$\min_{x \in \mathbb{R}^n} f(x) = \min_{x \in \mathbb{R}^n} \frac{1}{2}\langle Ax,\ x\rangle - \langle b,\ x\rangle, \tag{30}$$

where A is positive definite matrix and $1 \leq A_{ij} \leq 2$, for $i, j = 1, \dots, n$. From this we can conclude that:

$$\begin{aligned} L(f(x)) &= \lambda_{\max}(A) \geq \lambda_{\max}(1_n 1_n^T) = n, \\ L_i &= \sup_x \langle Ae_i,\ e_i\rangle = \langle Ae_i,\ e_i\rangle = A_{ii} \end{aligned} \tag{31}$$

The complexity of gradient descent depends on number of iterations (N) and complexity of one iteration (I). For fast gradient descent (FGD) this decomposition is expressed as [23]:

$$C_{FGD} = N_{FGD} \cdot I_{FGD} = \mathcal{O}\left(\sqrt{\frac{L(f)}{\varepsilon}}R\right) \cdot I_{FGD}, \tag{32}$$

where $R = \|x_0 - x^*\|$.

Algorithm 4. Accelerated Adaptive Stochastic Gradient-Free Descent

Require: Number of iterations N, accuracy ε, σ^2, starting point x^0.
1: **for** $k = 0, \dots, N-1$ **do**
2: generate $e_{k+1} \in E(1)$ independently
3: $L_i^{k+1} := \frac{L_i^k}{4}$
4: **repeat**
5: $L_i^{k+1} := 2L_i^{k+1}$.
6: $\alpha^{k+1} = (1 + \sqrt{1 + 4A^k L^{K+1}})/(2L^{k+1})$
7: $A^{k+1} = A^k + \alpha^{k+1}$
8: $r_{k+1} = \max\left\{\frac{n\sigma^2}{L_i^{k+1}\varepsilon}, 1\right\}$.
9: calculate $\widetilde{\nabla}^r_{i_{k+1}} f\left(x^k, \{\xi_l\}_{l=1}^r\right) = \frac{1}{r}\sum\limits_{l=1}^{r} \frac{f(x+te,\xi_l)-f(x,\xi_l)}{t} e_i$
10: $y^{k+1} = (\alpha^{k+1}u^k + A^k x^k)/A^{k+1}$.
11: $u^{k+1} = u^k - \alpha^{k+1}\widetilde{\nabla}^r_{i_{k+1}} f\left(y^{k+1}, \{\xi_l\}_{l=1}^r\right)$.
12: $x^{k+1} = (\alpha^{k+1}u^{k+1} + A^k x^k)/A^{k+1}$.
13: **until**

$$f(x^{k+1}) \le f(y^{k+1}) + \langle \widetilde{\nabla}^r_{i_{k+1}} f\left(y^{k+1}, \{\xi_l\}_{l=1}^r\right), x^{k+1} - y^{k+1}\rangle \tag{28}$$

$$+ L_i^{k+1}\left\|x^{k+1} - y^{k+1}\right\|^2 + \frac{\alpha^{k+1}}{2A^{k+1}}\varepsilon. \tag{29}$$

14: **end for**

By using coordinate-wise randomization we slightly change this bound. In fact, by randomly selecting the direction of gradient descent on each step, we change the constant L in complexity bound. Now it has the following form:

$$\sqrt{\overline{L}} = \frac{1}{n}\sum_{i=1}^{n}\sqrt{L_i} = \frac{1}{n}\sum_{i=1}^{n}\sqrt{A_{ii}} \le \sqrt{2} \tag{33}$$

So the complexity for coordinate-wise gradient descent (CGD) is expressed as:

$$C_{CGD} = N_{CGD} \cdot I_{CGD} = \mathcal{O}\left(n\sqrt{\frac{\overline{L}(f)}{\varepsilon}}R\right) \cdot I_{CGD} = \mathcal{O}\left(\frac{nR}{\sqrt{\varepsilon}}\right) \cdot I_{CGD} \tag{34}$$

The complexity of one iteration for fast gradient descent is:

$$I_{FGD} = \mathcal{O}\left(n^2\right), \tag{35}$$

while in coordinate-wise gradient descent it is only:

$$I_{CGD} = \mathcal{O}\left(n\right). \tag{36}$$

Overall the complexity of two methods for this task are expressed as:

$$C_{CGD} = \mathcal{O}\left(\frac{n^2 R}{\sqrt{\varepsilon}}\right) \le \mathcal{O}\left(\frac{n^{5/2} R}{\sqrt{\varepsilon}}\right) = C_{FGD} \tag{37}$$

3.3 Complexity and Oracle Calls

We assumed that upper bound for σ is known: $\sigma_0 \geq \sigma$.

It is shown [24] that number of oracle calls T in proposed algorithm is approximated as:

$$T = O\left(\frac{\sigma_0 R^2}{\varepsilon^2}\right) \tag{38}$$

and number of iterations N is:

$$N = O\left(\frac{LR^2}{\varepsilon}\frac{\sigma}{\sigma_0}\right) \tag{39}$$

It is natural that in practice our bound σ_0 is not approximated well. We have two cases of bad approximation: overestimation and underestimation. Using (38) and (39) we can analyze the behavior of the method in these two cases.

In case of overestimation we use bigger value for upper bound: $\overline{\sigma} > \sigma_0$. As a result, number of oracle calls is increased:

$$\overline{T} = O\left(\frac{\overline{\sigma} R^2}{\varepsilon^2}\right) > O\left(\frac{\sigma_0 R^2}{\varepsilon^2}\right) \tag{40}$$

If we underestimated upper bound we use smaller value: $\underline{\sigma} < \sigma_0$. In this case, the total number of iterations is increased:

$$\underline{N} = O\left(\frac{LR^2}{\varepsilon}\frac{\sigma}{\underline{\sigma}}\right) > O\left(\frac{LR^2}{\varepsilon}\frac{\sigma}{\sigma_0}\right) \tag{41}$$

4 Experiments

We conducted an experiments using propose method. Since we considering gradient free method applied to machine learning task, we want to analyze the performance on low-dimensional task with large number of terms in sum. Because of this we used MSD[1] dataset. The purpose of this dataset is a prediction of the release year of a song from audio features. We define this task as a binary classification problem (we predict that realize year is earlier or later than 2001) and use only 12 average features to describe samples. Number of train samples, i.e. number of terms in sum is set to 463,715 and number of test samples is set to 51,630. As a model we select logistic regression. During the training the following loss function is minimized:

$$f(w) = \sum_{i=1}^{m} \left[-y_i \log(h_w(x_i)) + (1 - y_i)\log(1 - h_w(x_i))\right], \tag{42}$$

[1] https://archive.ics.uci.edu/ml/datasets/YearPredictionMSD.

where x_i denotes the feature vector for i-th sample, y_i – binary label for this sample and $h_w(x_i)$ is the prediction of the model with parameters w:

$$h_w(x) = \frac{1}{1 + e^{-wx}} \tag{43}$$

As the measure of model quality we compute accuracy among the predictions and true answers:

$$\text{Accuracy} = \frac{\text{Number of correct predictions}}{\text{Total number of predictions}}, \tag{44}$$

The code for experiments can be found here: https://github.com/kamil-safin/AGFCW.

We implemented three described algorithms: coordinate-wise gradient free algorithm with accelerated and non-accelerated versions and algorithm with full gradient restoring. The per-iteration progress of this algorithms is illustrated on Fig. (1). As expected, full gradient restoration version takes more iterations to achieve the same level of target function. The time progress of algorithms is shown on Fig. (2).

As expected, the full restoration algorithm takes more time and steps to achieve the same level of the loss function as proposed method does. Moreover it requires more computational steps to restore the full gradient unlike the proposed method.

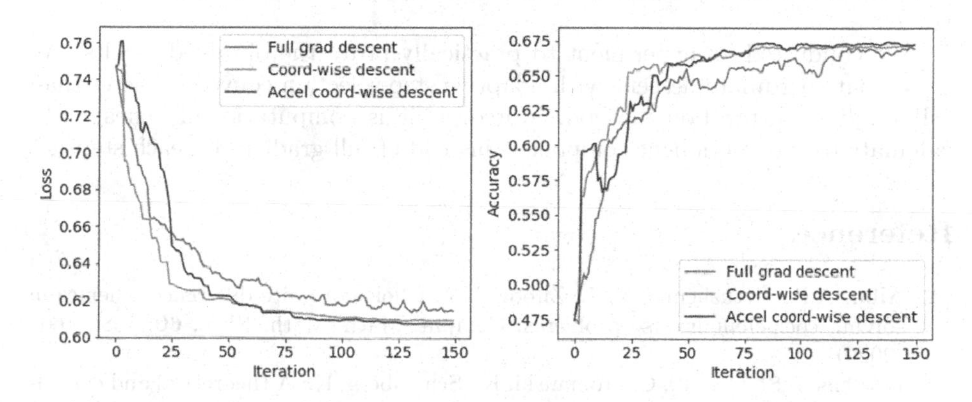

Fig. 1. Per-iteration progress of loss and accuracy

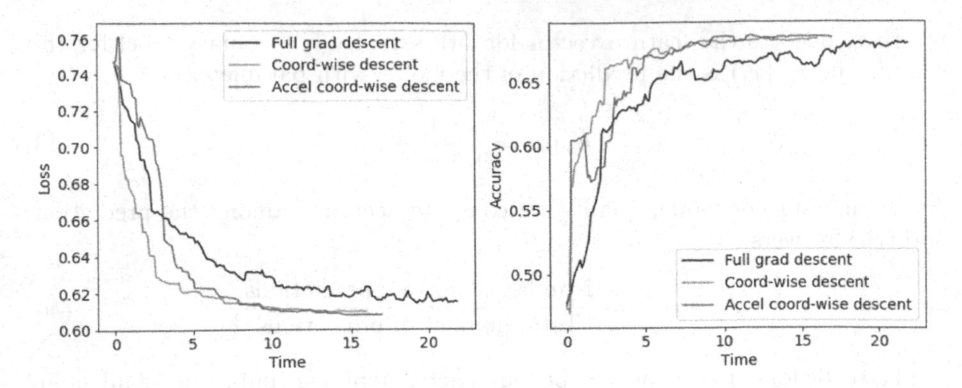

Fig. 2. Time progress of loss and accuracy

5 Conclusion

In this paper we proposed adaptive algorithm for zero-order stochastic optimization. We use coordinate descent with uniform selection of search direction. Adaptivity is based on adjusting Lipschitz constants for each direction independently. The iterative selection of Lipschitz constant is based on line-search procedure. Based on selected value of Lipschitz constant we calculate optimal batch size. In contrast to other works in this field, proposed method allows to decrease unknown Lipschitz constant which can potentially lead to faster convergence.

We conducted an experiment to practically prove the obtained results. We show that coordinate descent with adaptive step selection converges faster than full gradient restoration method. Moreover, it is computationally cheaper to calculate only one gradient component instead of full gradient on each step.

References

1. Albu, A.F., Evtushenko, Y.G., Zubov, V.I.: Choice of finite-difference schemes in solving coefficient inverse problems. Comput. Math. Math. Phys. **60**, 1589–1600 (2020)
2. Berahas, A.S., Cao, L., Choromanski, K., Scheinberg, K.: A theoretical and empirical comparison of gradient approximations in derivative-free optimization and Control (2019). arXiv: Optimization
3. Bollapragada, R., Wild, S.M.: Adaptive sampling quasi-Newton methods for derivative-free stochastic optimization. In: Beyond First Order Methods in Machine Learning (NeurIPS 2019 Workshop) (2019). sites.google.com/site/optneurips19/
4. Brent, R.: Algorithms for minimization without derivatives. Prentice-Hall (1973)
5. Chen, X., et al.: Zo-adamm: zeroth-order adaptive momentum method for black-box optimization. Adv. Neural Inf. Proc. Syst. **32** (2019)
6. Chen, Y., Orvieto, A., Lucchi, A.: An accelerated dfo algorithm for finite-sum convex functions. In: ICML (2020)

7. Choromanski, K., Iscen, A., Sindhwani, V., Tan, J., Coumans, E.: Optimizing simulations with noise-tolerant structured exploration. CoRR abs/1805.07831 (2018). arxiv.org/abs/1805.07831
8. Choromanski, K., Rowland, M., Sindhwani, V., Turner, R.E., Weller, A.: Structured evolution with compact architectures for scalable policy optimization. CoRR abs/1804.02395 (2018). arxiv.org/abs/1804.02395
9. Conn, A.R., Scheinberg, K., Vicente, L.N.: Introduction to derivative-free optimization. Soc. Ind. Appl. Math. USA (2009)
10. Croce, F., Rauber, J., Hein, M.: Scaling up the randomized gradient-free adversarial attack reveals overestimation of robustness using established attacks. CoRR abs/1903.11359 (2019). arxiv.org/abs/1903.11359
11. Cui, X., Zhang, W., Tüske, Z., Picheny, M.: Evolutionary stochastic gradient descent for optimization of deep neural networks. CoRR abs/1810.06773 (2018). arxiv.org/abs/1810.06773
12. Dereventsov, A., Webster, C., Daws, J.D.: An adaptive stochastic gradient-free approach for high-dimensional blackbox optimization. arXiv:2006.10887 (2020)
13. Duchi, J., Hazan, E., Singer, Y.: Adaptive subgradient methods for online learning and stochastic optimization. J. Mach. Learn. Res. **12**, 2121–2159 (2011)
14. Dvurechensky, P.E., Gasnikov, A.V., Tiurin, A.: Randomized similar triangles method: a unifying framework for accelerated randomized optimization methods (coordinate descent, directional search, derivative-free method). arxiv:1707.08486 (2017)
15. Ene, A., Nguyen, H.L., Vladu, A.: Adaptive gradient methods for constrained convex optimization. CoRR abs/2007.08840 (2020). arxiv.org/abs/2007.08840
16. Fazel, M., Ge, R., Kakade, S.M., Mesbahi, M.: Global convergence of policy gradient methods for linearized control problems. CoRR abs/1801.05039 (2018). arxiv.org/abs/1801.05039
17. Hanzely, F., Kovalev, D., Richtárik, P.: Variance reduced coordinate descent with acceleration: new method with a surprising application to finite-sum problems. arxiv:2002.04670 (2020)
18. Ji, K., Wang, Z., Zhou, Y., Liang, Y.: Improved zeroth-order variance reduced algorithms and analysis for nonconvex optimization. CoRR abs/1910.12166 (2019). arxiv.org/abs/1910.12166
19. Kingma, D.P., Ba, J.: Adam: a method for stochastic optimization. CoRR abs/1412.6980 (2015)
20. Leluc, R., Portier, F.: SGD with coordinate sampling: theory and practice (2021). arxiv:2105.11818
21. Liu, S., Kailkhura, B., Chen, P., Ting, P., Chang, S., Amini, L.: Zeroth-order stochastic variance reduction for nonconvex optimization. CoRR abs/1805.10367 (2018). arxiv.org/abs/1805.10367
22. Nazari, P., Tarzanagh, D.A., Michailidis, G.: Adaptive first-and zeroth-order methods for weakly convex stochastic optimization problems (2020)
23. Nesterov, Y., Stich, S.U.: Efficiency of the accelerated coordinate descent method on structured optimization problems. SIAM J. Optim. **27**, 110–123 (2017)
24. Ogaltsov, A., Dvinskikh, D., Dvurechensky, P.E., Gasnikov, A.V., Spokoiny, V.G.: Adaptive gradient descent for convex and non-convex stochastic optimization and control (2019). arXiv: Optimization
25. Ogal'tsov, A.V., Tyurin, A.I.: A heuristic adaptive fast gradient method in stochastic optimization problems. Comput. Math. Math. Phys. **60**(7), 1108–1115 (2020). https://doi.org/10.1134/s0965542520070088

26. Ruan, Y., Xiong, Y., Reddi, S.J., Kumar, S., Hsieh, C.: Learning to learn by zeroth-order oracle. CoRR abs/1910.09464 (2019). arxiv.org/abs/1910.09464
27. Spall, J.C.: Introduction to Stochastic search and optimization. (1st edn.) John Wiley and Sons Inc, USA (2003)
28. Strongin, R., Barkalov, K., Bevzuk, S.: Acceleration of global search by implementing dual estimates for Lipschitz constant, pp. 478–486 (2020). https://doi.org/10.1007/978-3-030-40616-5_46
29. Ye, H., Huang, Z., Fang, C., Li, C.J., Zhang, T.: Hessian-aware zeroth-order optimization for black-box adversarial attack. CoRR abs/1812.11377 (2018). arxiv.org/abs/1812.11377
30. Zhigljavsky, A., Zilinskas, A.: Stochastic global optimization, vol. 9 (2008). https://doi.org/10.1007/978-0-387-74740-8

Optimal Control

Synthesis of Power and Movement Control of Heating Sources of the Rod

K. R. Aida-zade(✉) and V.A. Hashimov

Institute of Control Systems of ANAS, B.Vahabzade 9, AZ1141 Baku, Azerbaijan
kamil_aydazade@rambler.ru
http://www.isi.az

Abstract. The article proposes an approach to solving the problem of synthesis of motion and power control of lumped sources with optimization of the locations of the points of the measurements. For specificity, the problem of feedback control of moving heat sources during rod heating is considered. The power and speed of point-wise sources are assigned depending on the state of the processes at the measurement points. The formulas for the gradient components of the objective functional, allowing for the numerical solution of the problem using of the first-order optimization methods are obtained.

Keywords: Rod heating · Feedback control · Moving sources · Temperature measurement points · Feedback parameters

1 Introduction

In this paper, we study the problem of synthesis of control of the rod heating process by lumped sources moving along the rod. The current values of the powers and speeds of movement of the sources are determined depending on the temperature at the points of measurement, the location of which is being optimized. The paper proposes to use the linear dependence of the control actions by the power and motion of the sources on the measured temperature values. Constant coefficients involved in these dependencies are the desired feedback parameters. Thus, the problem of synthesis of the control of moving heat sources is reduced to the problem of parametric optimal control.

Note that the problems of synthesis of control of objects described by both ordinary and partial differential equations are the most difficult both in the theory of optimal control and in the practice of their application [1–9].

For the problems of synthesis of control of objects with lumped parameters, there are certain, fairly general approaches to their solution, in particular, for linear systems [1,4–6,10]. There are no such approaches for objects with distributed parameters [2,3,5,7,8]. Firstly, this is due, to a wide variety of both mathematical models of such objects and possible variants of the corresponding formulations of control problems[2,3,5]. Secondly, the implementation of currently known methods for controlling objects with feedback in real time requires the use of expensive telemechanics, measuring and computing equipment [5,7,8].

N. N. Olenev et al. (Eds.): OPTIMA 2021, CCIS 1514, pp. 111–122, 2021.
https://doi.org/10.1007/978-3-030-92711-0_8

In general, despite the complexities of the mood of the feedback control systems, a large number of automatic control systems, automatic control of both objects with lumped and distributed parameters function [4,5,7,8] in practice.

It is clear that the further development of the use in practice of closed-loop control systems with feedback is essentially determined by the level of development of measuring and computing equipment, methods of computational mathematics, and control theory [6,10–14].

In this work, to determine the optimal values of the feedback parameters, the problem of parametric optimal control is solved using first-order numerical optimization methods. For this purpose, the necessary conditions for the optimal of the parameters are obtained, containing formulas for the components of the gradient of the objective functional for the optimized parameters. The formulas obtained make it possible to use effective first-order optimization methods for the numerical solution of the problem of determining the optimal values of the feedback parameters.

The proposed approach to the synthesis of control of moving sources can be easily extended to evolutionary processes described by other types of differential equations and initial-boundary conditions.

2 Formulation of the Problem

Consider the process of heating a rod by moving point-wise heat sources described by a second-order differential equation with partial derivatives of parabolic type [7]:

$$u_t(x,t) = a^2 u_{xx}(x,t) - \lambda_0[u(x,t) - \theta] + \sum_{i=1}^{N_c} q_i(t)\delta\left(x - z_i(t)\right), \tag{1}$$

$$x \in (0,l), \quad t \in (0,T],$$

with boundary conditions

$$\begin{aligned} u_x(0,t) &= \lambda_1(u(0,t) - \theta), \quad t \in (0,T], \\ u_x(l,t) &= -\lambda_2(u(l,t) - \theta), \quad t \in (0,T], \end{aligned} \tag{2}$$

Here $u(x,t)$ is the temperature of the rod at the point x at the moment of time t; l – rod length; T – heating process duration; a, λ_0, λ_1, λ_2 – given coefficients; $q_i(t)$ and $z_i(t)$ piece-wise continuous functions at t that determine the i-th source power and location on the rod; moreover

$$\underline{q_i} \le q_i(t) \le \overline{q_i}, \quad t \in [0,T], \quad i = 1,2,\ldots,N_c, \tag{3}$$

$$0 \le z_i(t) \le l, \quad t \in [0,T], \quad i = 1,2,\ldots,N_c. \tag{4}$$

θ – time constant temperature of the external environment, its exact the value at the time of the process is not specified, but the set is known possible values Θ and the distribution density function $\rho_\Theta(\theta)$ such that

$$\rho_\Theta(\theta) \geq 0, \quad \theta \in \Theta, \quad \int_\Theta \rho_\Theta(\theta) d\theta = 1.$$

$\underline{q_i}$, $\overline{q_i}$ – are given values; N_c – number of point sources; $\delta(\cdot)$ is a Dirac function such that for an arbitrarily continuous function $f(x)$ and points x, $\tilde{\xi} \in [0, l]$,

$$\int_0^l f(x)\delta(x - \tilde{\xi})dx = f(\tilde{\xi}).$$

It is assumed that the initial temperature of the rod at the initial moment of time $t = 0$ is not specified, but the set of its possible values is known, determined by parametrically specified functions, depending from s-dimensional vector of parameters.

$$u(x, 0) = \varphi(x; p), \quad x \in [0, l], \quad p \in P \subset \mathrm{R}^s. \tag{5}$$

Here P is a given set of values of the parameters of the initial function $\varphi(x; p)$, while the distribution density function $\rho_P(p) \geq 0$ of values is known such that

$$\rho_P(p) \geq 0, \quad p \in P, \quad \int_P \rho_P(p)dp = 1.$$

The trajectories of motion of point-wise sources $z_i(t)$ are controllable and determined by the equations

$$\ddot{z}_i(t) = a_i\dot{z}_i(t) + b_i z_i(t) + \vartheta_i(t), \quad t \in (0, T], \tag{6}$$

$$z_i(0) = z_i^0, \quad \dot{z}_i(0) = z_i^1, \quad i = 1, 2, \ldots, N_c. \tag{7}$$

Here a_i, b_i are the given parameters of the source motion; (z_i^0, z_i^1) – given initial positions of sources; $\vartheta_i(t)$ is a piece-wise continuous function that determines the motion control of the i-th source, satisfying the conditions:

$$\underline{\vartheta_i} \leq \vartheta_i(t) \leq \overline{\vartheta_i}, t \in [0, T], i = 1, 2, \ldots, N_c. \tag{8}$$

It is required to determine the functions controlling the process under consideration $q = q(t) = (q_1(t), q_2(t), \ldots, q_{N_c}(t))$, $\vartheta = \vartheta(t) = (\vartheta_1(t), \vartheta_2(t), \ldots, \vartheta_{N_c}(t))$, $w = w(t) = (q(t), \vartheta(t))$, minimizing the functional:

$$J(w) = \int_P \int_\Theta I(w; p, \theta)\rho_\Theta(\theta)\rho_P(p)d\theta dp, \tag{9}$$

$$I(w;p,\theta) = \int_0^l \mu(x)[u(x,T) - U(x)]^2 dx \quad (10)$$

$$+\varepsilon_1 \|q(t) - \hat{q}\|^2_{L_2^{N_c}[0,T]} + \varepsilon_2 \left\|\vartheta(t) - \hat{\vartheta}\right\|^2_{L_2^{N_c}[0,T]}.$$

Here $U(x)$ is a given piece-wise continuous function that determines the desired final temperature distribution on the rod; $\mu(x) \geq 0$, $x \in [0,l]$ – weight function; $u(x,t) = u(x,t;w,p,\theta)$ is a solution to the initial-boundary value problem (1), (2), (4) for admissible given control $w(t)$, parameters of the initial condition $\varphi(x;p)$ and ambient temperature θ.

It is easy to understand that the objective functional in the problem under consideration estimates the control functions $w(t) = (q(t), \vartheta(t))$ by the behavior of the heating process on average over all possible values of the parameters of the initial conditions $p \in P$ and ambient temperature $\theta \in \Theta$.

Let it be required to determine the current values of the control $w(t)$ from the results of continuous measurements of the temperature of the rod at N_o optimized observation points $\xi_j \in [0,l]$, $j = 1,2,\ldots,N_o$. The measured values are denoted by

$$u_j(t) = u(\xi_j, t), \quad t \in [0,T], \quad \xi_j \in [0,l], \quad j = 1,2,\ldots,N_o.$$

For the current values of the controls, we use the following linear dependencies on the measured temperature values

$$q_i(t) = \sum_{j=1}^{N_o} \alpha_i^j [u(\xi_j,t) - \tilde{\gamma}_i^j], \quad t \in [0,T], \quad i = 1,2,\ldots,N_c, \quad (11)$$

$$\vartheta_i(t) = \sum_{j=1}^{N_o} \beta_i^j [u(\xi_j,t) - \tilde{\gamma}_i^j], \quad t \in [0,T], \quad i = 1,2,\ldots,N_c. \quad (12)$$

Here α_i^j, β_i^j, $\tilde{\gamma}_i^j$, $i = 1,2,\ldots,N_c$, $j = 1,2,\ldots,N_o$ are synthesized constant feedback parameters. The parameter $\tilde{\gamma}_i^j$ characterizes the required value of the nominal temperature at the point $x = \xi^j$, which must be achieved by i-th point source. It is clear that this value must be close to the given desired value $U(\xi_j)$, $i = 1,2,\ldots,N_c$, $j = 1,2,\ldots,N_o$. The parameters α_i^j and β_i^j by analogy with synthesis problems for objects with lumped parameters will be called gain factors.

Let us write dependencies (11), (12) in the form

$$q_i(t) = \sum_{j=1}^{N_o} \alpha_i^j u(\xi_j,t) - \sum_{j=1}^{N_o} \alpha_i^j \tilde{\gamma}_i^j, \quad t \in [0,T], \quad i = 1,2,\ldots,N_c.$$

$$\vartheta_i(t) = \sum_{j=1}^{N_o} \beta_i^j u(\xi_j,t) - \sum_{j=1}^{N_o} \beta_i^j \tilde{\gamma}_i^j, \quad t \in [0,T], \quad i = 1,2,\ldots,N_c.$$

Introducing the notation

$$\gamma_i^q = \sum_{j=1}^{N_o} \alpha_i^j \tilde{\gamma}_i^j, \quad \gamma_i^\vartheta = \sum_{j=1}^{N_o} \beta_i^j \tilde{\gamma}_i^j,$$

we will get

$$q_i(t) = \sum_{j=1}^{N_o} \alpha_i^j u\left(\xi_j, t\right) - \gamma_i^q, \quad t \in [0, T], \quad i = 1, 2, \ldots, N_c, \tag{13}$$

$$\vartheta_i(t) = \sum_{j=1}^{N_o} \beta_i^j u\left(\xi_j, t\right) - \gamma_i^\vartheta, \quad t \in [0, T], \quad i = 1, 2, \ldots, N_c. \tag{14}$$

Combine the parameters α_i^j, β_i^j, γ_i^q, γ_i^ϑ, ξ_j into one $N = 2N_c(N_o + 1) + N_o$ – dimensional synthesized vector of feedback parameters and the coordinates of the measurement points: $y = \left(\alpha_i^j, \beta_i^j, \gamma_i^q, \gamma_i^\vartheta, \xi_j\right)$, $i = 1, 2, \ldots, N_c$, $i = 1, 2, \ldots, N_o$.

$$J(y) = \int_P \int_\Theta I(y; p, \theta)\rho_\Theta(\theta)\rho_P(p)d\theta dp, \tag{15}$$

$$I(y; p, \theta) = \int_0^l \mu(x)[u(x, T) - U(x)]^2 dx + \varepsilon \left\|y - \hat{y}\right\|_{R^N}^2. \tag{16}$$

Substituting dependencies (13), (14) into Eq. (1), (6), we obtain

$$u_t(x, t) = a^2 u_{xx}(x, t) - \lambda_0[u(x, t) - \theta] \tag{17}$$

$$+ \sum_{i=1}^{N_c} \delta\left(x - z_i(t)\right) \left(\sum_{j=1}^{N_o} \alpha_i^j u\left(\xi_j, t\right) - \gamma_i^q\right), \quad x \in (0, l), \quad t \in (0, T],$$

$$\ddot{z}_i(t) = a^i \dot{z}_i(t) + b_i z_i(t) + \sum_{j=1}^{N_o} \beta_i^j u\left(\xi_j, t\right) - \gamma_i^\vartheta, \quad t \in (0, T]. \tag{18}$$

The specificity of Eq. (17) lies, firstly, in the fact that it is point-loaded with respect to the spatial variable, which was investigated in such works as [15–18]. Secondly, Eq. (17), (18) with respect to the time variable must be solved simultaneously.

Consider constraints (3), (8) on controls $q = q(t) = (q_1(t), q_2(t), \ldots, q_{N_c}(t))$, $\vartheta = \vartheta(t) = (\vartheta_1(t), \vartheta_2(t), \ldots, \vartheta_{N_c}(t))$. It is clear that from technological considerations, the range of possible temperatures at the points of the rod during its heating can be considered known:

$$\underline{u} \le u(x, t) \le \overline{u}, \quad x \in [0, l], \quad t \in [0, T]. \tag{19}$$

Taking into account the linearity of dependencies (13), (14) with respect to $u(\xi_j, t)$ and condition (19), each of the constraints (3), (8) can be written in the form of two linear constraints with respect to the parameters of the inverse communication:

$$\underline{q_i} \le \overline{u} \sum_{j=1}^{N_o} \alpha_i^j - \gamma_i^q \le \overline{q_i}, \quad i = 1, 2, \ldots, N_c,$$

$$\underline{q_i} \le \underline{u} \sum_{j=1}^{N_o} \alpha_i^j - \gamma_i^q \le \overline{q_i}, \quad i = 1, 2, \ldots, N_c, \tag{20}$$

$$\underline{\vartheta_i} \le \overline{u} \sum_{j=1}^{N_o} \beta_i^j - \gamma_i^\vartheta \le \overline{\vartheta_i}, \quad i = 1, 2, \ldots, N_c,$$

$$\underline{\vartheta_i} \le \underline{u} \sum_{j=1}^{N_o} \beta_i^j - \gamma_i^\vartheta \le \overline{\vartheta_i}, \quad i = 1, 2, \ldots, N_c,$$

$$0 \le \xi_j \le l, \quad j = 1, 2, \ldots, N_o.$$

Above added restrictions on the location of the measurement points.

Thus, the source considered control problem for moving point-wise sources (1)–(10) with feedback (13), (14) is reduced to the parametric optimal control problem (15), (16), (2), (4) [9,11].

We note the following specific features of the investigated parametric optimal control problem.

First, the original control problem for moving sources (1)–(10) is generally not convex. This is due to the third term in Eq. (1), which contains the product of the optimized functions: ($q_i(t)$ and $\delta(x - z_i(t; \vartheta_i(t))$). The resulting problem of parametric optimal control is not convex in terms of the feedback parameters, as can be seen from the differential Eq. (17) and dependencies (13), (14).

Second, the system of partial differential Eq. (17) and ordinary derivatives (18) is specific. On the right-hand sides, they contain the values of the unknown phase state at the observation points.

Third, the problem is specific because of the objective functional (9), (10), which estimates the behavior of a bundle of phase trajectories with initial conditions from a parametrically given set.

In general, the obtained problem can also be classified as a class of finite-dimensional optimization problems with respect to the vector $y \in \mathrm{R}^N$. In this problem, to calculate the objective functional at any point, it is required to solve initial-boundary value problems for differential equations with partial and ordinary derivatives.

3 Approach to Determining Feedback

To minimize the functional (15), (16), taking into account the linearity of constraints (20), we use the gradient projection method [11]:

$$y^{n+1} = \mathcal{P}_{(20)}\left[y^n - \alpha_n \mathbf{grad} J\left(y^n\right)\right], \tag{21}$$
$$\alpha_n = \arg\min_{\alpha\geq 0} J\left(y^n - \alpha \mathbf{grad} J\left(y^n\right)\right), \quad n = 0, 1, \ldots$$

Here α_n is the one-dimensional minimization step, y^0 is an arbitrary starting point of the search from R^N; $\mathcal{P}_{(20)}[\cdot]$ – is the operator of projecting an arbitrary point $y \in \mathrm{R}^N$ onto the admissible domain defined by constraints (20). Taking into account the linearity of constraints (20), the operator $\mathcal{P}_{(20)}[\cdot]$ is easy to construct constructively [11]. It is known that iterative procedure (21) allows one to find only the local minimum of the objective functional closest to the point y^0. Therefore, for procedure (21), it is proposed to use the multistart method from different starting points. From the obtained local minimum points, the best functional is selected.

In the implementation of procedure (21), analytical formulas for the components of the gradient of the objective functional play an important role. Therefore, below we will prove the differentiability of the functional with respect to the optimized parameters and obtain formulas for its gradient, which make it possible to formulate the necessary optimal conditions for the synthesized feedback parameters y.

Theorem 1. *Under conditions on the functions and parameters involved in problem (15)–(18), (2), (5), (7), the gradient of functional (15), (16) with respect to the feedback parameters is differentiable, and its components are determined by the formulas:*

$$\frac{\partial J(y)}{\partial \alpha_i^j} = \int_P \int_\Theta \left\{ -\int_0^T \psi(z_i(t), t) u\left(\xi_j, t\right) dt + 2(\alpha_i^j - \hat{\alpha}_i^j) \right\} \rho_\Theta(\theta) \rho_P(p) d\theta dp,$$

$$\frac{\partial J(y)}{\partial \beta_i^j} = \int_P \int_\Theta \left\{ -\int_0^T \varphi_i(t) u\left(\xi_j, t\right) dt + 2(\beta_i^j - \hat{\beta}_i^j) \right\} \rho_\Theta(\theta) \rho_P(p) d\theta dp,$$

$$\frac{\partial J(y)}{\partial \xi_j} = \int_P \int_\Theta \left\{ -\sum_{i=1}^{N_c} \int_0^T \left(\alpha_i^j \psi\left(z^i(t), t\right) + \beta_i^j \varphi^i(t)\right) u_x(\xi^j, t) dt \right. \tag{22}$$
$$\left. +2\left(\xi_j - \hat{\xi}_j\right) \right\} \rho_\Theta(\theta) \rho_P(p) d\theta dp,$$

$$\frac{\partial J(y)}{\partial \gamma_i^q} = \int\limits_P \int\limits_\Theta \left\{ \int\limits_0^T \psi(z_i(t), t) dt + 2\varepsilon(\gamma_i^q - \hat{\gamma}_i^q) \right\} \rho_\Theta(\theta) \rho_P(p) d\theta dp,$$

$$\frac{\partial J(y)}{\partial \gamma_i^\vartheta} = \int\limits_P \int\limits_\Theta \left\{ \int\limits_0^T \varphi_i(t) dt + 2\varepsilon(\gamma_i^\vartheta - \hat{\gamma}_i^\vartheta) \right\} \rho_\Theta(\theta) \rho_P(p) d\theta dp.$$

$i = 1, 2, \ldots, N_c$, $j = 1, 2, \ldots, N_o$. Functions $\psi(x,t)$ and $\varphi_i(t)$, $i = 1, 2, \ldots, N_c$ for every given parameters $\theta \in \Theta$ and $p \in P$ are solutions of the following conjugate initial-boundary value problems:

$$\psi_t(x,t) = -a^2 \psi_{xx}(x,t) + \lambda_0 \psi(x,t)$$

$$-\sum_{i=1}^{N_c} \sum_{j=1}^{N_o} \left(\alpha_i^j \psi(z_i(t), t) + \beta_i^j \varphi_i(t) \right) \delta(x - \xi_j), \quad x \in \Omega, \quad t \in [0, T),$$

$$\psi(x, T) = -2\mu(x) \left(u(x, T) - U(x) \right), \quad x \in \Omega, \tag{23}$$

$$\psi_x(0, t) = \lambda_1 \psi(0, t), \quad t \in [0, T),$$

$$\psi_x(l, t) = -\lambda_2 \psi(l, t), \quad t \in [0, T),$$

$$\ddot{\varphi}_i(t) = -a_i \dot{\varphi}_i(t) + b_i \varphi_i(t) + \psi_x(z_i(t), t) \sum_{j=1}^{N_o} \alpha_i^j u(\xi_j, t) - \gamma_i^q, \quad t \in [0, T),$$

$$\dot{\varphi}_i(T) = -a_i \varphi_i(T), \quad \varphi_i(T) = 0, \quad i = 1, 2, \ldots, N_c.$$

Proof. To prove the differentiable of the functional $J(y)$ with respect to y, we use the increment method.

¿From the obvious interdependence of all parameters of the initial conditions $p \in P$ and the temperature of the environment $\theta \in \Theta$, the validity of the formula follows:

$$\mathbf{grad} J(y) = \mathbf{grad} \int\limits_P \int\limits_\Theta I(y; p, \theta) \rho_\Theta(\theta) \rho_P(p) d\theta dp \tag{24}$$

$$= \int\limits_P \int\limits_\Theta \mathbf{grad} I(y; p, \theta) \rho_\Theta(\theta) \rho_P(p) d\theta dp.$$

Therefore, it is sufficient to obtain formulas for $\mathbf{grad} I(y; p, \theta)$ for arbitrarily given admissible values of $p \in P$ and $\theta \in \Theta$.

Let us introduce a notation for the third term on the right-hand side of (17), which depends on all optimized parameters of the feedback y:

$$W(t; y) = \sum_{i=1}^{N_c} \delta(x - z_i(t)) \left(\sum_{j=1}^{N_o} \alpha_i^j u(\xi_j, t) - \gamma_i^q \right), \quad t \in [0, T].$$

Let $u(x,t) = u(x,t; y, p, \theta)$, $z(t) = z(t; y, p, \theta)$ are solutions, respectively, initial boundary-value problem (17), (2), (5) and Cauchy problem (18), (7) for given values of the parameters p and θ. Let the feedback parameters y be incremented Δy: $\tilde{y} = y + \Delta y$. It is clear that the corresponding solutions of problems (17), (2), (5) and (18), (7) also receive increments, which we denote as follows:

$$\tilde{u}\,(x,t;\tilde{y},p,\theta) = u(x,t;y,p,\theta) + \Delta u(x,t;y,p,\theta),$$

$$\tilde{z}\,(t;\tilde{y},p,\theta) = z(t;y,p,\theta) + \Delta z(t;y,p,\theta).$$

The increments $\Delta u(x,t;y,p,\theta)$ and $\Delta z(t;y,p,\theta)$ are solutions of the following initial-boundary-value problems:

$$\Delta u_t(x,t) = a^2 \Delta u_{xx}(x,t) - \lambda_0 \Delta u(x,t) + \Delta W(t;y), \quad x \in (0,l), \quad t \in (0,T], \tag{25}$$

$$\Delta u(x,0) = 0, \quad x \in [0,l], \tag{26}$$

$$\Delta u_x(0,t) = \lambda_1 \Delta u(0,t), \quad t \in (0,T], \tag{27}$$

$$\Delta u_x(l,t) = -\lambda_2 \Delta u(l,t), \quad t \in (0,T].$$

$$\Delta \ddot{z}_i(t) = a_i \Delta \dot{z}_i(t) + b_i \Delta z_i(t) + \Delta \vartheta_i(t), \quad t \in (0,T], \tag{28}$$

$$\Delta z_i(0) = 0, \quad \Delta \dot{z}_i(0) = 0, \quad i = 1,2,\ldots,N_c. \tag{29}$$

The functional $\Delta I(y;p,\theta)$ will be incremented

$$\Delta I(y) = I(y + \Delta y;p,\theta) - I(y;p,\theta) \tag{30}$$

$$= 2\int_0^l \mu(x)\,(u(x,T) - U(x))\,\Delta u(x,T)dx + 2\varepsilon\,\langle y - \widehat{y}, \Delta y\rangle\,.$$

Move the right-hand sides of differential Eq. (25) and (28) to the left, multiply both sides of the obtained qualities by so far arbitrary functions $\psi(x,t)$ and $\varphi_i(t)$, respectively. We integrate over $t \in (0,T)$ and $x \in (0,l)$. The resulting left-hand sides equal to zero are added to (30). Will have:

$$\Delta I(y) = 2\int_0^l \mu(x)\,(u(x,T) - U(x))\,\Delta u(x,T)dx + 2\varepsilon_1\,\langle y - \hat{y}, \Delta y\rangle +$$

$$\int_0^T\!\!\int_0^l \psi(x,t)\,\big(\Delta u_t(x,t) - a^2 \Delta u_{xx}(x,t) + \lambda_0 \Delta u(x,t) - \Delta W(t;y)\big)dxdt$$

$$+\sum_{i=1}^{N_c}\int_0^T \varphi_i(t)\,(\Delta \ddot{z}_i(t) - a_i \Delta \dot{z}_i(t) - b_i \Delta z_i(t) - \Delta \vartheta_i(t))\,dt.$$

Integrating by parts, grouping and taking into account conditions (26), (27), (29), we obtain the following expression for the increment of the functional:

$$\Delta I(y) = 2\int_0^l \mu(x)\,(u(x,T) - U(x))\,\Delta u(x,T)dx + \int_0^l \psi(x,T)\Delta u(x,T)dx$$

$$+\int_0^T\int_0^l \left(-\psi_t(x,t) - a^2\psi_{xx}(x,t) + \lambda_0\psi(x,t)\right) \Delta u(x,t)dxdt$$

$$-\sum_{i=1}^{N_c}\sum_{j=1}^{N_o}\int_0^T\int_0^l \left\{\alpha_i^j\psi\left(z_i(t),t\right) + \beta_i^j\varphi_i(t)\right\}\delta\left(x-\xi_j\right)\Delta u(x,t)dxdt$$

$$+a^2\int_0^T \left(\psi_x(l,t) + \lambda_2\psi(l,t)\right)\Delta u(l,t)dt - a^2\int_0^T \left(\psi_x(0,t) - \lambda_1\psi(0,t)\right)\Delta u(0,t)dt$$

$$+\sum_{i=1}^{N_c}\sum_{j=1}^{N_o}\Delta\alpha_i^j\left\{-\int_0^T \psi\left(z_i(t),t\right)u\left(\xi_j,t\right)dt + 2\varepsilon\left(\alpha_i^j - \hat{\alpha}_i^j\right)\right\}$$

$$+\sum_{i=1}^{N_c}\sum_{j=1}^{N_o}\Delta\beta_i^j\left\{-\int_0^T \varphi_i(t)u\left(\xi_j,t\right)dt + 2\varepsilon\left(\beta_i^j - \hat{\beta}_i^j\right)\right\}$$

$$+\sum_{i=1}^{N_c}\Delta\gamma_i^q\left\{\int_0^T \psi\left(z_i(t),t\right)dt + 2\varepsilon\left(\gamma_i^q - \hat{\gamma}_i^q\right)\right\}$$

$$+\sum_{i=1}^{N_c}\Delta\gamma_i^\vartheta\left\{\int_0^T \varphi_i(t)dt + 2\varepsilon\left(\gamma_i^\vartheta - \hat{\gamma}_i^\vartheta\right)\right\}$$

$$+\sum_{j=1}^{N_o}\Delta\xi_j\left\{-\sum_{i=1}^{N_c}\int_0^T\left(\alpha_i^j\psi\left(z_i(t),t\right) + \beta_i^j\varphi_i(t)\right)u_x(\xi_j,t)dt + 2\left(\xi_j - \hat{\xi}_j\right)\right\}$$

$$-\sum_{i=1}^{N_c}\left(\dot{\varphi}_i(T) + \varphi_i(T)a_i\right)\Delta z_i(T) + \sum_{i=1}^{N_c}\varphi_i(T)\Delta\dot{z}_i(T)$$

$$+\sum_{i=1}^{N_c}\int_0^T\left\{\ddot{\varphi}_i(t) + a_i\dot{\varphi}_i(t) - b_i\varphi_i(t) - \psi_x\left(z_i(t),t\right)\sum_{j=1}^{N_o}\alpha_i^j u\left(\xi_j,t\right) - \gamma_i^q\right\}\Delta z_i(t)dt.$$

Using the arbitrariness of the choice of the functions $\psi(x,t)$ and $\varphi_i(t)$, we require them to satisfy conditions (23).

Then the components of the gradient of the functional $I(y;p,\theta)$, determined by the linear parts of the functional increment with the corresponding feedback parameters [11], are defined by the following formulas:

$$\frac{\partial I(y;p,\theta)}{\partial\alpha_i^j} = -\int_0^T \psi(z_i(t),t)u\left(\xi_j,t\right)dt + 2(\alpha_i^j - \hat{\alpha}_i^j),$$

$$\frac{\partial I(y;p,\theta)}{\partial \beta_i^j} = -\int_0^T \varphi_i(t) u\left(\xi_j, t\right) dt + 2(\beta_i^j - \hat{\beta}_i^j),$$

$$\frac{\partial I(y;p,\theta)}{\partial \xi^j} = -\sum_{i=1}^{N_c} \int_0^T \left\{\alpha_i^j \psi\left(z_i(t), t\right) + \beta_i^j \varphi_i(t)\right\} u_x(\xi_j, t) dt + 2\left(\xi_j - \ddot{\xi}_j\right), \quad (31)$$

$$\frac{\partial I(y;p,\theta)}{\partial \gamma_i^q} = \int_0^T \psi\left(z_i(t), t\right) dt + 2\varepsilon\left(\gamma_i^q - \hat{\gamma}_i^q\right),$$

$$\frac{\partial I(y;p,\theta)}{\partial \gamma_i^\vartheta} = \int_0^T \varphi^i(t) dt + 2\varepsilon\left(\gamma_i^\vartheta - \hat{\gamma}_i^\vartheta\right).$$

Taking into account formula (24) from (31), we obtain the required formulas (22) given in Theorem 1.

Based on the variational form of the necessary optimality conditions, the following theorem can be proved.

Based on the variational form of the necessary optimality conditions, taking into account the convexity of the set of admissible values of the optimized parameters determined by the linear functions participating in (20) the following theorem can be proved.

Theorem 2. *Let the feedback parameters y^* satisfy the conditions of problem (20) and provide a minimum to functional (15), (16). Then, for arbitrary admissible $y \in R^N$ satisfying conditions (20), the following inequality holds:*

$$\langle \boldsymbol{gradJ}(y^*), y - y^* \rangle \leq 0.$$

4 Conclusion

The article proposes an approach to solving the problem of synthesis of motion and power control of lumped sources in systems with distributed parameters. For concreteness, the problem of feedback control of moving heat sources during rod heating is considered. The powers and velocities of point sources are assigned depending on the measured measuring points. Formulas for the linear dependence of the synthesized parameters on the measured temperature values are proposed. The differentiable of the functional with respect to the parameters of the feedback is shown, formulas for the gradient of the functional with respect to the synthesized parameters are obtained. The formulas make it possible to use effective first-order numerical optimization methods and available standard software packages to solve the problem of synthesizing the control of lumped sources.

We note that the proposed approach to the synthesis of control of lumped sources can be used in automatic control systems and automatic control of lumped sources for many other technological processes and technical objects. The objects themselves can be described by other partial differential equations and types of boundary conditions.

References

1. Antipin, A.S., Khoroshilova, E.V.: Feedback synthesis for a terminal control problem. Comput. Math. and Math. Phys. **58**, 1903–1918 (2018)
2. Butkovskiy, A.G.: Methods of control of systems with distributed parameters. Nauka, Moscow (1984).(In Russian)
3. Deineka, V.S., Sergienko, I.V.: Optimal control of non-homogeneous distributed systems. Naukova Dumka, Kiev (2003).(In Russian)
4. Utkin, V.I.: Sliding Modes in Control and Optimization. Springer, Heidelberg (1992). https://doi.org/10.1007/978-3-642-84379-2
5. Ray W.H.: Advanced process control. McGraw-Hill Book Company (2002)
6. Yegorov, A.I.: Bases of the control theory. Fizmatlit, Moscow (2004).(In Russian)
7. Butkovskiy, A.G., Pustylnikov, L.M.: The theory of mobile control of systems with distributed parameters. Nauka, Moscow (1980).(In Russian)
8. Sirazetdinov, T.K.: Optimization of systems with distributed parameters. Nauka, Moscow (1977).(In Russian)
9. Sergienko, I.V., Deineka, V.S.: Optimal control of distributed systems with conjugation conditions. Kluwer Acad. Publ, New York (2005)
10. Polyak, B.T., Khlebnikov, M.V., Rapoport, L.B.: Mathematical theory of automatic control. LENAND, Moscow (2019)
11. Vasilyev, F.P.: Optimization methods, 824. Faktorial Press, Moscow (2002).(In Russian)
12. Guliyev, S.Z.: Synthesis of zonal controls for a problem of heating with delay under non-separated boundary conditions. Cybern. Syst. Analysis. **54**(1), 110–121 (2018)
13. Aida-zade, K.R., Abdullaev, V.M.: On an approach to designing control of the distributed-parameter processes. Autom. Remote Control **73**(9), 1443–1455 (2012)
14. Aida-zade, K.R., Hashimov, V.A., Bagirov, A.H.: On a problem of synthesis of control of power of the moving sources on heating of a rod. Proc. Inst. Math. Mech. ANAS **47**(1), 183–196 (2021)
15. Nakhushev, A.M.: Loaded equations and their application. Nauka, Moscow (2012)
16. Alikhanov, A.A., Berezgov, A.M., Shkhanukov-Lafishev, M.X.: Boundary value problems for certain classes of loaded differential equations and solving them by finite difference methods. Comp. Math. Math. Phys. **48**(9), 1581–1590 (2008)
17. Abdullaev, V.M., Aida-zade, K.R.: Numerical method of solution to loaded nonlocal boundary value problems for ordinary differential equations. Comp. Math. Math. Phys. **54**(7), 1096–1109 (2014)
18. Abdullayev, V.M., Aida-zade, K.R.: Finite-difference methods for solving loaded parabolic equations. Comp. Math. Math. Phys. **56**(1), 93–105 (2016)

Evolutionary Algorithms for Optimal Control Problem of Mobile Robots Group Interaction

Sergey Konstantinov[1,2](✉) and Askhat Diveev[1]

[1] Federal Research Center "Computer Science and Control" of the Russian Academy of Sciences, Moscow 119333, Russia
[2] RUDN University, 117198 Moscow, Russia

Abstract. An optimal control problem of mobile robots group interaction on a plane with hourglass-shaped phase constraints is presented. Hourglass-shaped phase constraints can be represented as checkpoints that must be traversed by any or all of controlled objects. The modern evolutionary algorithms are used for searching the control that provides passage of all checkpoints by all robots of the group in minimum time. In a computational experiment the performance of hybrid evolutionary algorithm for solving this task is considered for mobile robots being launched simultaneously.

Keywords: Evolutionary algorithms · Optimal control problem · Group interaction

1 Introduction

The optimal control problem belongs to the class of optimization problems in infinite-dimensional space. The solution of the optimal control problem is a control function of time, which provides the control object to achieve the terminal states optimally by some quality criterion. The generally known method to solve optimal control problem is a method based on the Pontryagin maximum principle. According to this method to obtain an analytical solution to the optimal control problem, it is necessary to construct a system of differential equations for the conjugate variables. Next the optimal control function has to be found taking into account the maximum of the Hamiltonian and the constraints on the control values to be fulfilled. The joint solution of differential equations for basic variables and conjugate variables with the found control allows to construct an analytical solution to the optimal control problem. Here it is necessary to obtain general solutions for two systems of differential equations of basic and conjugate variables. Partial solution of these systems is associated with the difficulty of finding the initial values for the system of conjugate variables. Practically,

The research was supported by the Russian Science Foundation (project No 19-11-00258).

N. N. Olenev et al. (Eds.): OPTIMA 2021, CCIS 1514, pp. 123–136, 2021.
https://doi.org/10.1007/978-3-030-92711-0_9

in most cases it is impossible to receive an analytical solution for the applied optimal control problems.

This problem is also complicated by the fact that most applied optimal control problems in addition to the restrictions on the control values, have restrictions on state variable values. These restrictions are called phase constraints. Phase constraints can be interpreted as obstacles on the control object movement trajectory. Indirect methods and in particular Pontryagin maximum principle are mostly unable to solve such problems. This type of optimal control problem is well researched in [1,2].

To solve complex applied optimal control problem it is suggested to use direct method that reduces the initial problem to the nonlinear programming problem [3]. The solution of the received nonlinear programming problem provides general numerical solution to the optimal control problem. Researches in [4,5] showed that none of the classic optimization methods including the most commonly used gradient methods are not capable of solving nonlinear programming problem received from optimal control problem. It is suggested to use modern evolutionary algorithms that show good functionality even for rather complex allied problems.

Evolutionary algorithms appeared at the end of the twentieth century [6]. New evolutionary methods appear every year. From a computational point of view, all evolutionary algorithms have two common features. Firstly, they use the initial set of possible solutions generated randomly. Secondly, at a stage of evolution this set is being modified based on the information about previously calculated optimization criterion values for elements of a set of possible solutions.

The reason why evolutionary algorithms perform better than other well-known methods is that the information about topology of the function being optimized is unknown. If the function contains a large number of local extrema, then classic methods applied to solve nonlinear programming problem will most likely find local extrema, the value of which does not provide information about the location of the global optimum.

Another challenge that evolutionary algorithms cope with is high dimension of the space of the unknown parameters. The simplest reduction of optimal control problem to a nonlinear programming problem consists in discretizing the components of control vector in time. With this approach, the more sampling points, the more accurate the numerical solution of optimal control problem, but the larger dimension of a space of the required parameters.

When solving an optimal control problem by reducing it to a nonlinear programming problem and further applying evolutionary algorithms, it is required to calculate the value of the quality criterion for each possible solution. To calculate it, it is necessary to integrate the system of differential equations for each moment of time, and also to check the fulfillment of phase constraints.

In such a problem, the issue of group interaction of two control objects is of particular interest. In this paper the problem of group interaction of two tracked robots moving in Euclidean space is considered. The classical form of phase constraints in the form of obstacles in this problem is replaced by a more complex form of constraints in the form of an "hourglass". With this type of restrictions the control object needs to ensure the passage of special points on the state space—checkpoints in addition to reaching the terminal state. At the same time, dynamic phase constraints are imposed on the moving robots which consist in avoiding collisions with each other.

2 Optimal Control Problem

Consider optimal control problem in its classical formulation.

A mathematical model of any control object is given by

$$\dot{\mathbf{x}}(t) = \mathbf{f}(\mathbf{x}(t), \mathbf{u}(t)),$$

where $\mathbf{x}(t)$ represents state variables, $\mathbf{x}(t) \in \mathbb{R}^n$, $\mathbf{u}(t)$ represents control variables, $\mathbf{u}(t) \in \mathrm{U} \subseteq \mathbb{R}^m$, U is bounded closed set, $m \leq n$.

Initial conditions are

$$\mathbf{x}(0) = \mathbf{x}^0,$$

where $\mathbf{x}^0$ is a given vector of initial state.

Terminal conditions are

$$\mathbf{x}(t_f) = \mathbf{x}^f \in \mathbb{R}^n, \tag{1}$$

where $\mathbf{x}^f$ is a given vector of a terminal state, t_f is terminal time,

$$t_f = \begin{cases} t, \text{ if } t < t_{max} \text{ and } \left\|\mathbf{x}^f - \mathbf{x}(t)\right\| \leq \varepsilon_0 \\ t_{max}, \text{ otherwise} \end{cases},$$

ε_0 is a small positive value, t_{max} is given time limit to reach terminal conditions (1) in free terminal time t_f problem. The norm $\left\|\mathbf{x}^f - \mathbf{x}(t)\right\|$ is selected depending on the features of the problem being solved.

The quality criterion depends on the terminal time t_f, the accuracy of reaching terminal state and violation of phase constraints

$$J = \alpha_1 \left\|\mathbf{x}^f - \mathbf{x}(t_f)\right\| + \int_0^{t_f} f_0(\mathbf{x}(t), \mathbf{u}(t))dt + J_p \rightarrow \min, \tag{2}$$

where

$$J_p = \alpha_2 \sum_{j=1}^{K} \int_0^{t_f} \vartheta\left(h_j(\mathbf{x}(t))\right) dt \tag{3}$$

is a penalty for violating phase constraints, $h_j(\mathbf{x})$ is condition of satisfying phase constraints

$$h_j(\mathbf{x}) \leq 0,\ j = \overline{1, K},$$

K is a given number of phase constraints, $\vartheta(h_j(\mathbf{x}))$ is the Heaviside function, which adds an extra value proportional to the time the object is in the state of phase constraints violation,

$$\vartheta(h_j(\mathbf{x})) = \begin{cases} 1, \text{ if } h_j(\mathbf{x}) > 0 \\ 0, \text{ otherwise} \end{cases},$$

α_1, α_2 are penalty coefficients. In case of phase constraints in the form of an "hourglass" $h_j(\mathbf{x})$ is condition of passing through each of K checkpoints. The penalty (3) for such case will take the following form

$$J_p = \alpha_2 \left(K - \sum_{j=1}^{K} \vartheta \int_0^{t_f} (h_j(\mathbf{x}(t)))\, dt \right). \tag{4}$$

If the phase constraints in the form (3) or (4) are met and the terminal conditions (1) are satisfied, the quality criterion (2) takes the form

$$J_1 = \alpha_1 \varepsilon_0 + \int_0^{t_f} f_0(\mathbf{x}(t), \mathbf{u}(t))dt \to \min.$$

The solution to the optimal control problem is a control vector $\mathbf{u} = \mathbf{v}(t)$. Its components are piecewise-continuous functions of time.

In order to apply the optimal control of nonlinear programming methods to the solution of the initial problem, it is necessary to approximate the required components of the control vector $\mathbf{u} = \mathbf{v}(t)$ with the functional dependencies on a finite number of parameters. For this purpose, polynomials, orthogonal series, or piecewise-functional approximations are often used [7]. Based on the authors experience a piecewise-linear approximation gave the best result among the listed types of approximations for the considered applied problem taking into account results and computational complexity.

Further to reduce the optimal control problem to the nonlinear programming problem a piecewise-linear approximation is used. For this a small interval $\Delta t > 0$ is defined the number of intervals is determined

$$M = \left\lceil \frac{t_{max}}{\Delta t} \right\rceil.$$

The control value $\tilde{\mathbf{u}}(t)$ at the time t is determined from the following relation

$$\tilde{u}_j(t) = \begin{cases} u_j^-, \text{ if } q(t, j, i, \Delta t) < u_j^- \\ u_j^+, \text{ if } q(t, j, i, \Delta t) > u_j^+ \\ q(t, j, i, \Delta t), \text{ otherwise} \end{cases}, \tag{5}$$

$$q(t, j, i, \Delta t) = q_{(j-1)M+i} + (q_{(j-1)M+i+1} - q_{(j-1)M+i}) \frac{(t - (i-1)\Delta t)}{\Delta t},$$

where $i\Delta t \leq t < (i+1)\Delta t$, $i = 1, \ldots, M$, $j = 1, \ldots, m$.

As a result, the search for the control is replaced by the search for the vector of constant parameters

$$\mathbf{q} = [q_1 \dots q_p]^T, \tag{6}$$

where $p = m(M+1)$. The parameter values are limited as follows

$$q_i^- \le q_i \le q_i^+,\ i = 1, \dots, p,$$

where q_i^- and q_i^+ are the given values for the constraints for parameters, $q_i^- \le u^-_{\lfloor i/(M+1) \rfloor + 1}, u^+_{\lfloor i/(M+1) \rfloor + 1} \le q_i^+$, $i = 1, \dots, p$. The control constraint does not coincide with the constraints on the values of the parameters in order to increase the Lipschitz property of the unknown function.

3 Evolutionary Algorithms

Evolutionary algorithms appeared at the end of the 20th century. Genetic algorithm is known as the first of evolutionary algorithms [8]. At the moment there are more than 160 unique evolutionary algorithms [6].

All evolutionary algorithms use a set of possible solutions with a given number of elements, in which the evolution of the possible solutions is performed the given number of times at each iteration. The solution of the problem is the best possible solution for the value of the objective function in the resulting set.

Nowadays the most rapidly developing the most interesting branches of the evolutionary algorithms are algorithms based on swarm intelligence [9]. The term swarm intelligence originates in 1993. The first and most known method based on swarm intelligence is Particle swarm optimization (PSO) proposed in 1995 [10]. The main features of swarm-based evolutionary algorithms are the following [6,9]:

- Multiple candidate solutions are considered on each iteration;
- Each solution takes a part in avoiding locally optimal solutions and navigating to promising areas of search space;
- The best solution obtained so far and other information about the search space is preserved over the iterations;
- Evolution of solutions in evolutionary algorithms is based on the succession property. This means that with a given probability the new and old solutions are resemble to each other with some rate value and the better the current solution the higher the value of this rate;
- There are fewer parameters to adjust.

The difference between evolutionary algorithms lies in the difference in the transformations of possible solutions at the stage of evolution.

It should also be noted that one of the new and popular approaches to create more effective evolutionary algorithms is to use the hybridization of some known effective evolutionary algorithms [11].

Hybridization allows to improve the efficiency and the scope of evolutionary algorithms. Hybrid algorithms combine two or more evolutionary algorithms. In this scheme the weakness in efficiency of one algorithm is compensated by the another algorithm and vice versa.

Generally hybridization can be divided into two types. The first type is cooperation. This is a high-level of hybridization, in which the combining algorithms retain own modification strategies of possible solution. The second type is coupling. This type is a low-level hybridization with high degree of algorithms integration, which allows one to speak of creating a practically new method.

One of the features of using evolutionary algorithms for solving the optimal control problem is that after a significant number of search iterations, the process possible solutions modification may cease to bring any significant improvement to the current best solution. In this case it is appropriate to change the evolutionary modification strategy for a given set of possible solutions. Hybrid algorithms can be used to provide such variation of modification strategies.

The paper presents a hybrid algorithm based on the use of three highly efficient and popular evolutionary algorithms: Particle swarm optimization [10], Bees algorithm (BA) [12] and Gray wolf optimizer (GWO) [13]. These algorithms showed a good efficiency for solving optimal control problem [4]. To provide the better search convergence, it would be enough to use a high-level hybridization.

Let us present the evolutionary modification strategies for three mentioned above algorithms. These strategies were combined in a hybrid algorithm.

Each possible solution in PSO is called particle. Apart from the current solution each particle stores the best own solution from previous iterations. The evolution process of each possible solution in PSO takes into account current solution, the best possible solution found so far and previous best own solution.

In PSO the evolution of the i-th solution $\mathbf{q}^i$ in the set of all possible solutions is performed by the following formula

$$q_{i,j}^{new} = q_{i,j} + \alpha v_{i,j}^{new},$$

where

$$v_{i,j}^{new} = \beta v_{i,j} + \xi_1 (q_{b,j} - q_{i,j}) + \xi_2 (q_{b_i,j} - q_{i,j}),$$

$\mathbf{q}^b$ is the best possible solution found so far, $\mathbf{q}^{b_i}$ is best own solution on previous iterations, ξ_1 and ξ_2 are random variables, $\xi_1, \xi_2 \in [0;\gamma]$, α, β and γ are the given algorithm parameters.

Each possible solution in Bees algorithm is called bee [12]. Such terminology can sometimes be embarrassing, but some scientists believe that this promotes understanding and popularization of algorithms. According to the authors, the algorithm is based on the simulation of the behavior of honey bees when collecting nectar.

In the BA each possible solution is ranked according to the value of the quality criterion. A certain number of subdomains of the search space is determined around the best solutions. These subdomains are investigated more intensively, while the radius of the selected subdomains decreases with each iteration.

Each possible solution in Grey wolf optimizer is called wolf [13], which also refers not to mathematical terminology, but to the special terminology for understanding and popularization.

According to the authors, the algorithm is based on the simulation of the behavior of a pack of wolves when hunting for prey. In GWO at each iteration three best possible solutions are determined. The evolution of each possible solution uses the information about these three best solutions.

In GWO the evolution of the i-th solution $\mathbf{q}^i$ in the set of all possible solutions is performed by the following formula

$$q_{i,j}^{new} = \frac{\alpha_{i,j} + \beta_{i,j} + \gamma_{i,j}}{3},$$

where

$$\alpha_{i,j} = q_{\alpha,j} - 2\xi_1|(2\xi_2 - 1)aq_{\alpha,j} - q_{i,j}|$$
$$\beta_{i,j} = q_{\beta,j} - 2\xi_3|(2\xi_4 - 1)aq_{\beta,j} - q_{i,j}|$$
$$\gamma_{i,j} = q_{\alpha,j} - 2\xi_5|(2\xi_6 - 1)aq_{\gamma,j} - q_{i,j}|,$$

$\mathbf{q}^\alpha$, $\mathbf{q}^\beta$ and $\mathbf{q}^\gamma$ are correspondingly the first, the second and the third best possible solutions found so far, $\xi_1, \ldots, \xi_6$ are randomly distributed variables in the interval $[0; 1]$,

$$a = 2 - \frac{2w}{W},$$

w is the current iteration number, W is the maximum number of search iterations.

4 A Computational Experiment

Consider an optimal control problem of a group of two mobile two-track robots moving on the plane.

The mathematical model of a group of mobile two-track robots is given in the form of a system of differential equations [14]

$$\begin{aligned} \dot{x}_{i,1} &= 0.5(u_{i,1} + u_{i,2})\cos(x_{i,3}), \\ \dot{x}_{i,2} &= 0.5(u_{i,1} + u_{i,2})\sin(x_{i,3}), \\ \dot{x}_{i,3} &= 0.5(u_{i,1} - u_{i,2}), \end{aligned} \tag{7}$$

where $\mathbf{x}_i = [x_{i,1}x_{i,2}x_{i,3}]^T$ is a state space vector of i-th robot, $x_{i,1}$ and $x_{i,2}$ are coordinates on the plain, $x_{i,3}$ is a rotation angle, $\mathbf{u}_i = [u_{i,1}u_{i,2}]^T$ is a control vector of i-th robot, $i = \overline{1,N}$, $N = 2$ is the number of robots.

Restriction on control are given

$$-10 = u^- \leq u_{i,j} \leq u^+ = 10, i = 1, 2, j = 1, 2.$$

Each robot has its own initial and terminal states. Initial states for the first and second robots, respectively

$$\mathbf{x}^{0,1} = [4 \quad 1.5 \quad 0]^T, \qquad \mathbf{x}^{0,2} = [4 \quad 0.5 \quad 0]^T.$$

Terminal states for the first and second robots, respectively

$$\mathbf{x}^{f,1} = [4\ \ 0.5\ \ 0]^T, \quad \mathbf{x}^{f,2} = [4\ \ 1.5\ \ 0]^T.$$

Before reaching the respective terminal states robots have to cross all checkpoints. Checkpoints are phase constraints in the form of "hourglass". For each checkpoint the allowed zone of the robot's passage is set by two coordinates on the plane. In the computational experiment robots have to cross $K = 4$ checkpoints with the following coordinates

$$\mathrm{C}^i = \left(\mathbf{l}^i, \mathbf{r}^i\right),$$
$$\mathbf{l}^i = \left[l_{i,1}\ l_{i,2}\right]^T, \mathbf{r}^i = \left[r_{i,1}\ r_{i,2}\right]^T, i = \overline{1, K},$$

where

$$\text{checkpoint 1:}\quad \mathbf{l}^1 = \left[5\ \ 0\right]^T,\ \mathbf{r}^1 = \left[5\ \ 2\right]^T,$$
$$\text{checkpoint 2:}\quad \mathbf{l}^2 = \left[8\ \ 5\right]^T,\ \mathbf{r}^2 = \left[10\ \ 5\right]^T,$$
$$\text{checkpoint 3:}\quad \mathbf{l}^3 = \left[5\ \ 8\right]^T,\ \mathbf{r}^3 = \left[5\ \ 10\right]^T,$$
$$\text{checkpoint 4:}\quad \mathbf{l}^4 = \left[0\ \ 5\right]^T,\ \mathbf{r}^4 = \left[2\ \ 5\right]^T.$$

If robot reaches the terminal state without passing checkpoint then the value of the quality criterion is being increased by the value of the corresponding penalty.

For the model under consideration the quality criterion should take into account the accuracy of reaching the terminal state, the time to reach the terminal state, as well as the penalty for failing to pass checkpoints and the penalty for colliding robots with each other. Such the quality criterion has the following form

$$J = t_f + \alpha_1 \sum_{i=1}^{N} \left\|\mathbf{x}^{f,i} - \mathbf{x}^i(t_f)\right\| + J_p + J_c \to \min, \tag{8}$$

where J_p is a penalty for not passing checkpoints

$$J_p = \alpha_2 \left(NK - \sum_{i=1}^{N} \sum_{j=1}^{K} \vartheta \int_0^{t_f} \left(h_j(\mathbf{x}^i(t))\right) dt \right),$$

J_c is a penalty for colliding robots with each other

$$J_c = \alpha_3 \sum_{i=1}^{N-1} \sum_{j=i+1}^{N} \int_0^{t_f} \vartheta \left(r_0 - \sqrt{(x_{1,1}(t) - x_{1,2}(t))^2 + (x_{2,1}(t) - x_{2,2}(t))^2}\right) dt,$$

α_1, α_2 and α_3 are penalty coefficients, r_0 is a minimum distance between two robots which ensures they are not colliding, $h_j(\mathbf{x}^i(t))$ is condition of passing i-th robot through j-th checkpoint. It returns 1 if robot has passed the corresponding checkpoint and 0 otherwise.

Experiments have shown that this form of quality criterion is not enough to provide the attraction of robots to the area with checkpoints. During the modeling process of robots movement, checkpoints should act as attractors making robots to approach and cross them. To implement this the quality criterion (8) was modified in the following form

$$J = t_f + \alpha_1 \sum_{i=1}^{N} \left\| \mathbf{x}^{f,i} - \mathbf{x}^i(t_f) \right\| + J_a + J_p + J_c \rightarrow \min, \tag{9}$$

where

$$J_a = \sum_{i=1}^{N} \sum_{j=1}^{K} \left(\alpha_4 \left(1 - \vartheta \int_0^{t_f} (h_j(\mathbf{x}^i(t)))\, dt \right) \right.$$
$$\left. \min \left(\sqrt{ \left(x_{i,1}(t) - \frac{l_{j,1} - r_{j,1}}{2} \right)^2 + \left(x_{i,2}(t) - \frac{l_{j,2} - r_{j,2}}{2} \right)^2 } \right) \right)$$

is a penalty which value is the lower, the closer the robot to the corresponding checkpoint, α_4 is an attraction coefficient.

The modeling was performed with the following values of given parameters: $r_0 = 0.5$, $\alpha_1 = 1$, $\alpha_2 = 3$, $\alpha_3 = 10$, $\alpha_4 = 1$, $\varepsilon_0 = 0.01$, $t_{max} = 4.5$, $\Delta t = 0.3$, $M = \left\lceil \frac{t_{max}}{\Delta t} \right\rceil = 15$, $p = 2N(M+1) = 64$, $q_i^- = -12$, $q_i^+ = 12$, $i = \overline{1,p}$.

To find a solution, the original optimal control problem was reduced to a non-linear programming problem, and then evolutionary algorithms were applied. A hybrid evolutionary algorithm based on three high-performance algorithms—Particle swarm optimization, Grey wolf optimizer and Bees algorithm—was used. The hybrid algorithm used had the following values of parameters: the size of population $H = 64$, the number of search iterations $W = 10000$, the values of algorithm-specific parameters were chosen according to the recommended ones [10,12,13].

The computational experiments were performed independently 10 times. Each experiment provided a solution in the form of vector $\mathbf{q}$ (6), which being substituted into expression (5) gave the discrete values of control. The best of 10 experiments solution had the value of quality criterion (9) $J = 2.8237$. The deviation of the best found solution from others was no more than 0.03. Components of best solution vector $\mathbf{q}$ are presented below.

$$\mathbf{q} = \begin{bmatrix} 9.9823 & 9.9151 & 9.9991 & 9.9546 & 9.9953 & 9.9908 & 7.8571 & 9.9897 \\ 9.9765 & 8.3731 & 0 & 0 & 0 & 0 & 0 & 0 \\ 9.7580 & 3.5776 & 6.3722 & 4.8915 & 5.3755 & 3.0772 & 6.9895 & 3.4072 \\ 9.9989 & -1.1137 & 0 & 0 & 0 & 0 & 0 & 0 \\ 9.9495 & 9.9838 & 9.9984 & 9.9751 & 9.9998 & 9.9816 & 10 & 9.9691 \\ 9.9675 & 9.9487 & 0 & 0 & 0 & 0 & 0 & 0 \\ 8.6749 & 5.0724 & 4.6589 & 8.8469 & 3.0949 & 5.9391 & 6.6820 & 1.9519 \\ 9.9145 & 0.7711 & 0 & 0 & 0 & 0 & 0 & 0 \end{bmatrix} \tag{10}$$

Results of simulation the system (7) with control obtained by using (10) are presented in the Figs. 1, 2, 3, 4, 5 and 6.

In the Fig. 1 the best found trajectories of two-track robots movement on the plane $\{x_1, x_2\}$ are presented. Figure 1 demonstrates that both robots achieve the terminal condition with sufficient accuracy after making a circle movement through all 4 checkpoints. The first robot trajectory is showed with red line, the second robot trajectory is showed with blue line. Checkpoints are indicated by green triangles. The passage of robots between these triangles means that corresponding "hourglass" constraint is fulfilled.

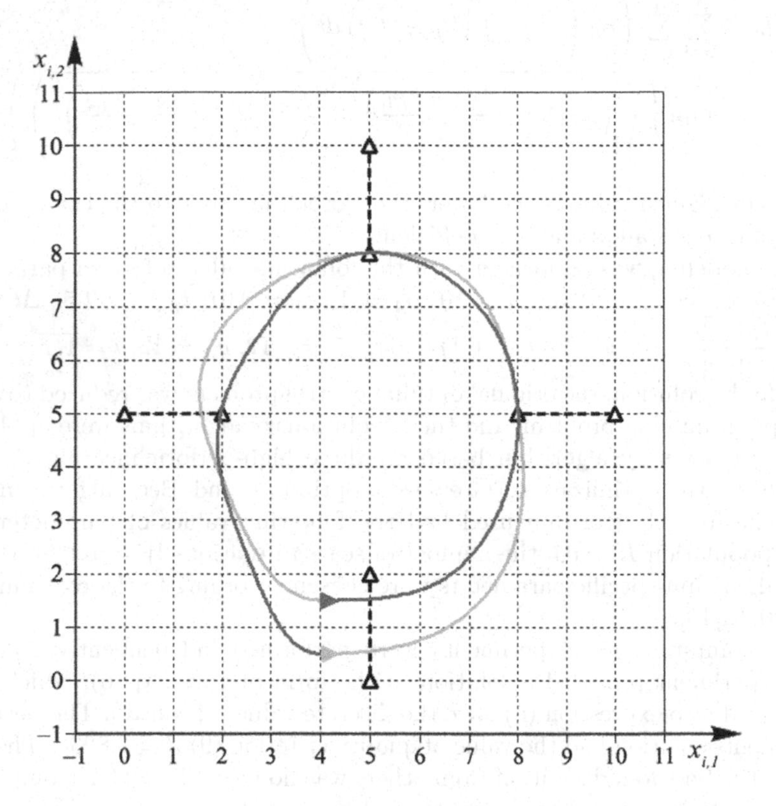

Fig. 1. Best found trajectories of two robots on the plane (robot 1 – red line; robot 2 – blue line; checkpoints – green dashed line) (Color figure online)

Figures 2, 3 and 4 show the graphs of state components over time. Since the states $x_{1,3}$ and $x_{2,3}$ are the angles of robots position, the final values on the Fig. 4 which both are approximately equal to $6.28 \approx 2\pi$ correspond to the terminal values $x_{f,1,3} = 0$ and $x_{f,2,3} = 0$ respectively.

Figures 5 and 6 show the graphs of control components over time.

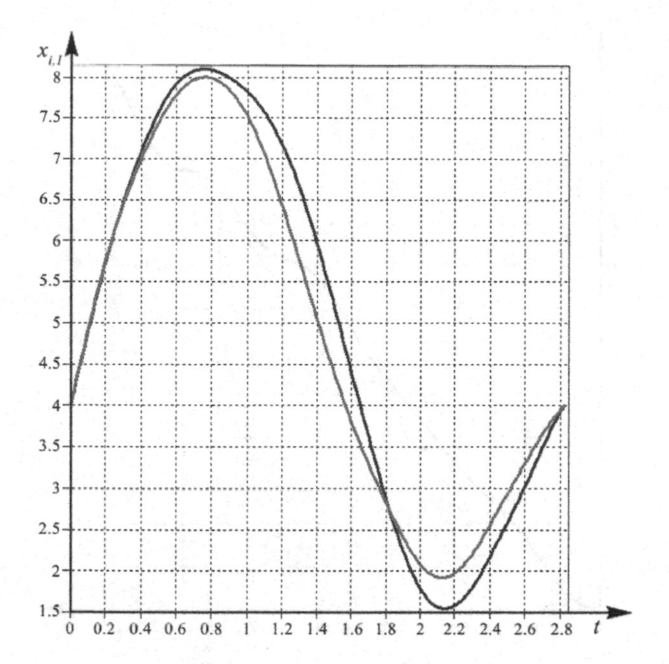

Fig. 2. State components $x_{1,1}$ and $x_{2,1}$ over time

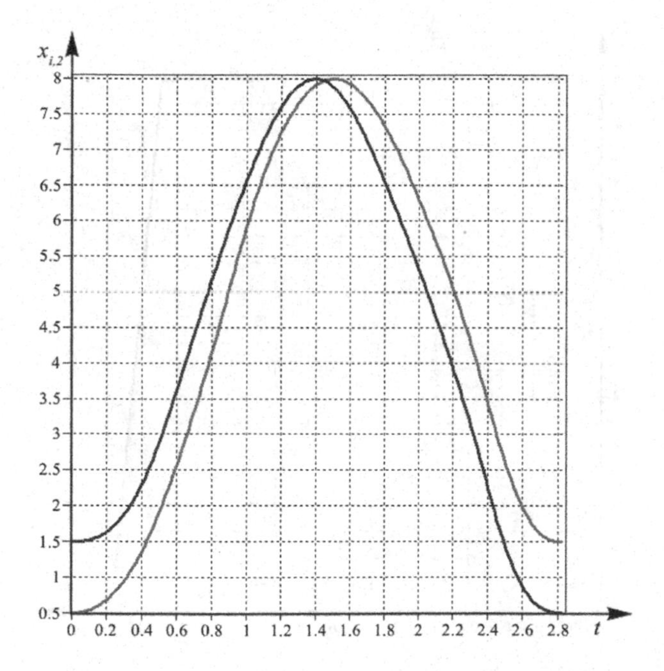

Fig. 3. State components $x_{1,2}$ and $x_{2,2}$ over time

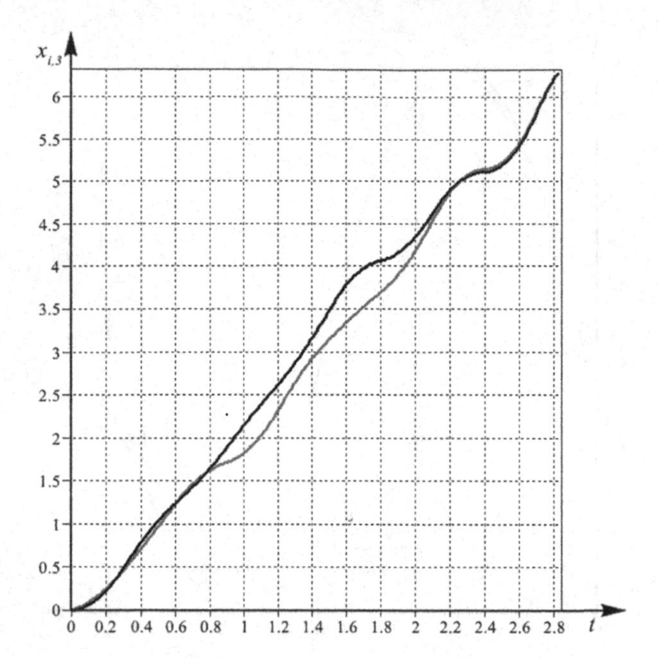

Fig. 4. State components $x_{1,3}$ and $x_{2,3}$ over time

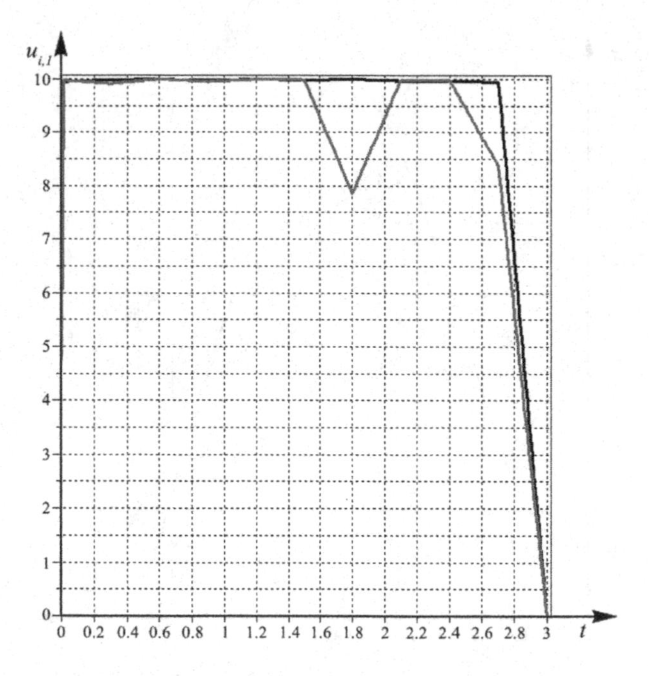

Fig. 5. State components $u_{1,1}$ and $u_{2,1}$ over time

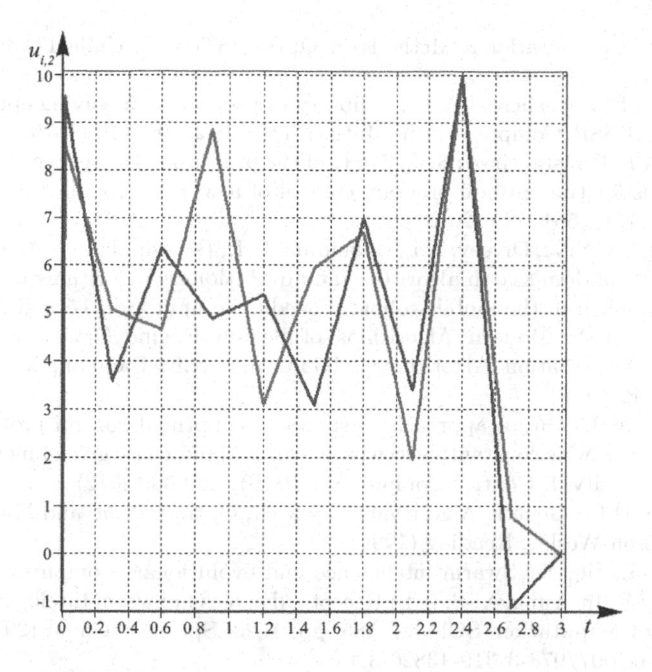

Fig. 6. Control components $u_{1,2}$ and $u_{2,2}$ over time

5 Conclusion

To solve applied optimal control problems in the presence of phase constraints, indirect methods are usually not applicable due to the practical impossibility of solving two systems of differential equations of basic and conjugate variables. The main features and challenges of numerical solution of optimal control problem reduced to a nonlinear programming problem lie in the uncertainty of the function topology and the high dimension of the search space. Evolutionary algorithms prove their efficiency for the numerical solution of optimal control problem. The use of a hybrid evolutionary algorithm based on the Particle swarm optimization, Grey wolf optimizer and Bees algorithm in the computational experiment made it possible to solve a complex applied problem of group interaction of two-track robots moving on a plane with constraints in the form of "hourglass". To solve this problem it was necessary to construct a quality criterion that takes into account the passage of the "hourglass" constraints by robots, as well as the fact of their collisions with each other. The simulation results have shown the high efficiency of this approach for the problem considered.

References

1. Evtushenko, Y.G.: Optimization and Rapid Automatic Differentiation. Computing Center of RAS, Moscow (2013). (in Russian)

2. Polak, E.: Computational Methods in Optimization. A Unified Approach, New York (1971)
3. Grachev, I.I., Evtushenko, Y.G.: A library of programs for solving optimal control problems. USSR Comput. Math. Math. Phys. **19**(2), 99–119 (1979)
4. Diveev, A.I., Konstantinov, S.V.: Study of the practical convergence of evolutionary algorithms for the optimal program control of a wheeled robot. J. Comput. Syst. Sci. Int. **57**(4), 561–580 (2018)
5. Konstantinov, S.V., Diveev, A.I., Balandina, G.I., Baryshnikov, A.A.: Comparative research of random search algorithms and evolutionary algorithms for the optimal control problem of the mobile robot. Procedia Comput. Sci. **150**, 462–470 (2019)
6. Karpenko, A.P.: Modern Algorithms of Search Engine Optimization. Nature-Inspired Optimization Algorithms, MGTU n.a. N.E. Bauman, Moscow (2014), 446p. (in Russian)
7. Rahimov, A.B.: On an approach to solution to optimal control problems on the classes of piecewise constant, piecewise linear, and piecewise given functions. Tomsk State University J. Control Comput. Sci. **2**(19), p20-30 (2012)
8. Goldberg, D.E.: Genetic Algorithms in Search, Optimization, and Machine Learning. Addison-Wesley, Reading (1989)
9. Yang, X.-S., He, X.: Swarm intelligence and evolutionary computation: overview and analysis. In: Yang, X.-S. (ed.) Recent Advances in Swarm Intelligence and Evolutionary Computation. SCI, vol. 585, pp. 1–23. Springer, Cham (2015). https://doi.org/10.1007/978-3-319-13826-8_1
10. Kennedy, J., Eberhart, R.: Particle swarm optimization. In: Proceedings of ICNN'95 - International Conference on Neural Networks, pp. 1942–1948 (1995)
11. Raidl, G.R.: A unified view on hybrid metaheuristics. In: Almeida, F., et al. (eds.) HM 2006. LNCS, vol. 4030, pp. 1–12. Springer, Heidelberg (2006). https://doi.org/10.1007/11890584_1
12. Pham, D.T., et al.: The Bees Algorithm – A Novel Tool for Complex Optimisation Problems. In: Intelligent Production Machines and Systems - 2nd I*PROMS Virtual International Conference, 3–14 July 2006, pp. 454–459. Elsevier Ltd. (2006)
13. Mirjalili, S., Mirjalili, S.M., Lewis, A.: Grey wolf optimizer. Adv. Eng. Softw. **69**, 46–61 (2014)
14. Šuster, P., Jadlovska, A.: Tracking trajectory of the mobile robot Khepera II using approaches of artificial intelligence. Acta Electrotechnica et Informatica **11**(1), 38–43 (2011)

Mathematical Economics

Model LSFE with Conjectural Variations for Electricity Forward Market

Natalia Aizenberg(✉)

Melentiev Energy Systems Institute SB RAS, Lermontov Street, 130, Irkutsk, Russia

Abstract. The goal of our study is to analyze the theoretical basis of the forward electricity market effectiveness. In Russian practice, it is represented by the free long-term contracts market. This is one of the recognized tools to reduce price volatility in the electricity market. In addition, forward contracts also reduce the market power of large generation companies, consisting of the ability to systematically raise prices relative to their costs. It is known that the efficiency of this tool fundamentally depends on the electricity market properties. One of the important characteristics of the market reaction is described by the conjectural variations (the concept from game theory) and is related to a competitor's reaction to changes in the supply volumes of other market participants. Our model is a modification of the supply functions equilibrium model with forward contracts. We describe the reaction of each competitor to the change of forward contract of another competitor through a conjectural variation. We show that firms with high slope of marginal costs relative to others enter first to the forward market. The example of modeling an electric power system, similar to the real one, confirms this result.

Keywords: Electricity markets · Conjectural variations model · Forward contracts · Supply functions equilibrium

1 Introduction

Electricity markets have their own specifics. The traded goods are homogeneous, not stored, the elasticity of demand for it is quite low, the barriers to entry into the industry are high due to high initial investment costs. These characteristics contribute to the fact that large generating companies are effective, which can neutralize the risks of a shortage of electricity to the consumer, including against the background of underinvestment in production. Thus, almost all electricity markets have an oligopoly structure. It is known that market power in such markets is pronounced without additional regulatory efforts. The state tries to

This work is carried out as part of the State Assignment Project (no. FWEU- 2021-0001) of the Fundamental Research Program of Russian Federation 2021–2030, and with the support of the Russian Foundation for Basic Research, grant 019-010-00183.

N. N. Olenev et al. (Eds.): OPTIMA 2021, CCIS 1514, pp. 139–153, 2021.
https://doi.org/10.1007/978-3-030-92711-0_10

control the processes taking place in these markets. Forward contracts are one of the possible instruments to reduce market power.

The theoretical foundations of the effectiveness of the introduction of the forward contracts in the electric power wholesale markets were considered in [1]. The conclusion of preliminary agreements for the supply of electricity increases the elasticity of demand in the main market, stabilizing prices [2].

In modeling, the most important characteristic that determines the feasibility of introducing forwards is the degree of competitors' reaction to changes in the supply volumes of other market participants. This can be defined through the concept of reflection from game theory when competitors can assume and then react to possible actions of each other. In [3], the author suggests a model that takes into account a similar reflection. It is based on the well-known supply function equilibrium model (SFE). The modeling result shows that if competitors have a weak reaction to changes in the forward supply volumes of other participants (following Cournot behavior), then the introduction of the forward market is not optimal. And efficiency increases as competitors react to each other. Forward markets based on the SFE model [4–7,13] also was developed in [8–11]. The model we have considered is a modification of the supply function equilibrium model, taking into account the actions of generating companies [12]. Generation companies (GK) are a collection of several capacities, each of which must bid an application to the market on a "day ahead". This statement is caused by the rules of the Russian electric power market.

The research question is whether and in what form there will be a forward electricity market for an electric power system similar to the system "Siberia", which features large hydro generation, as well as cogeneration of heating and power plants with differing costs of generating an electricity unit. The paper is organized as follows. In the first part, the problem of the spot and the forward market is formulated as the Stackelberg leader-follower problem type. Further, a solution is considered in the spot market under the assumption of fixed forward sales. The next part examines the forward market in the formulation of hypothetical variations when players assume the degree of competitors' reaction to changes in the volume of supply. All solutions are given for the well-known linear supply function equilibrium (LSFE) model for the case when companies contain several capacities. The last part contains an example of calculating a system similar to the electric power system "Siberia".

2 Model

2.1 Basic Forward Market Model

Our model is based on the well-known SFE model [4,8,13,15,16]. Here we briefly describe what it consists of and give a modification for the forward markets [1,17]. Pricing in the spot electricity market is organized as a bilateral auction. Generating companies give their bids (supply functions) to the market depending on the anticipated actions of their competitors. Consumers also give their bids as a demand function. After that, the price and volumes of electricity (supplies) are determined based on the equality of supply and demand. Power plants that are part of generating companies differ in their technological capabilities and type of costs [12].

All consumers are aggregated by a single nonincreasing demand function $D(p)$. The demand function is continuous, infinitely, and differentiable

$$D(0) \geq 0, \, D(\infty) \leq 0.$$

Supply functions of individual firms: $q_k(p)$ is the output of company k, $k = \overline{1,m}$, $m \geq 2$ is the number of generating power in the market and

$$\sum_{k=1}^{m} q_k(p) = D(p).$$

The inverse demand function is $p = D^{-1}\left(\sum_{k=1}^{m} q_k\right)$. Each generating company k, $k = \overline{1,m}$ has several capacities, in its composition. First, we consider a situation where each generating company consists of only one power plant.

We will define the outputs of competitors for company k as $q_{-k}(p) = \sum_{j \neq k} q_j$, this is the total output excluding k. Thus, the residual demand of the generation company k is

$$q_k(p) = D(p) - q_{-k}(p). \tag{1}$$

Here $p \in R_+^1$ is the price formed through the interaction of agents in the market. Cost function $C_k(q_k)$, $k = \overline{1,m}$ is continuous, continuously differentiable, convex and increasing, $q_k \geq 0$, $k = \overline{1,m}$. Generation companies maximize the profit on the residual demand provided in equilibrium the demand equal the total output of the companies

$$\pi_k(p, q_k) = p \cdot (D(p) - q_{-k}(p)) - \sum_{k=\overline{1,m}} C_k(q_k(p)). \tag{2}$$

The company k assumes the supply functions of all other companies and uses this information. The scale of the reactions to changes in the output of other players (generating companies) is assumed by firm k itself. Hence, these answers can differ from the actual reactions of competitors. A simplified form of this model is the SFE model [5] where all stakeholders have information about competitors.

Adding a forward market transforms the problem into a model where players make decisions in two related markets in two steps. In this case, the problem is similar to the Stackelberg game: the decision in the spot market is made taking into account the sales of the forward one. Thus, it is intuitively clear that the players in the spot market are followers, and they are leaders in the forward market.

The strategy of the k-th company in two markets is to define its own supply function $q_k(p)$ which is given to the operator in real markets as the bid. It is assumed here that the type of the supply function is set (for example, linear), and as a result of maximization (2), the company determines specific coefficients. In the forward market, the strategy is to determine the trading volume x_k.

The solution algorithm is as follows: first, the equilibrium in the spot market is obtained, taking into account the fact that residual demand on which the k-th company maximizes profit is dependent on the forward contracts (x_k, x_{-k}), and the supply functions of other companies $q_{-k}(p)$ in the spot market. Then the equilibrium is searched on the forward market on the variable x_k, $k = \overline{1,m}$.

General Problem of the Spot Market for the k-th Company. We maximize the profit of k-th company in the spot market with fixed-volume forward sales:

$$\pi_k(p) = p \cdot (q_k(p) - x_k) + f \cdot x_k - C_k(q_k(p)) \to \max_p, \ k = \overline{1,m}, \quad (3)$$

$$q_k \geq 0, \quad k = \overline{1,m}.$$

Here

$$q_k(p) - x_k = D(p) - q_{-k}(p) - x_k \quad (4)$$

is the residual demand for company k on the spot market, selling on the forward market x_k are fixed.

Proposition 1. The equilibrium in the spot market exists and is unique if the demand function $D(p)$ is concave with respect to the price p, and the supply function of each supplier is convex with respect to the price p.

Proof. The market equilibrium is unique if all profit functions of the companies are continuous and concave [18]. Let us define the conditions when this is fulfilled for the described market.

Show that $\pi_k(x)$ is a continuous piecewise differentiable function. It follows simply from the definition of the market-clearing price function p^*, and cost function

Let FOC has form, assuming that q_j of each supplier is some function of p:

$$\frac{d\pi_k(p)}{dp} \equiv (p - C_k'(q_k)) \cdot \left(\frac{dD(p)}{dp} - \sum_{j \neq k} \frac{dq_j(p)}{dp} \right) + D(p) - q_{-k}(p) - x_k = 0.$$

Rewrite taking into account (4):

$$q_k(p) = -(p - C_k'(q_k)) \cdot \left(\frac{dD(p)}{dp} - \sum_{j \neq k} \frac{dq_j(p)}{dp} \right). \quad (5)$$

The generating company sells electricity on the market, therefore $q_k(p) \geq 0$. We assume the standard form of the sentence function. It is an increasing function, therefore, for each k, $j \neq k$ $\frac{dq_j(p)}{dp} \geq 0$. We have standard assumptions for the demand function. It is a decreasing function, therefore $\frac{dD(p)}{dp} \leq 0$. So, $\frac{dD(p)}{dp} - \sum_{j\neq k} \frac{dq_j(p)}{dp} \leq 0$.

Then we have that the price in such a market for all companies is always greater than or equal to the marginal costs $C'_k(q_k)$:

$$p - C'_k(q_k) \geq 0. \tag{6}$$

The SOC is as follows

$$\frac{d^2\pi_k(p)}{dp^2} = (p - C'_k(q_k)) \cdot \left(\frac{d^2 D(p)}{dp^2} - \sum_{j\neq k} \frac{d^2 q_j(p)}{dp^2} \right) +$$

$$2\left(\frac{dD(p)}{dp} - \sum_{j\neq k} \frac{dq_j(p)}{dp} \right) - C''_k(q_k) \cdot \left(\frac{dD(p)}{dp} - \sum_{j\neq k} \frac{dq_j(p)}{dp} \right)^2.$$

Let us define the conditions when $\frac{d^2\pi_k(p)}{dp^2} \leq 0$.

(i) In the first term of the expression, we have $p - C'_k(q_k) \geq 0$ proved above. Then in order to $\frac{d^2 D(p)}{dp^2} - \sum_{j\neq k} \frac{d^2 q_j(p)}{dp^2} \leq 0$ it suffices that $D(p)$ is concave, including the standard linear case, and the supply function $q_j(p)$ is convex, and $\frac{dq_j(p)}{dp} \geq 0$.

(ii) The second term is negative as proved above for FOC.

(iii) The third term is negative because the cost function is increasing and convex, and therefore $C''_k(q_k) \geq 0$, and it is included in the SOC with the '-' sign. ■

So, all profit functions of companies are concave, continuous, and, therefore, the Nash equilibrium exists and is unique.

Discussion. The result is determined by the concavity of the demand function and the convexity of the supply function. Both assumptions are quite natural. The convexity of the supply function may follow from the fact that it may be similar to the marginal cost function. These are standard assumptions from firm theory, corresponding to cubic or quadratic total cost functions [19]. The concave demand function describes the demand for normal goods, the demand for which has a saturation limit: if the saturation limit is reached, a decrease in price does not give an increase in the volume of demand. This function graph has a vertical closing segment.

The difficulty of finding equilibrium in this problem lies in the fact that $q_j(p)$, and p is an implicit function of forward output $q_k(p(x_k, x_{-k}))$ defined below.

General Problem of the Forward Market for the k-th Company. Consider *the forward market problem*, the leader problem. Suppose the forward market price is close to the spot market price and the players have near-perfect foresight $f = p$. This is the standard assumption in such models. Then each company looks for the maximum of its profit by the volumes of trade x_k in the forward market:

$$\pi_k(p) = p(x_k, x_{-k}) \cdot q_k(x_k, x_{-k}) - C_k(q_k(x_k, x_{-k})) \to \max_{x_k}, \ k = \overline{1,m}, \quad (7)$$

$$q_k \geq 0, \quad k = \overline{1,m}.$$

FOC has the form

$$\frac{\partial p}{\partial x_k} \cdot q_k + (p - MC_k(q_k)) \cdot \frac{\partial q_k}{\partial x_k} + \quad (8)$$

$$\sum_{j \neq k} \left(\frac{\partial p}{\partial x_j} \cdot q_k + (p - MC_k(q_k)) \cdot \frac{\partial q_k}{\partial x_j} \right) \cdot \frac{\partial x_j}{\partial x_k} = 0,$$

where the concept of marginal cost is used $MC_k(q_k) = C_k'(q_k)$. The main points of action in the two markets, which are described here by the SFE model, are associated with the reflection of the participants in the interaction regarding each other's actions. For the forward market, these reactions are described by $w_{jk} = \partial x_k / \partial x_k$. This denotes the conjectural variation, which shows the generator i's belief about the extent to which the generator j's offered quantity will change in response to a change in its own offer in the forward market. These coefficients satisfy standard conditions: $w_{jk} \in [-1, 0]$, $-1 \leq \sum_{j=1,m,j\neq k} w_{jk} \leq 0$. For interaction of the Cournot type $w_{jk} = 0$, competitors react with changes in each other's output by the elasticity of demand. For the Bertrand model, we have the maximum response (or changes of output): $\sum_{j=1,m,j\neq k} w_{jk} = -1$.

Let us consider in more detail the situation when

1) the demand function is linear:

$$D(p) = N - \gamma \cdot p, \quad (9)$$

where $\gamma > 0$;

2) generation powers are heterogeneous in convex quadratic operating cost functions and constraints on power generation:

$$C_k(q_k) = \frac{1}{2} c_k q_k^2 + a_k q_k + d_k, \quad (10)$$

$c_k > 0$, $a_k \neq 0$, $k = \overline{1,m}$. For each generation power is $q_k \leq V_k$, $k = \overline{1,m}$. The marginal costs are $MC(q_k) = c_k q_k + a_k$;

3) in the classical formulation of the SFE model, it is assumed that the supply function of each generating company is linear, the parameters of which β_k, α_k we determine in the process of solving the problem [14]:

$$q_k(p) = \beta_k p + \alpha_k, \quad k = \overline{1,m}. \quad (11)$$

For the above assumptions, it was shown in [20,21] that the total output $q_k(x_k, x_{-k})$ of a company trading in two consecutive markets, where the decision is made as a leader-follower, increases in x_k if $w_{jk} \neq 0$. For a linear demand and a convex cost function, equilibrium exists if $q_k(x_k, x_{-k})$ is concave in x_k. The following is a solution for the forward market of the well-known LSFE model, for which all of the above conditions hold. In particular, the function $q_k(x_k, x_{-k})$ is linear (a special case of concave) in x_k.

2.2 Equilibrium in the LSFE Model for the Forward Market

Decision in the Spot Market Taking into Account Forward Sales. In problem (3) with variables β_k, α_k in the positive domain $q_k \geq 0, \beta_k \geq 0, k = \overline{1,m}$, the solution taking into account fixed sales x_k in the forward market is unique [6]. Maximizing the profit with respect to p on residual demand $q_k = N - \gamma \cdot p - \sum_{j \neq k} q_j(p) - x_k$, we obtain a system of non-linear equations describing the equilibrium with the variables β_k, α_k:

$$\beta_k = (1 - c_k \beta_k)\left(\gamma + \sum_{j \neq k} \beta_j\right), k = \overline{1,m}; \tag{12}$$

$$\alpha_k = x_k \cdot (1 - c_k \beta_k) - a_k \left(\gamma + \sum_{j \neq k} \beta_j\right), k = \overline{1,m}.$$

Let $B = 1/\left(\gamma + \sum_j^m \beta_j\right)$, $\sigma_k(c_k, \beta_k) = 1 - c_k \beta_k$. Solving it with respect to β_k, α_k , $k = \overline{1,m}$, we get the following for p and q_k:

$$p(x_k, x_{-k}) = B \cdot \left(N - \sum_j^m (\sigma_j x_j - a_j \beta_j)\right), \tag{13}$$

$$q_k(x_k, x_{-k}) = \beta_k B \cdot \left(x_k + N - \sum_{j, j \neq k}^m (\sigma_j x_j - (a_j - a_k)\beta_j) - a_k \gamma\right) \tag{14}$$

The function $q_k(x_k, x_{-k}), k = \overline{1,m}$ is linear concerning all variables. Therefore, the conditions under which equilibrium in the spot market exists and is unique [20,21] hold.

Decision in the Forward Market. The next step is to substitute (13) and (14) in (8):

$$\frac{\partial \pi_k}{\partial x_k} \equiv \sum_j \left(\frac{\partial p}{\partial x_j} \cdot q_k + (p - a_k - c_k q_k) \cdot \frac{\partial q_k}{\partial x_j}\right) \cdot \frac{dx_j}{dx_k}, k = \overline{1,m} = 0. \tag{15}$$

Here are

$$\frac{\partial p}{\partial x_j} = \frac{\sigma_j}{B}, j = \overline{1,m}, \frac{\partial q_k}{\partial x_j} = \frac{\beta_k \sigma_j}{B}, j \neq k, \frac{\partial q_k}{\partial x_k} = \frac{\beta_k}{B}. \tag{16}$$

An important characteristic is $-1 \le \sum_{j \ne k} \sigma_j w_{jk} \le 0$. Exactly it affords the cumulative reaction to changes in the company's sales in the forward market and the outputs in the spot market depend on it.

Equilibrium in the forward market exists if the payoff function for each k-th company, formed based on of continuous functions (13) and (14), is concave in x_k. Let us define the SOC for the problem (7) using known (13) and (14), and proof the Proposition.

Proposition 2. Profit functions π_k of each k-th company (7) with (13), and (14) are concave with respect to sales x_k in the forward market for the case $w_{jk} \in [-1, 0]$, $-1 \le \sum_{j=1,m,j \ne k} w_{jk} \le 0$.

Proof. See appendix. ■

Using the results of Proposition 2 we find the solution in the forward market from FOC (15), defining it through the general output of the company q_k. FOC has form

$$B \cdot \left(-\sigma_k \cdot x_k - \sum_{j, j \ne k} \left(q_k \sigma_j + \sigma_k \left(q_k - x_k \right) \sigma_j \right) \cdot w_{jk} \right) = 0,$$

hence

$$x_k = q_k \frac{(1 + \sigma_k) \cdot \sum_{j \ne k} \sigma_j w_{jk}}{\sigma_k \cdot \left(\sum_{j \ne k} \sigma_j w_{jk} - 1 \right)}. \tag{17}$$

The ratio (17) is similar to one obtained by Green for two companies [3]. Based on (17), it is possible to modeling related sales in the forward and spot markets. There are some important conclusions for modeling from (17).

Forward Market Size. 1. The existence of the forward market is possible only if competition differs from Cournot competition in the direction of a perfect market, that is if $w_{jk} \ne 0$.

2. The forward market volume depends on the ratio of the aggregate reaction of competitors $\sum_{j \ne k} \sigma_j w_{jk}$ to the change in the forward sales x_k of company k and how much the slope of the supply function of the company k exceeds the slope of its marginal cost function σ_k. If $\sum_{j \ne k} \sigma_j w_{jk} > -\sigma_k$, then the firm enters both the spot and forward markets $q_k > x_k$.

The most important characteristic here is σ_k. In the case of perfect competition, if the company reports exactly its marginal costs, equilibrium will be obtained through the relationship $p = a_k + c_k q_k$. In this model, $\beta_k \le \frac{1}{c_k}$, and $\sigma_k \left(c_k, \beta_k \right)$ shows how much the slope of the supply function can deviate from the slope of the marginal cost function.

Proposition 3. The function $\sigma_k \left(c_k, \beta_k \right)$ is decreasing with respect to c_k, which is the slope of the marginal cost of function (10).

Proof. Let us define β_k from (12):

$$\beta_k = \frac{1}{\left(1/\left(\gamma + \sum\limits_{j \neq k} \beta_j\right) - c_k\right)},$$

find $\phi(c_k) = c_k \beta_k$ and determine the value of its derivative with respect to c_k:

$$\phi(c_k) = \frac{c_k}{\left(1/\left(\gamma + \sum\limits_{j \neq k} \beta_j\right) - c_k\right)},$$

$$\phi'_{c_k}(c_k) = \frac{1/\left(\gamma + \sum\limits_{j \neq k} \beta_j\right)}{\left(1/\left(\gamma + \sum\limits_{j \neq k} \beta_j\right) - c_k\right)^2} \geq 0.$$

Therefore, $\phi(c_k)$ is increasing, and $\sigma_k(c_k, \beta_k)$ is decreasing in c_k. ■

Discussion. The case of the smallest slope among all marginal costs corresponds to the largest σ, which means that the player has competitive advantages in the markets. He can deviate more from the marginal cost and influence the final price more. Therefore, such a company is more inclined to trade on the spot market. And vice versa. Forward trading primarily benefits companies with low marginal cost elasticity (or high MC slope) relative to other players. This conclusion implies that in electricity markets, for systems consisting primarily of capacities with low marginal cost elasticities, the forward market is small.

3 Spot and Forward Market Modeling on the Electricity Market Example

Consider a system similar in some parameters to the electric power system "Siberia". The features that should be taken into account in the calculations are the generating companies including several power plants.

3.1 Initial Data

We use a scheme with 15 nodes (Fig. 1) and model formation of the price taking into account the strategic interaction of producers in the market. We use the main characteristics of generation and consumption at the nodes of the system, average hourly consumption, and average annual costs of producers in the system like the electric power system "Siberia". We discuss the modeling of the interaction on spot and forward without network constraints, based on the models described in Sect. 2.

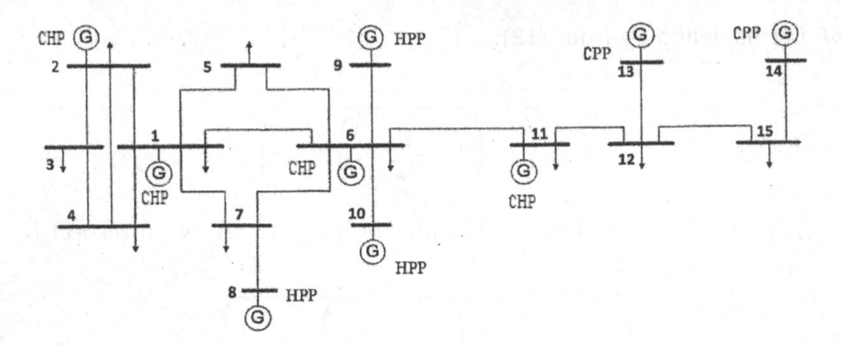

Fig. 1. Electric power system diagram.

The demand function in our model is linear. As usual, we assume that the electricity demand has low elasticity close to zero. The demand elasticity in [22] is −0.165 for the urban population and −0.28 for the rural population. These data have been calculated for Siberia (Novosibirskenergo). In this case, the parameter of the slope of the demand function corresponding to the elasticity of −0.3 is $\gamma = 40$.

Table 1. Demand and supply characteristics of the electricity market.

N of nodes generation	Type GP	q_{max}	a_k	c_k	N of nodes consumption	q_d
1	CHP	4297	20	0.08	2	2439
2	CHP	3214	12.4	0.09	3	1586
6	CHP	7906	21	0.042	4	1829
11	CHP	4762	18	0.064	5	1336
13	CPP	2020	6.8	0.2	6	6487
14	CPP	665	9.12	0.588	7	8023
8	HPP	4781	–	–	11	8268
9	HPP	3873	–	–	12	795
10	HPP	8657	–	–	15	1087

Producers (CHP[1], CPP[2] and HPP[3]) are divided into strategic producers that significantly influence the price, the price makers, and the price-takers. The second group consists of HPP that are assumed to have zero marginal costs; they are present in the market by providing information only on the volumes of generated power. All plants have constraints on generation. We consider a planning problem and assume that transmission networks do not impose constraints on market equilibrium. It should be said that the addition of network

[1] Co-generation of heating and power.
[2] Condensation electric power-plant.
[3] Hydropower plants.

restrictions to the problem greatly complicates the formulation and subsequent analysis. Forward trading problems with network restrictions are solved using EPEC approaches, simplifying the supply functions [17].

Table 1 presents the characteristics of generating companies competing in the forward and spot markets. For HPP, production limits are given.

3.2 Results of Modeling

We have $m = 6$ generating stations in our model for which it is necessary to set the values of conjectural variations $w_{jk} \in [-1, 0], \forall j \neq k$, $\sum_{j=1,m,j\neq k} w_{jk} \in [-1, 0]$. In our case, there are 5 competitors. This amount is determined from the characteristics of the electricity system used for testing in the forward market. There are 9 generating stations in total. Hydroelectric power plants do not compete in the electricity market, since they have "almost zero" variable costs, and in all electricity markets, according to the rules, they are considered price-taking. The remaining 6 stations as for the stations located in the 13-th and 14-th nodes belong to the same company.

If everyone believes that reactions of other competitors to changes in sales in the forward market are symmetrical, the value of $w_{jk} \in [-0.2, 0]$. The low bound -0.2 corresponds with the reaction in the Bertrand model and should get the maximum forward value of x_k.

The results are presented in Tables 2, 3. Table 2 contains output values of the spot market competitors 1, 2, 6, 11, 13, 14. The power plants in nodes 13 and 14 are owned by the same company. We have performed several types of models: Cournot model (K), linear supply function equilibrium (LSFE), Bertrand model (B), the model with the forward market (LSFE+F). The last line of Table 2 contains market prices. Prices in nodes of the model are the same. It is assumed that there are no transmission restrictions and losses. Restrictions on power plant output in models LSFE and LSFE+F are performed by iteration procedure [23]. The demand is partially covered by hydroelectric generation companies, which have zero marginal costs and are price-takers in the electricity market.

Table 2. Comparison of generation volumes q_k (MWh) of companies in the electricity market for different models.

Node/Model	B	LSFE+F $w_{jk} = -0.2$	LSFE+F $w_{jk} = -0.1$	LSFE+F $w_{jk} = -0.01$	LSFE	K
1	2974	3055	2986	2919	2912	2766
2	2728	2843	2776	2711	2595	2595
6	5924	5205	5118	5036	5027	4403
11	3749	3711	3626	3548	3539	3680
13	1255	1407	1369	1333	1328	1362
14	423	492	478	466	464	499
Sum $\sum_k^m q_k$	17053	16713	16353	16013	15974	14902
Price $/MWh	36.7	38.1	39.4	40.8	41	45.1

The highest consumption and lowest prices correspond to the Bertrand model or the perfect competition model. This model does not work for real electricity markets and we provide it here as a reference point. The highest prices and lowest generation rates correspond to the Cournot model. The model with the forward market improves the competitive situation relative to the LSFE model, bringing the result closer to a perfectly efficient one. The introduction of preliminary sales leads to an increase in total output and a decrease in the elasticity of demand in the spot market, which contributes to falling prices. Figure 2 shows the dependence of the company's total output on the volume of the forward trading, confirming the previous thesis: an increase in x_k leads to an increase in q_k.

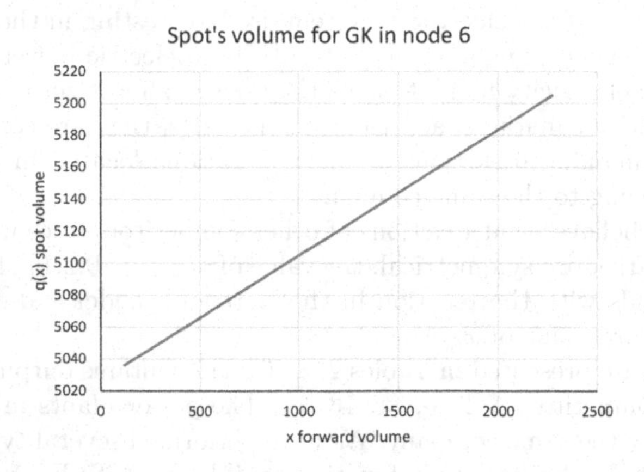

Fig. 2. Dependence of total electricity sales volumes on the size of forward trade for GK located in node 6.

Table 3 shows company-by-company results for several cases of the model with conjectural variations. The size of the forward market decreases with the decrease in the reaction on the output changes of each other's w_{jk}.

Primarily companies with high c_k trade on the forward market, which confirms the conclusions from Sect. 3 about the forward market size. For high values of conjectural variations in the forward market $w_{jk} = -0.2$, we have the largest volumes of forwarding trade of companies relative to other values of w_{jk} (Table 3). Some companies, namely nodes 2, 13, and 14, prefer to supply the entire volume of electricity to the forward market. It is these companies that have the highest slope of marginal costs relative to others. The smaller the assumed coefficient of conjectural variations w_jk, the smaller electricity volumes x_k is supplied by companies with the low slope of marginal costs supply to the forward market. Such companies supply as much as possible electricity volumes to the spot market. The last company to start trading in the spot market is the company at node 14. This station has the highest slope of marginal cost.

The power system in the illustrative example is similar to the Siberia power system, which has a large share of cheap electricity. With these advantages,

Table 3. Changes in the supply volumes of companies in the forward x_k market when the value of the conjectural variations changes w_{jk}.

Node	1	2	6	11	13	14
$w_{jk} = -0.2$, 75% in forward						
$x_k$forward	2829	2843	2276	2749	1407	492
$q_k - x_k$spot	226	0	2929	962	0	0
$q_k$sum	3055	2843	5205	3711	1407	492
$w_{jk} = -0.15$, 60% in forward						
x_k	2155	2241	1730	2084	1388	485
$q_k - x_k$	866	569	3432	1585	0	0
q_k	3021	2810	5162	3669	1388	485
$w_{jk} = -0.1$, 45% in forward						
x_k	1459	1517	1169	1405	1369	478
$q_k - x_k$	1527	1259	3949	2221	0	0
q_k	2986	2776	5118	3626	1369	478
$w_{jk} = -0.05$, 26% in forward						
x_k	741	770	592	710	877	471
$q_k - x_k$	2209	1971	4481	2873	442	0
q_k	2950	2741	5073	3583	1349	471
$w_{jk} = -0.01$, 6% in forward						
x_k	150	156	118	143	177	185
$q_k - x_k$	2769	2555	4917	3405	1156	281
q_k	2919	2711	5036	3548	133	466

generating companies tend to trade on the spot market. The main conclusion is that for power systems similar to the system "Siberia", if the volume of the forward trade is obtained only by market mechanisms, then it will be small, and, therefore, it will be poorly to reduce the market power.

4 Conclusion

The paper considers the forward electric power market model. Using the supply function equilibrium model, we work out the ratio of supply volumes in two markets: spot and forward. In the case of interaction between companies with different costs, not all players enter the forward market. This is also true for the spot market. The size of the trade is determined by the elasticity of the demand function, costs and the response to changes in competitor's supply volumes. The model is designed and analyzed for the case with generating companies that own several capacities.

Our finding is that the companies with a more increasing cost curve and therefore the faster-growing cost of sales per unit of production will prefer the forward market because they would face extra expenses on the residual demand in the spot market and would be less competitive. This is illustrated by an example where companies with a lower MC slope sell relatively less volume in the forward market than generator companies with a high MC slope. Therefore, in power systems where there are a lot of cheap large generating capacities, the forward market is small.

Appendix

*ProofProposition*2. Let us write FOC taking into account (13)–(16) and $(p - a_k - c_k q_k) = \sigma_k (q_k - x_k) / \beta_k$

$$\frac{\partial \pi_k}{\partial x_k} = -B\sigma_k q_k + \beta_k B \sigma_k (q_k - x_k) / \beta_k - B \sum_{j, j \neq k} \left(q_k \sigma_j + \sigma_k (q_k - x_k) \sigma_j \frac{\beta_k}{\beta_k} \right) w_{jk}.$$

Hence

$$\frac{\partial \pi_k}{\partial x_k} = B \cdot \left(-\sigma_k x_k + (q_k + \sigma_k (q_k - x_k)) \cdot \sum_{j, j \neq k} \sigma_j w_{jk} \right).$$

Then

$$\frac{\partial^2 \pi_k}{\partial x_k^2} = -B \cdot \left[\sigma_k \left(1 + \sum_{j, j \neq k} \sigma_j w_{jk} \right) \right.$$

$$\left. -\beta_k B (1 + \sigma_k) \cdot \sum_{j, j \neq k} \sigma_j w_{jk} + \beta_k B \cdot \left(\sum_{j, j \neq k} \sigma_j w_{jk} \right)^2 \right].$$

Since $-1 \leq \sum_{j \neq k} \sigma_j w_{jk} \leq 0$, then the first, second, and third terms in brackets are positive. Accordingly, $\frac{\partial^2 \pi_k}{\partial x_k^2} \leq 0$. The payoff function of the company π_k is concave with respect to x_k.

References

1. Green, R.: The electricity contract market in England and Wales. J. Ind. Econ. **47**(1), 107–124 (1999)
2. Giulietti, M., Grossi, L., Waterson, M.: Price transmission in the UK electricity market: was NETA beneficial? Energy Econ. **32**(5), 1165–1174 (2010)
3. Green, R.: Retail competition and electricity contracts. Faculty Econ. (2004)
4. Klemperer, P., Meyer, M.: Supply Function Equilibria in Oligopoly under uncertainty. Econometrica **57**, 1243–1277 (1989)
5. Newbery, D.: Analytic solutions for supply function equilibria: uniqueness and stability. In: Cambridge Working paper in Economics 0824 (2008)

6. Rudkevich, A.: Supply function equilibrium in power markets: learning all the way. TCA technical paper 1299, N 1702 (1999)
7. Vasin, A.A., Kartunova, P.A.: Auctions of homogeneous goods: game-theoretic analysis. Contrib. Game Theory Manag. **8**(10), 315–335 (2015)
8. Anderson, E.J., Hu, X.: Forward contracts and market power in an electricity market. Int. J. Ind. Organ. **26**(3), 679–694 (2008)
9. Bushnell, J.: Oligopoly equilibria in electricity contract markets. J. Regul. Econ. **32**(3), 225–245 (2007)
10. Yu, C.W., Zhang, S.H., Wang, X., Chung, T.S.: Modeling and analysis of strategic forward contracting in transmission constrained power markets. Electr. Power Syst. Res. **80**(3), 354–361 (2010)
11. Vasin, A., Vasina, P.: The impact of the forward market on producer's market power. Working Paper # WP/2008/085 - Moscow, New Economic School, 28p. (2008). (in Russian)
12. Aizenberg, N.: Application of supply function equilibrium model to describe the interaction of generation companies in the electricity market. In: Kochetov, Y., Khachay, M., Beresnev, V., Nurminski, E., Pardalos, P. (eds.) DOOR 2016. LNCS, vol. 9869, pp. 469–479. Springer, Cham (2016). https://doi.org/10.1007/978-3-319-44914-2_37
13. Ayzenberg, N., Kiseleva, M., Zorkaltsev, V.: Models of imperfect competition in analysis of Siberian electricity market. J. New Econ. Assoc. **2**, 88–98 (2013). (in Russian)
14. Baldick, R., Grant, R., Kahn, E.: Theory and application of linear supply function equilibrium in electricity markets. J. Regul. Econ. **25**, 143–167 (2004)
15. Vasin, A., Dolmatova, M., Gao, H.: Supply function auction for linear asymmetric oligopoly: equilibrium and convergence. Procedia Comput. Sci. **55**, 112–118 (2015)
16. Aizenberg, N.I.: Analysis of the mechanisms of functioning of wholesale electricity markets. ECO **6**, 97–112 (2014). (in Russian)
17. Zhang, S.X., Chung, C.Y., Wong, K.P., Chen, H.: Analyzing two-settlement electricity market equilibrium by coevolutionary computation approach. IEEE Trans. Power Syst. **24**(3), 1155–1164 (2009)
18. Rosen, J.B.: Existence and uniqueness of equilibrium points for concave N-person games. Econometrica **33**, 347–351 (1965)
19. Mas-Colell, A., Whinston, M.D., Green, J.R.: Microeconomic Theory. Oxford University Press, New York (1995)
20. Sherali, H.D.: A multiple leader Stackelberg model and analysis. Oper. Res. **32**(2), 390–404 (1984)
21. Su, C.-L.: Equilibrium problems with equilibrium constraints: stationarities, algorithms, and applications. Ph.D Dissertation. Stanford University (2005)
22. Nahata, B., Izyumov, A., Busygin, V., Mishura, A.: Application of Ramsey model in transition economy: a Russian case study. Energy Econ. **29**, 105–125 (2007)
23. Anderson, E., Hu, X.: Finding supply function equilibria with asymmetric RMS. Oper. Res. **56**, 697–711 (2008)

Meta Algorithms for Portfolio Optimization Using Reinforcement Learning

Yury Kolomeytsev(✉)

Lomonosov Moscow State University, Leninskie Gory,
Moscow 119991, Russian Federation
yury.kolomeytsev@gmail.com

Abstract. We explore the effectiveness of various machine learning algorithms, especially deep reinforcement learning, for solving the portfolio optimization problem. The investigated algorithms can be divided into the following groups: "Follow-the-Winner" (using trend follow principle), "Follow-the-Loser" (using mean reversion principle) "Pattern-Matching" (for example, using correlations between current prices and historical prices), "Meta-Learning Algorithms" (combination of several algorithms) and, finally, strategies based on reinforcement learning. We propose a novel meta-learning algorithm based on deep reinforcement learning, which we call MetaRL. It utilizes algorithms from different groups and optimizes their combination. Reinforcement learning (RL) technique is well suited to solve portfolio selection tasks because it can learn hidden dependencies in financial data and trading patterns by trial and error, directly predicts trading actions, and allows to consider delayed rewards. Moreover, RL naturally takes into account transaction costs and can model real trading when the strategy can impact the market. As agents, we use deep neural networks that decide which of the algorithms will be more profitable at a particular moment. We provide extensive experiments using financial data and show that our new MetaRL algorithm substantially outperforms other portfolio optimization methods.

Keywords: Portfolio optimization · Reinforcement learning · Deep learning · Quantitative finance · Meta learning

1 Introduction

Online portfolio selection (OLPS), which aims to maximize cumulative wealth by sequentially rebalancing a portfolio, is an important task for asset portfolio management. Earlier, all trades on exchanges were made by human traders and were executed in open outcry markets on a trading floor. It involved different forms of communication like shouting and using hand signals to buy or sell orders. However, with the development of computers and technology, the open outcry method has been replaced by electronic trading systems which have made trading faster, more convenient and easier. Nowadays we are able to get a huge

N. N. Olenev et al. (Eds.): OPTIMA 2021, CCIS 1514, pp. 154–168, 2021.
https://doi.org/10.1007/978-3-030-92711-0_11

amount of different information that could be used for trading decisions: asset prices, news, economic conditions, fundamental information about companies. These new market conditions require the development of new algorithms that will find hidden dependencies in data. That is why portfolio selection has been broadly studied by many researchers in finance, statistics, machine learning, data mining, and artificial intelligence. More and more trades nowadays are made by decisions of trading robots. And this is reasonable because, unlike humans, robots can instantly process large amounts of information, constantly follow a specified strategy and have no emotion, do not have any needs and problems that can negatively affect the performance of trading. Using algorithmic trading and steadily following a specified strategy can ensure to potentially receive income from the market and avoid human errors.

The main focus of this paper is devoted to deep learning and reinforcement learning and their applications for solving portfolio management problem. Deep learning (DL) models allow to find complex non-linear hidden dependencies in data by using deep artificial neural networks containing multiple layers. Reinforcement learning (RL) allows an agent to learn close to optimal behavior in an environment by interacting with it and improves its performance by trial and error. One attractive feature of reinforcement learning is considering delayed rewards, which helps us to take into account transaction costs. It allows an agent to learn situations when it is not worth rebalancing a portfolio if the immediate reward is lower than the transaction costs that are paid for this rebalancing.

The main contributions of this paper are the following. We investigate the effectiveness of a wide class of machine learning algorithms, especially deep reinforcement learning for solving online portfolio selection problem. These include the research of existing OLPS algorithms, understanding their drawbacks and limitations, and improving them. Moreover, we propose a novel meta-learning OLPS algorithm that was called MetaRL. It is based on deep reinforcement learning and utilizes different algorithms by their combination. Apart from that, we developed a trading platform with a user-friendly graphical user interface for backtesting in order to compare the performance of various strategies. We provide extensive backtesting of different OLPS methods including our new MetaRL algorithm using stock data from Moscow Exchange (MOEX) [12] from 2014 till 2019. It appears that our new MetaRL algorithm outperforms all other portfolio selection algorithms using all main metrics of financial performance such as Sharpe ratio, final cumulative return and maximum drawdown.

2 Related Work

Over the years, a lot of research has been performed in portfolio management and online portfolio selection. In this section we discuss works that are most relevant to our research. In [9] Bin Li *et al.* made an extensive survey of different online portfolio selection strategies including benchmarks and state-of-the-art algorithms, and implemented OLPS toolbox in Matlab. The portfolio management methods can be classified into four categories: "Follow-the-Winner", "Follow-the-Loser", "Pattern-Matching", and "Meta-Learning".

When using Follow-the-Winner strategy, we increase the weights of more successful assets hoping that they will continue to rise in price. This group includes such strategies as Universal Portfolios [2] and others.

When using Follow-the-Loser strategy, we transfer wealth from winners to losers under the mean reversion assumption, which assumes that if there is no trend, price will return to its average computed for some recent past. So, we expect that very good-performing assets will perform poorly in the future, and vice versa, poorly-performing assets will perform well.

Pattern-Matching based strategies may utilize both winners and losers. These approaches consider the non-independent identically distributed market and maximize the conditional expectation of the return using past observations. Nonparametric Nearest Neighbor Log-optimal Strategy (BNN), proposed in [5], combines the nearest neighbor sample selection and log-optimal utility function. Correlation-driven nonparametric learning approach (CORN) was proposed by Li *et al.* [10]. This method combines correlation-driven sample selection and log-optimal utility function.

Meta-Learning is another promising group of algorithms for OLPS that is closely related to expert learning [1]. Generally, meta-agents assume several base experts from different classes of algorithms. Each expert predicts a portfolio vector for the next period, and a meta-agent combines these portfolios to form a final portfolio. This final portfolio is used for the next rebalancing. In [3] authors propose two meta optimization algorithms, named Online Gradient Update and Online Newton Update, which are extensions of Exponential Gradient and Online Newton Step.

In [4,7] the authors investigate the Recurrent Reinforcement Learning approach for solving OLPS model for one asset. The performance functions that they consider for reinforcement learning are profit or wealth, economic utility, the Sharpe Ratio (SR), and also they propose Differential Sharpe Ratio (DSR). However, they provide the solution of the OLPS problem only for one asset. More recent study [6] use deterministic policy gradient method to fit RL agent to trade on cryptocurrency market. In [11] authors also use similar approach to trade on China stock market. Apart from policy gradient they implement Deep Deterministic Policy Gradient (DDPG) and Proximal Policy Optimization (PPO) algorithms. Although DDPG and PPO algorithms are more complicated and have nice theoretical properties, it appears that in the experiments DDPG and PPO algorithms work worse than PG. Inspired by the results of these studies, we also use reinforcement learning for solving OLPS. However, instead of directly managing assets, we develop a Meta-Learning approach and manage other strategies.

3 Problem Statement

Now we will mathematically formulate online portfolio selection task. Consider trading of N assets for T periods. The closing prices of all assets form a price vector $\mathbf{p}_t = \{p_{t,n}\}_{n=1}^{N}$, where $p_{t,n}$ is the closing price of n-th asset at time t.

In order to trade successfully in situations when most assets are going down, we should be able to move investment money into risk-free asset. So it simply indicates that we do not invest this amount of money. We assume that one of the N assets is risk-free. The price change is represented by a N-dimensional price relative vector $\mathbf{x}_t \in \mathbb{R}_+^N, t = 1, \ldots, T$, where the n-th element of t-th prices relative vector $x_{t,n}$ denotes the ratio of t-th closing price to last closing price for the n-th assets: $x_{t,n} = \frac{p_{t,n}}{p_{t-1,n}}$. The price relative of risk-free asset is always equal to 1. So we have

$$\mathbf{x}_t = \Big(x_{t,1}, \ldots, x_{t,N}\Big) = \Big(\frac{p_{t,1}}{p_{t-1,1}}, \ldots, \frac{p_{t,N}}{p_{t-1,N}}\Big).$$

So, on period t an investment in asset n increases by $x_{t,n}$ factor. Denote the market price changes from period t_1 to t_2 ($t_2 > t_1$) by a market window. It consists of a sequence of price relative vectors $\mathbf{x}_{t_1}^{t_2} = \{\mathbf{x}_{t_1}, \ldots, \mathbf{x}_{t_2}\}$, where t_1 denotes the beginning period and t_2 denotes the ending period. Our whole trading period will be $\mathbf{x}_1^T = \mathbf{x}_1, \ldots, \mathbf{x}_T$.

At the beginning of the t-th period, an investment is specified by a portfolio vector $\mathbf{w}_t = (w_{t,1}, \ldots, w_{t,N}), t = 1, \ldots, T$. The n-th element of t-th portfolio, $w_{t,n}$, represents the proportion of capital invested in the n-th asset. Typically, it is assumed a portfolio is self-financed and no margin/short is allowed. Thus, a portfolio satisfies the constraint that each entry is non-negative and all entries sum up to one, that is, $\mathbf{w}_t \in \Delta_N$, where $\Delta_N = \{\mathbf{w} : \mathbf{w} \geq 0, \mathbf{w}^\top \mathbf{1} = 1\}$. Here, $\mathbf{1}$ is the N-dimensional vector of all 1, and $\mathbf{w}^\top \mathbf{1}$ denotes the inner product. Let's assume that at the very beginning everything is in risk-free asset. Then a portfolio strategy is a sequence of mappings:

$$\mathbf{w}_1 = (\frac{1}{N}, \ldots, \frac{1}{N}), \quad \mathbf{w}_t : \mathbb{R}_+^{N(t-1)} \longrightarrow \Delta_N, t = 2, 3, \ldots, T,$$

where $\mathbf{w}_t = \mathbf{w}_t(\mathbf{x}_1^{t-1})$ denotes the portfolio computed from the past market window $\mathbf{x}_1^{t-1}$. Let us denote the portfolio strategy for T periods as $\mathbf{w}_1^T = (\mathbf{w}_1, \ldots, \mathbf{w}_T)$.

Later we will reformulate this problem considering short positions.

For the t-th period, a portfolio manager divides its capital according to portfolio $\mathbf{w}_t$ at the opening time, and holds the portfolio until the closing time. Thus, the portfolio wealth will increase by a factor of $\mathbf{w}_t^\top \mathbf{x}_t = \sum_{n=1}^N w_{t,n} x_{t,n}$. Since this model uses price relatives and reinvests the capital, the portfolio wealth will increase multiplicatively. From period 1 to T, a portfolio strategy $\mathbf{w}_1^T$ increases the initial wealth S_0 by a factor of $\prod_{t=1}^T \mathbf{w}_t^\top \mathbf{x}_t$ that is, the final cumulative wealth after a sequence of T periods is

$$S_T(\mathbf{w}_1^T) = S_0 \prod_{t=1}^T \mathbf{w}_t^\top \mathbf{x}_t = S_0 \prod_{t=1}^T \sum_{n=1}^N w_{t,n} x_{t,n}. \tag{1}$$

In practice very often investment funds that have a lot of money under management face a problem with liquidity and market impact because of high trading volumes. That is why there is an upper bound on the invested money into each asset and each asset has its own bound. Thus in such cases using cumulative wealth metric may be not suitable. In this situation we can use not cumulative wealth which means that we always invest fixed amount of money S_0 and do not reinvest the money we earned. The formula for this type of final wealth takes the following form

$$\hat{S}_T(\mathbf{w}_1^T) = S_0\big(1 + \sum_{t=1}^{T}(\mathbf{w}_t^\top \mathbf{x}_t - 1)\big) = S_0\big(1 + \sum_{t=1}^{T}\sum_{n=1}^{N}(w_{t,n}x_{t,n} - 1)\big). \quad (2)$$

It should be mentioned that apart from final cumulative wealth there are some other utility functions that measure trading performance. Let us consider Sharpe ratio which in addition to returns takes into account the risk of the strategy that is the standard deviation of the returns. Denote R_t as the profitability of a strategy for period t calculated by

$$R_t = \mathbf{w}_t^\top \mathbf{x}_t - 1,$$

then we can calculate Sharpe Ratio formula

$$SR_T(\mathbf{w}_1^T) = \frac{\hat{R}_T}{\sigma(R_T)}, \quad (3)$$

where $\hat{R}_T = \frac{1}{T}\sum_{t=1}^{T} R_T$, $\sigma(R_T) = \sqrt{\frac{1}{T}\sum_{t=1}^{T}(R_t - \hat{R}_T)^2}$.

Summing up, the goal of OLPS task is to make portfolio strategy $\mathbf{w}_1^T$ that maximizes some utility function that can be, for example, the portfolio cumulative wealth $S_n(\mathbf{w}_1^T)$ or Sharpe Ratio $SR_T(\mathbf{w}_1^T)$.

4 Algorithms for OLPS

In this section we will briefly describe the algorithms that we study in this research project. They can be divided into 8 categories. The grouping of considered algorithms is outlined in Table 1.

The first category is "Baselines" and contains baseline strategies. The second category is "Follow-the-Winner" also known as momentum investing, which follows the current market trend and invests into stocks that perform better at the current moment. The second category is named "Follow-the-Loser" and it, on the contrary, moves wealth from winning assets to losers following the principle of mean reversion. The "Pattern-Matching" approach constructs a portfolio by finding similar patterns of price changes of assets in history. The forth approach named "Meta-Learning" combines multiple strategies into one. The next category is for strategies based on reinforcement learning which is one of the main topic of this research, thus the description of this approach can be found in a separate chapter. Another category is for strategies based on Modern Portfolio Theory (MPT), also known as mean-variance analysis, that was developed by

Harry Markowitz. These strategies consider 2 factors of a portfolio: expected return and risk. They construct portfolios that are located on the efficient frontier. Efficient frontier is the set of portfolios that satisfy the condition that no other portfolio exists with a higher expected return but with the same standard deviation of return. The last category is for strategies that could not be classified into any of previous categories.

Table 1. Portfolio selection algorithms.

Category	Algorithm
Baselines	Buy And Hold (BAH)
	Best Stock (BS)
	Constant Rebalanced Portfolios (CRP)
Follow-the-Winner	Universal Portfolios (UP)
	Exponential Gradient (EG)
	Online Newton Step (ONS)
	Trend Follow (TF)
	Bollinger
	Maximum Return
Follow-the-Loser	Anti Correlation (Anticor)
	Passive Aggressive Mean Reversion (PAMR)
	Weighted Moving Average Mean Reversion (WMAMR)
	Confidence Weighted Mean Reversion (CWMR)
	Online Moving Average Reversion (OLMAR)
	Robust Median Reversion (RMR)
	Combination Forecasting Reversion (CFR)
Pattern-Matching	Nonparametric Kernel-based Log-optimal Strategy (BK)
	Nonparametric Nearest Neighbor Log-optimal Strategy (BNN)
	Correlation-driven Nonparametric Learning Strategy (CORN)
Meta-Learning	Online Gradient Updates (OGU)
	Online Newton Updates (ONU)
	Boosting Moving Average Reversion (BMAR)
	Reinforcement Learning based Meta-Learning (MetaRL)
Reinforcement Learning	Policy Gradient
Modern Portfolio Theory	Markowitz Portfolio
	Minimum Variance Portfolio
	Maximum Sharpe Portfolio
Other	Online ARMA
	Online ARIMA
	Peak Price Tracking-Based Learning (PPT)
	Short-term Sparse Portfolio Optimization (SSPO)

5 Reinforcement Learning Background

Reinforcement Learning is a machine learning technique when a system (agent) is trained by interacting with an environment, and uses the received feedback from its own actions and experiences to learn the optimal behavior. An action moves an agent to a new state and agent gets some reward that can be positive or negative. The goal is to find a suitable action model that would maximize the total cumulative reward of the agent.

Formally, we can define the reinforcement learning problem as follows. At discrete time steps $t \geq 0$ the agent observes information about the environment and finds itself in a state $s_t \in \mathcal{S}$ where $\mathcal{S}$ is a space of all system's states. Based on s_t and its previous experience, the agent makes an action $a_t \in \mathcal{A}$ where $\mathcal{A}$ is a space of all available actions. After making the action agent moves to a new state s_{t+1} and receives a reward $r_{t+1} \in \mathbb{R}$ from the environment.

Agent's behavior is represented by its policy π which is a mapping from states to actions $\pi : \mathcal{S} \to \mathcal{A}$. The policy may be deterministic or stochastic. A deterministic policy gives a unique action given a state: $a = \pi(s)$. A stochastic policy $\pi(a|s)$ can be interpreted as a probability of taking an action a in a given state s. To behave optimally the agent should use deterministic policies. However, at the beginning of learning process, when the agent does not know enough information about the environment, it may be helpful explore the environment by using stochastic policies and then reduce the stochasticity while learning optimal behavior.

After making T steps the agent gets cumulative reward for this episode, which is also called the return:

$$R_t = \sum_{k=1}^{T} r_{t+k}.$$

In case of continuing tasks when $T = \infty$ the return can be infinite. To solve this problem the discounted return is used

$$R_t = \sum_{k=1}^{T} \gamma^{k-1} r_{t+k},$$

where $\gamma \in [0, 1]$ is a discount rate. This means that a reward received in the future is worth less than the immediate reward. If $\gamma < 1$ and the sequence of rewards is bounded then infinite sum will have a finite value. When $\gamma = 0$ the agent is only interested in maximizing the immediate rewards. When γ approaches to 1, the agent considers future rewards more strongly.

Finally, the main goal of reinforcement learning agent is to find a policy that maximizes the expected return:

$$\mathbb{E}_\pi(R) \to \max_\pi.$$

6 A Novel Meta-learning Algorithm Using Reinforcement Learning

We propose a novel Meta-Learning algorithm for OLPS based on reinforcement learning that will be further called MetaRL agent. Unlike previous works in which RL agents managed the assets directly and produced actions whether to buy or sell a certain amount of the particular asset, in our approach RL agent manages other strategies. These strategies under management will be called base strategies. Base strategies may come from different OLPS categories, operate with different number of assets and use different information to predict portfolio vectors. Although these strategies may be weak and volatile when used separately, their proper combination can lead to high and stable returns.

The main goal of MetaRL agent is to find out what base strategies will be more profitable for the current trading period given the previous returns of base strategies. At each period it should assign a positive weight to each base strategy. All weights should finally sum to 1. These weights can be interpreted as the degree of confidence of our agent that the particular base strategy will be profitable or not: the greater the weight of the strategy, the more confident is our meta agent that the base strategy will be profitable.

The key point of combining base strategies is their diversity in order to capture more opportunities of gaining returns from the market in different periods. To achieve that one should try to consider a lot of base strategies that have small correlations between each other. In the first part of the research we investigated and implemented different approaches for solving OLPS problem that can be used in our new Meta-Learning strategy. All in all we have 31 different strategies. Some of these strategies work differently depending on their hyper-parameters and can capture different market dependencies. For example, depending on the size of windows of trend follow strategy it can capture short-term trends or long-term trends. Thus we can run all of our strategies with different parameters, pick the most profitable and stable ones on a train set and feed them into our MetaRL agent.

6.1 Proposed Algorithm

Now let us formally describe our MetaRL algorithm step by step. First of all we will describe training part.

Assume that we have N assets and K_0 base strategies. First of all we should filter base strategies and take only those that show good performance on a train set. As metrics of good performance we take Sharpe ratio, Maximum drawdown and Final return. Assume that after filtering we have K base strategies left which will be combined.

Denote $r_{t,k}$ as the return of k-th strategy for t-th period. We can compute the cumulative return $c_{t,k}$ of this strategy as

$$c_{t,k} = \prod_{i=0}^{t} r_{i,k}.$$

Denote $\mathbf{R}$ as a matrix of size $T \times K$ of cumulative returns of K base strategies for T periods. Its t-th row looks the following way:

$$\mathbf{R}_t = (c_{t,1}, c_{t,2}, \ldots, c_{t,K}) = (\prod_{i=0}^{t} r_{i,1}, \prod_{i=0}^{t} r_{i,2}, \ldots, \prod_{i=0}^{t} r_{i,K}). \quad (4)$$

At time t our Meta agent observes previous cumulative returns of strategies on a window of size M which is a hyper-parameter, and at a particular step t we fed our neural network with a sub-matrix

$$\hat{\mathbf{R}}_t = (\mathbf{R}_{t-M}, \ldots, \mathbf{R}_{t-1})^\top.$$

In order to successfully train a neural network this input sub-matrix is normalized by the last value at time $t-1$:

$$\hat{\mathbf{R}}_t^{\text{norm}} = (\mathbf{R}_{t-M} \oslash \mathbf{R}_{t-1}, \mathbf{R}_{t-M+1} \oslash \mathbf{R}_{t-1}, \ldots, \mathbf{I})^\top,$$

where $\oslash$ denotes an element-wise division operator and $\mathbf{I} = (1, 1, \ldots, 1)$.

Denote weights of i-th base strategy at time t as a vector $\mathbf{b}_i^t = (b_{1,i}^t, \ldots, b_{N,i}^t)^\top$.

To deal with transaction costs at each period apart from previous historical returns of base strategies at each time t we give our Meta agent a matrix $\mathbf{B}_t$ of predicted weights of base strategies which has size $N \times K$:

$$\mathbf{B}_t = \begin{pmatrix} b_{1,1}^t & \cdots & b_{1,K}^t \\ \vdots & \ddots & \vdots \\ b_{N,1}^t & \cdots & b_{N,K}^t \end{pmatrix}.$$

The i-th column of that matrix represents the portfolio vector of i-th base strategy at time t.

Assume that for period $t-1$ the prediction of our MetaRL agent was $\mathbf{w}_{t-1}$. At the beginning of t period our portfolio vector due to price changes became $\mathbf{w}'_t = (w'_{t,1}, \ldots, w'_{t,N}) = \frac{\mathbf{x}_t \odot \mathbf{w}_{t-1}}{\mathbf{x}_t \mathbf{w}_{t-1}}$ and we should construct $\mathbf{w}_t$. We compute a vector of costs of rebalancing current portfolio vector to weights of particular strategy

$$C_t = (c_1^t, \ldots, c_K^t) = \Big(\sum_{n=1}^{N} |b_{n,1}^t - w'_{t,n}|, \ldots, \sum_{n=1}^{N} |b_{n,K}^t - w'_{t,n}| \Big).$$

Then we normalize this vector

$$\hat{C}_t = \Big(\frac{1 - c_1^t}{\sum_{k=1}^{K} c_k^t}, \ldots, \frac{1 - c_K^t}{\sum_{k=1}^{K} c_k^t} \Big). \quad (5)$$

Obtained vector $\hat{C}_t$ means how profitable for us is to choose a particular base strategy in terms of transaction costs (more profitable strategies have higher value). It is fed to our neural network along with cumulative returns of base strategies.

Our neural network predicts the weight to every base strategy which forms a vector

$$\mathbf{p}_t = (p_{t,1}, \ldots, p_{t,K}),$$

which satisfies the conditions: $\sum_{k=1}^{K} p_{t,k} = 1$ and $p_{t,k} \geq 0$, $k = 1, \ldots, K$.

Then we can calculate the final portfolio vector of our strategy

$$\mathbf{w}_t = (w_{t,1}, \ldots, w_{t,N}) = \Big(\sum_{i=1}^{K} p_{t,i} b_{1,i}^t, \ldots, \sum_{i=1}^{K} p_{t,i} b_{N,i}^t\Big). \tag{6}$$

After we finally obtained $\mathbf{w}_t$ we should calculate the reward function using the actual prices of assets and make a gradient step.

Denote θ as the parameters that define our policy $a_t = \pi_\theta(s_t)$. The goal of the agent is to maximize the return calculated for an episode of size $\hat{t}$. In our experiments, we maximize the average logarithmic cumulative return taking into account transaction costs c:

$$\begin{aligned} R(s_1, a_1, \ldots, s_{\hat{t}}, a_{\hat{t}}) &= \frac{1}{t} \log \prod_{t=1}^{\hat{t}} \Big(\mathbf{w}_t^\top \mathbf{x}_t - c \sum_{n=1}^{N} |w_{t-1,n} - w_{t,n}|\Big) \\ &= \frac{1}{t} \sum_{t=1}^{\hat{t}} \log(\mathbf{w}_t^\top \mathbf{x}_t - c \sum_{n=1}^{N} |w_{t-1,n} - w_{t,n}|). \end{aligned} \tag{7}$$

The performance metric of our policy for an episode of size $\hat{t}$ will be

$$J_{\hat{t}}(\pi_\theta) = R(s_1, \pi_\theta(s_1), \ldots, s_{\hat{t}}, \pi_\theta(s_{\hat{t}})).$$

Then we should update the parameters of our agent along the gradient direction:

$$\theta_{\text{new}} \leftarrow \theta_{\text{old}} + \lambda \nabla_\theta J_{\hat{t}}(\pi_\theta),$$

where λ is a learning rate.

We also tried to use Sharpe ratio as a reward. Both reward functions consider transaction costs. It turned out that using logarithmic cumulative return, our agent trained better than when using Sharpe ratio and had much better performance.

Denote θ as the parameters that define our policy $\pi_\theta(\mathbf{s}_t)$ we can formally describe the reward function.

When training a neural network we should use a loss function that we are to minimize instead of the reward that is maximized. However, this is solved by simply using negative reward as a loss function.

As an optimizer for neural network, we use adaptive moment estimation (Adam) [8] because it has some good benefits in comparison with simple stochastic gradient descent optimizer and in many cases finds a better optimum.

7 Neural Networks for Meta-learning Approach

We experiment with different architectures of our Meta-Learning agents' neural networks. Four architectures were fitted and tested with different parameters, namely CNN, RNN, LSTM and GRU.

In Fig. 1 we can see the architecture of CNN network that we use in our algorithm.

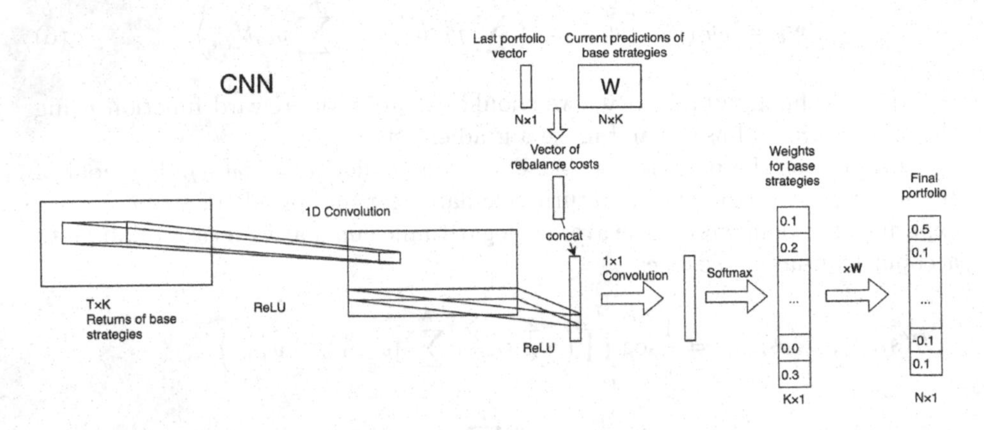

Fig. 1. Convolutional neural network for Meta-Learning algorithm.

In Fig. 2 we can see the architecture of RNN networks (RNN, LSTM, GRU) that we use in our algorithm.

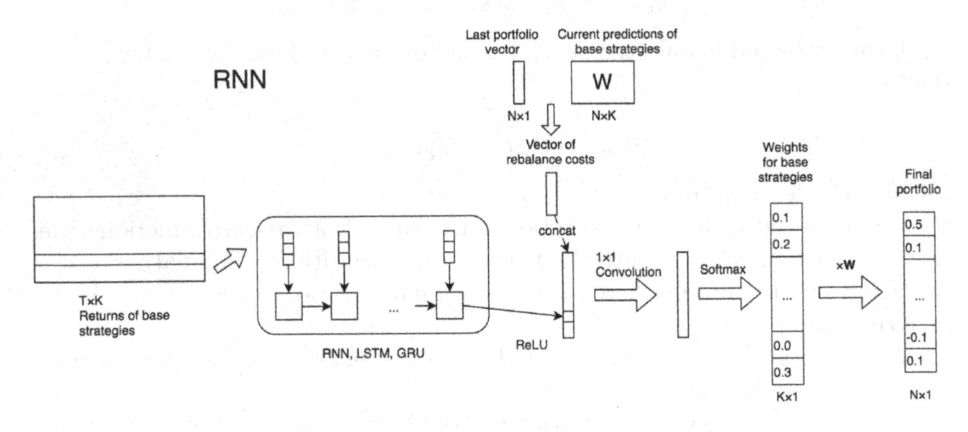

Fig. 2. Recurrent neural network for Meta-learning algorithm.

8 Experiments

8.1 Data

In our experiments we trade on the stock market on Moscow Exchange (MOEX). The frequency of trading is 8 h (2 trades per day at 10 and 18 o'clock). As data we use aggregated trades in the OHLCV candlesticks format. OHLCV stands for Open, High, Low, Close, Volume. Open, High, Low, Close are the first, highest, lowest and last prices of trades during the specified period. Volume is the overall volume that was traded during this period. The historical data for the last 5.5 years, starting from the beginning of 2014 year and ending on May 2019, for top 30 companies that are traded on MOEX is downloaded freely from the internet. Then for our experiments we take only those companies that have large traded volume.

8.2 Evaluation Methodology

In our experiments we evaluate the performance using the following metrics: cumulative wealth (Return), Sharpe ratio, Annualized Percentage Yield (APY), Maximum Drawdown (MDD), turnover and volatility.

When measuring the performance of trading algorithms, it is important to take transaction costs into account. One of the most popular ways of taking transaction costs that is used by many brokers when trading on the stock market is to take proportional transaction costs to the amount of traded money. For example, if we want to buy some stocks for S \$ and transaction costs are γ then our broker will charge us γS \$ for the services. In our experiments we compute the performance considering the transaction costs equal to 0.03% which is the average fee of BCS broker that provides trading on stock market of Moscow Exchange.

9 Results

In this section we provide the results of using our novel Meta-Learning algorithm based on reinforcement learning (MetaRL) for solving OLPS tasks. We extensively test it in various settings with different parameters and neural network's architectures and see in what situations we can achieve good performance and how the parameters influence on it. We conduct these experiments on MOEX exchange using 30 stocks that have the highest traded volume at the beginning of the train period. We fit our MetaRL agents on the train set from 2014-02-01 till 2018-10-30. The agents' comparison is conducted on the test set from 2018-10-31 till 2019-05-03. The traiding frequency is 8 h which means that we make 2 trades per day at 10 and 18 o'clock.

As base algorithms we all the considered strategies in these research as well as their improved versions with different parameters. All in all we got 156 algorithms. For training our Meta-Learning algorithm we took 30 best algorithms based on Sharpe ratio metric on train set. Also we used all considered 30 assets as simple strategies. All in all we got 60 base strategies for combination. In our experiments we tested several different architectures of agent's neural network, namely: CNN, RNN, LSTM and GRU. In each case we used several window sizes which indicates the number of historical periods in the past that uses the agent for making its prediction. In case of CNN we tried different sizes of 1-D convolutions. In case of RNN, LSTM and GRU we fit them using different number of units and dropout probability. All agents were trained during 100000 steps with Adam optimizer. The batch size was taken 100.

In Fig. 3 we can see the final comparison of our novel Meta-Learning algorithms based on CNN, RNN, LSTM and GRU with other strategies on the test set. We can see that all 4 models highly outperform other strategies.

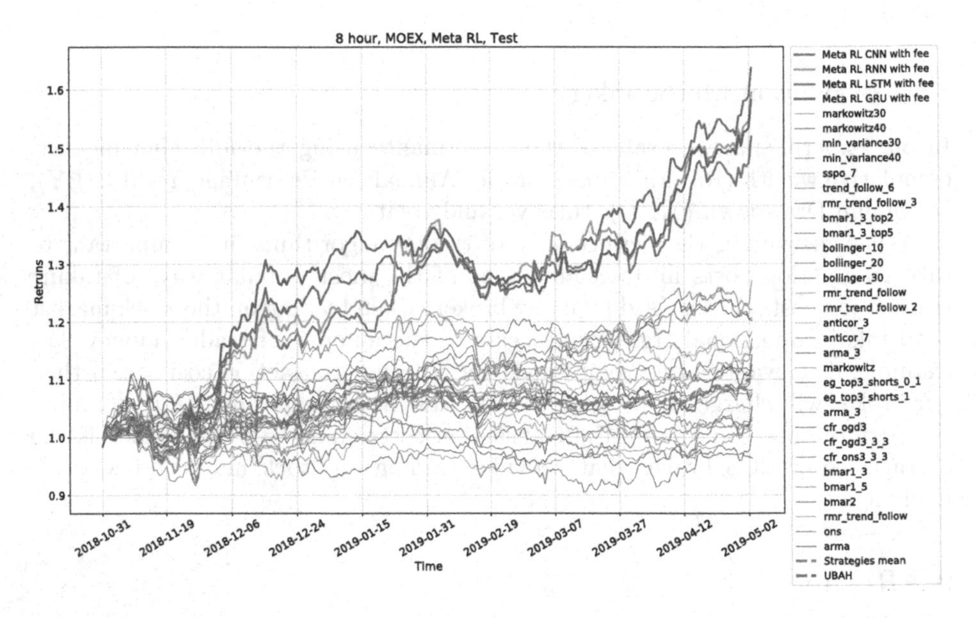

Fig. 3. Cumulative wealth of MetaRL agents in comparison with base algorithms on the test set.

We can see that the best results were obtained using CNN network. Although when using RNN, LSTM and GRU we can achieve pretty high performance, it is still a little bit lower than when using CNN and the training is not so stable.

In Table 2 we can see the metrics of our MetaRL algorithms in comparison with other strategies on the test set. We can see that all 4 models outperform other strategies by Sharpe ratio, Return, and APY metrics. Also, they have an average drawdown and relatively low volatility. The best results by Sharpe metric gets CNN agent, then GRU, third place gets RNN and the worst RL agent is LSTM. We should say that looking at Sharpe ratio metric is more important than Return and APY as it measures risk-adjusted returns. Considering risks is one of the main priorities of traders. If the agent is not stable and has low Sharpe but finally gets high returns then it cannot be considered as a good strategy.

Table 2. Comparison of algorithms with Meta-learning algorithms based on RL on MOEX exchange on the test set.

	Sharpe	Return	APY	MDD	Volatility
MetaRL CNN	**4.59**	**1.64**	**1.68**	0.08	0.0098
MetaRL GRU	**4.22**	**1.58**	**1.50**	0.12	0.0100
MetaRL RNN	**3.77**	**1.59**	**1.54**	0.14	0.0114
MetaRL LSTM	**3.66**	**1.52**	**1.31**	0.10	0.0105
markowitz40	3.02	1.21	0.46	0.05	0.0057
e.g. top3 shorts 0 1	2.73	1.21	0.47	0.06	0.0065
e.g. top3 shorts 1	2.56	1.19	0.43	0.06	0.0064
bmar1 3	2.13	1.25	0.57	0.10	0.0100
bmar1 5	2.11	1.25	0.56	0.11	0.0099
bmar1 3 top2	2.11	1.14	0.31	0.12	0.0059
anticor 7	2.04	1.16	0.35	0.09	0.0067
min variance30	1.95	1.08	0.16	0.03	0.0035
anticor 3	1.92	1.13	0.27	0.08	0.0058
min variance40	1.74	1.07	0.15	0.04	0.0038
ons	1.69	1.07	0.15	0.04	0.0039
bollinger 30	1.55	1.12	0.26	0.08	0.0069
rmr trend follow 3	1.36	1.13	0.27	0.10	0.0085
rmr trend follow	1.25	1.10	0.22	0.09	0.0075
cfr ons3 3 3	1.25	1.14	0.29	0.10	0.0100
arma 3	1.11	1.12	0.27	0.12	0.0106
arma	1.11	1.12	0.27	0.12	0.0106
arma 3	1.11	1.12	0.27	0.12	0.0106
bollinger 20	1.05	1.08	0.17	0.08	0.0071
sspo 7	1.01	1.11	0.24	0.11	0.0106
trend follow 6	0.72	1.07	0.15	0.16	0.0105
bmar1 3 top5	0.72	1.03	0.06	0.08	0.0038
markowitz	0.57	1.04	0.07	0.12	0.0063
rmr trend follow	0.52	1.02	0.05	0.07	0.0042
rmr trend follow 2	0.42	1.03	0.06	0.12	0.0076
bmar2	0.35	1.03	0.06	0.15	0.0111
cfr ogd3 3 3	0.35	1.03	0.06	0.12	0.0102
markowitz30	0.19	1.01	0.02	0.07	0.0058
cfr ogd3	0.07	0.99	−0.01	0.16	0.0105
bollinger 10	−0.15	0.98	−0.04	0.12	0.0071

10 Conclusions

The main aim of this work was to investigate the effectiveness of different machine learning algorithms, especially deep reinforcement learning for solving online portfolio selection task. We implemented, tested and compared existing OLPS algorithms, investigated their drawbacks and limitations. We developed a novel Meta-Learning OLPS approach based on deep reinforcement learning that utilizes various OLPS algorithms by their combination.

The main contributions of this paper are the following:

- We developed a novel Meta-Learning OLPS algorithm based on reinforcement learning and compared it with state-of-the-art algorithms. The experimental results on stock trading on Moscow Exchange show that our method outperforms all other considered algorithms in key trading performance metrics such as Sharpe Ratio, Cumulative wealth and APY.
- We developed a trading platform for testing and comparing strategies with convenient, user-friendly graphical user interface which has a library that consists of more than 30 OLPS algorithms.

References

1. Cesa-Bianchi, N., Lugosi, G.: Prediction, Learning, and Games. Cambridge University Press, New York (2006)
2. Cover, T.M.: Universal portfolios. Math. Financ. **1**(1), 1–29 (1991). https://doi.org/10.1111/j.1467-9965.1991.tb00002.x
3. Das, P., Banerjee, A.: Meta optimization and its application to portfolio selection. In: Proceedings of the ACM SIGKDD International Conference on Knowledge Discovery and Data Mining, pp. 1163–1171. San Diego (2011). https://doi.org/10.1145/2020408.2020588
4. Gold, C.: FX trading via recurrent reinforcement learning. In: 2003 IEEE International Conference on Computational Intelligence for Financial Engineering, Proceedings, pp. 363–370, March 2003. https://doi.org/10.1109/CIFER.2003.1196283
5. Györfi, L., Udina, F., Walk, H.: Nonparametric nearest neighbor based empirical portfolio selection strategies. Stat. Decisions **26**, 145–157 (2008)
6. Jiang, Z., Xu, D., Liang, J.: A deep reinforcement learning framework for the financial portfolio management problem. ArXiv e-prints (2017)
7. John, M., Lizhong, W., Yuansong, L., Matthew, S.: Performance functions and reinforcement learning for trading systems and portfolios. J. Forecast. **17**(5–6), 441–470 (1998)
8. Kingma, D.P., Ba, J.: Adam: a method for stochastic optimization. CoRR abs/1412.6980 (2015)
9. Li, B., Hoi, S.C.H.: Online portfolio selection: a survey. ACM Comput. Surv. **46**(3), 35:1–35:36 (2014). https://doi.org/10.1145/2512962
10. Li, B., Hoi, S.C., Gopalkrishnan, V.: CORN: correlation-driven nonparametric learning approach for portfolio selection. ACM Trans. Intell. Syst. Technol. **2**(3) (2011). https://doi.org/10.1145/1961189.1961193
11. Liang, Z., Chen, H., Zhu, J., Jiang, K., Li, Y.: Adversarial deep reinforcement learning in portfolio management. ArXiv e-prints (2018)
12. Moscow exchange. https://www.moex.com

Optimization in Data Analysis

Simultaneous Detection and Discrimination of the Known Number of Non-Linearly Extended Alphabet Elements in a Quasiperiodic Sequence

Liudmila Mikhailova(✉)

Sobolev Institute of Mathematics, 4 Koptyug Avenue, 630090 Novosibirsk, Russia
mikh@math.nsc.ru

Abstract. The paper is devoted to one unexplored discrete optimization problem. This problem arises when applying the non-traditional a posteriori approach to the applied problem of simultaneous detection and discrimination of the given number of subsequences-fragments coinciding with non-linearly extended elements of the given sequence alphabet in a quasiperiodic sequence. It is constructively proved that this optimization problem is polynomial-time solvable. The numerical simulation results are presented.

Keywords: Discrete optimization problem · Quasiperiodic sequence · Detection · Discrimination · Polynomial-tyme solvability · Non-linear extension

1 Introduction

The current paper continues a series [1–3] of articles dealing with a posteriori processing of one specific type of quasiperiodic numerical sequences. It adds the new previously unstudied problem to this family. The subject of the study is an unexplored discrete optimization problem arising in the framework of a posteriori approach. The goal of the study is to analyze the algorithmic complexity of this optimization problem and to substantiate an algorithm for its solution.

A sequence is said to be a quasiperiodic one if it includes disjoint subsequences-fragments (subsequences formed by consecutive elements of the sequence and having some predefined characteristic properties) such that the interval between the initial positions of two consecutive fragments is not greater than the given constant and the fragments are included in the sequence as a whole, i.e., the first and the last fragments in the sequence are not cut by the sequence borders. The second restriction is not necessary and is added for the sake of simplicity.

The study was supported by the Russian Foundation for Basic Research, project no. 19-07-00397, the study was carried out within the framework of the state contract of the Sobolev Institute of Mathematics (project no. 0314-2019-0015).

N. N. Olenev et al. (Eds.): OPTIMA 2021, CCIS 1514, pp. 171–183, 2021.
https://doi.org/10.1007/978-3-030-92711-0_12

Numerical quasiperiodic sequences arise, for example, when sampling a quasiperiodic pulse train using a uniform grid. In these pulse trains, we allow two types of variability: fluctuations in the interval between pulses and admissible modifications in the shape of every pulse. Examples of such pulse trains can be found among the data obtained as a monitoring result for natural objects having quasiperiodically repeating states (bio-medical, underwater, underground, aerospace, etc.). By quasiperiodic repeatability of states, we mean that an object is not in two different states simultaneously, the distances between two successive state repetitions are limited from above by the given constant, and a typical state allows some fluctuations from one repetition to another. These two types of variability (see the description of quasiperiodic state repeatability) are reflected in the properties of a numerical quasiperiodic sequence. The constraints on the fragment beginnings correspond to the variability of pulse-to-pulse intervals, and the predefined properties of a fragment describe the variability of the object state.

There are two different approaches to solving sequence processing problems: sequential and a posteriori. The most traditional widely-used approach is the sequential one. This approach involves a breakdown of the problem into separate stages. These stages are, for example, optimal noise filtration [6,7], hypothesis testing and classification [8,9], change-point detection [10,11], and so on, used in various combinations. There are plenty of well-studied algorithms and methods to implement these stages. This approach has two main undoubted advantages: the reuse of available mathematical tools and the low complexity as a rule of the resulting algorithms. The main disadvantage of the approach is the lack of the optimality guarantee of the final solution, even if the solution at each stage is optimal.

Contrary to the sequential approach, a posteriori one allows obtaining optimal solutions or solutions with proven quality estimates. The essence of this approach is to reduce the data processing problem to an optimization one. There are no separate stages when solving the problem. The reason why this approach is less common is that the arising optimization problem is unique and tends to be unexplored. Examples of applying a posteriori approach to problems of quasiperiodic sequence processing (with various characteristic properties of a single fragment) can be found in [1–3,12,13], and the works cited there.

In this paper, as in [1–3], a characteristic property of an individual fragment is the fragment coincidence with a sequence obtained as a non-linear extension (by duplicating components) of some element of the given sequence set (alphabet). This characteristic property describes well the fluctuations of various natural signals. Clear examples of the described repeatability of states can be found among biomedical processes. Below, in Sect. 5, the sequence processing example, in the case when the alphabet includes ECG-like pulses, is used for illustration (see, for example [4,5], for a detailed description of their shape and characteristic waves).

The essence of the simultaneous detection and discrimination problem is as follows. Having the observable sequence (the unobservable quasiperiodic sequence distorted by additive noise) as a part of the problem input, we should get a collection of subsequences-fragments as the output. In other words, we have to find the collection of initial indices and lengths of subsequences (the detection) and the collection of corresponding alphabet elements (the identification). The output collections allow us to restore the unobservable sequence. In the particular case of the problem, when the alphabet consists of only one sequence [2,3], the identification is not needed, and the problem reduces to the detection problem solely.

The difference between the current problem and the close in formulation previously studied simultaneous detection and discrimination problem [1] is the presence of apriori information about the number of the fragments to be found: here, this number is known, contrary to [1], where it is unknown. This modification leads, in the framework of a posteriory approach, to the previously unexplored optimization problem. It turned out that the set of admissible solutions to this problem grows exponentially with the length of the processed sequence. Despite this polynomial growth, it is constructively proved that this optimization problem is polynomial-time solvable.

In addition to mathematical interest, the constructed new algorithm for noise-resistant processing of noisy quasiperiodic sequences is also interesting from an applied point of view. In the case of intensive noise, the algorithm [1] tends to make mistakes when determining the number of fragments. These mistakes lead to inaccuracies in detecting and discriminating fragments. If information about the number of fragments is available, the new algorithm allows us to obtain better results (in terms of restoring an unobservable sequence) in conditions of intensive noise.

2 Detection and Discrimination Problem

A sequence $U = (u_1, \dots, u_q) \in \Re^q$ is said to be a **reference sequence** of length q. Any finite set $V = \{U_1, \dots, U_K\}$, where U_k, $k = 1, \dots, K$, is a reference sequence, is said to be an **alphabet of reference sequences** or an **alphabet**. For any $U \in V$ denote the length of the alphabet element by $q(U)$, i.e., $q(U) = \dim U$, and denote by $q_{\max}(V)$ and $q_{\min}(V)$ the lengths of the longest and the shortest alphabet element:

$$q_{\max}(V) = \max_{U \in V} q(U), \quad q_{\min}(V) = \min_{U \in V} q(U).$$

The model of quasiperiodic sequence containing subsequences-fragments coinciding with non-linearly extended elements of the given alphabet of reference sequences is described in detail in [1]. This model allows us to describe quasiperiodic sequences with the structure we are interested in, regardless of whether the number of fragments is known or not. But despite the use of one model when describing a sequence, the availability of information about the number of fragments significantly affects the algorithmic processing of sequences. In the current work, this number is known; denote it by M.

Let $T_{\max}$ be the maximal admissible interval between the beginnings of two neighboring fragments, and $\ell \geq q_{\max}(V)$ limits from above the length of any extension of the alphabet element. The general formula for the n-th term of the quasiperiodic sequence $X = (x_1, \ldots, x_N)$ of length N is as follows

$$x_n = \sum_{m=1}^{M} u^{(m)}_{J^{(m)}(n-n_m+1)}, \quad n = 1, \ldots, N, \tag{1}$$

where $(u_1^{(m)}, \ldots, u_{q_m}^{(m)}) \in V$, $m = 1, \ldots, M$, $u_i^{(m)} = 0$, if $i = 0, -1, -2, \ldots$ or $i = q_m + 1, q_m + 2, \ldots$, $m = 1, \ldots, M$, the variables n_m, p_m, $m = 1, \ldots, M$, are constrained by the inequalities:

$$\begin{gathered} q_m \leq p_m \leq \ell \leq T_{\max} \leq N, \;\; m = 1, \ldots, M, \\ p_{m-1} \leq n_m - n_{m-1} \leq T_{\max}, \;\; m = 2, \ldots, M, \\ p_M \leq N - n_M + 1, \end{gathered} \tag{2}$$

and every mapping $J^{(m)} : \{1, \ldots, p_m\} \longrightarrow \{1, \ldots, q_m\}$, $p \in \{q_m, \ldots, \ell\}$, $m = 1, \ldots, M$, satisfies the following restrictions

$$\begin{gathered} J^{(m)}(1) = 1, \; J^{(m)}(p_m) = q_m, \\ 0 \leq J^{(m)}(i+1) - J^{(m)}(i) \leq 1, \; i = 1, \ldots, p-1. \end{gathered} \tag{3}$$

In this notation, variables n_m and p_m correspond to the initial index of the m-th fragment in X and its duration, respectively. The form of the fragment is described by $U^{(m)} = (u_1^{(m)}, \ldots, u_{q_m}^{(m)}) \in V$ and $J^{(m)}$. So, the inequality system (2) guarantees that the conditions of quasiperiodicity are valid for the fragments and the restrictions (3) define the admissible extension of $(u_1^{(m)}, \ldots, u_{q_m}^{(m)}) \in V$.

It follows from (1) that the quasiperiodic sequence X is uniquely defined by four collections: a collection $\mathcal{U}_M = (U^{(1)}, \ldots, U^{(M)}) \in V^M$ of alphabet elements, collections $\mathcal{M}_M = (n_1, \ldots, n_M)$ and $\mathcal{P}_M = (p_1, \ldots, p_M)$ of positive integers, and a collection $\mathcal{J}_M = (J^{(1)}, \ldots, J^{(M)})$ of mappings, so

$$x_n = x_n(\mathcal{U}_M, \mathcal{M}_M, \mathcal{P}_M, \mathcal{J}_M), \; n = 1, \ldots, N,$$

and

$$X = X(\mathcal{U}_M, \mathcal{M}_M, \mathcal{P}_M, \mathcal{J}_M).$$

In such a way, with given M, N, ℓ, and $T_{\max}$, alphabet V induces the set $\mathcal{X}_M$ of all possible quasiperiodic sequences, which include the known number M of fragments having the above-described property. At that, every element of this set is uniquely determined by $\mathcal{U}_M$, $\mathcal{M}_M$, $\mathcal{P}_M$, and $\mathcal{J}_M$.

Assume a sequence $Y = (y_1, \ldots, y_N)$, which is the element-wise sum of an unobservable sequence $X \in \mathcal{X}_M$ and some sequence reflecting possible noise distortion, be available for processing. **The problem of simultaneous detection and discrimination** is to find the collection of M fragments when processing the observable sequence Y. This is equivalent to the calculation of 4 collections: $\mathcal{U}_M$, $\mathcal{M}_M$, $\mathcal{P}_M$, and $\mathcal{J}_M$, on the base of Y.

Interpreting X and Y as vectors in N-dimensional space, the problem of simultaneous detection and discrimination can be formulated as the following optimization problem

$$\|Y - X(\mathcal{U}_M, \mathcal{M}_M, \mathcal{P}_M, \mathcal{J}_M)\|^2 \longrightarrow \min_{\mathcal{U}_M, \mathcal{M}_M, \mathcal{P}_M, \mathcal{J}_M} . \tag{4}$$

This problem is equivalent to the problem of searching for the best approximation of Y by $X \in \mathcal{X}_M$.

It is shown [2], that except for the trivial cases $M = 1$ or $T_{\max} = \ell = q$, the cardinality of $\mathcal{X}_M$, as well as the cardinality of the set of all admissible collections $(\mathcal{U}_M, \mathcal{M}_M, \mathcal{P}_M, \mathcal{J}_M)$, grows exponentially as a function of N even in the particular case of the problem, when $|V| = 1$. So, the set of admissible solutions to problem (4) grows exponentially with the length of the processed sequence, and constructing a polynomial-time algorithm for its solution is of undoubted mathematical interest.

3 Optimization Problem and Related Problems

Taking into account (1) and constraints (2), when expanding the squares of the norm in (4), we obtain by simple calculations that

$$\|Y - X\|^2 = \sum_{n=1}^{N} y_n^2 + \sum_{m=1}^{M} \sum_{i=1}^{p_m} \{(u^{(m)}_{J^{(m)}(i)})^2 - 2y_{n_m+i-1} u^{(m)}_{J^{(m)}(i)}\}.$$

Here the first term on the right-hand side is constant and, so, approximation problem (4) is equivalent to the following discrete optimization problem.

Problem 1. *Given:* the numerical sequence $Y = (y_1, \ldots, y_N)$, the alphabet V, $|V| = K$, of reference sequences, the positive integers $T_{\max}$, $\ell \geq q_{\max}(V)$, and $M \leq \lfloor N/q_{\min}(V) \rfloor$. *Find:* the collection $\mathcal{U}_M = (U^{(1)}, \ldots, U^{(M)})$ of numerical sequences, where $U^{(m)} = (u^{(m)}_1, \ldots, u^{(m)}_{q_m}) \in V$, the collection $\mathcal{M}_M = \{n_1, \ldots n_M\}$ of indices of the sequence Y, the collection $\mathcal{P}_M = \{p_1, \ldots, p_M\}$ of positive integers, the collection $\mathcal{J}_M = \{J^{(1)}, \ldots, J^{(M)}\}$ of mappings, where $J^{(m)} : \{1, \ldots, p_m\} \longrightarrow \{1, \ldots, q_m\}$, that minimize the objective function

$$F(\mathcal{U}_M, \mathcal{M}_M, \mathcal{P}_M, \mathcal{J}_M) = \sum_{m=1}^{M} \sum_{i=1}^{p_m} \{(u^{(m)}_{J^{(m)}(i)})^2 - 2y_{n_m+i-1} u^{(m)}_{J^{(m)}(i)}\}, \tag{5}$$

under the constraints (2) on the elements of the sought collections $\mathcal{M}_M$, $\mathcal{P}_M$, and under the constraints (3) on the mappings from the collection $\mathcal{J}_M$.

The modification of Problem 1, where the number M of the sought subsequences is unknown, was studied in [1]. The optimal solution to this modification can be found in $\mathcal{O}(T_{\max}^3 K^2 N)$ time. Below, in Sect. 5, it is illustrated that the information about the number of subsequences is really important when processing quasiperiodic sequences.

Problem 1 has two previously studied particular cases. In the first one [2], the alphabet V consists of only one sequence. The restriction $|V| = 1$ in the conditions of Problem 1 means that there is no identification at all, and all the fragments coincide with the extensions of the same alphabet element. So, the problem can be treated as a detection one exclusively. The modification of this particular case, when the number M of the sought subsequences is unknown, was studied in [3]. The algorithms suggested in [2] and [3] allow obtaining the optimal solutions to this particular case in time $\mathcal{O}(T_{\max}^3 MN)$ and to its modification in time $\mathcal{O}(T_{\max}^3 N)$.

In another particular case [12] of Problem 1, the assumption $|V| \geq 1$ is left unchanged, but the supplementary restrictions $q_1 = \ldots = q_K = q$, $p_m = q$, $J^{(m)}(i) = i$, $m = 1, \ldots, M$, are added. These restrictions mean that the fragments coincide with alphabet elements that are not distorted by stretching. There is the modification of this particular case in [13]. In it, the number M of the fragments is not a part of the problem input. The algorithms justified in [12] and [13] find the exact solutions to this particular case and its modification in time $\mathcal{O}((T_{\max} + (K-1)q)MN)$ and $\mathcal{O}((T_{\max} + (K-1)q)N))$, respectively.

4 Algorithm

To write down the algorithm solving Problem 1, we need the following lemma, which defines the sets of admissible values for interconnected variables of Problem 1.

Lemma 1. *Let the conditions of Problem 1 holds. Then*

(1) the set of possible values for n_m is

$$\omega_m = \{1 + (m-1)q_{\min}(V), \ldots, N - (M - m + 1)q_{\min}(V) + 1\}, \ m = 1, \ldots, M;$$

(2) if $n_m = n$, $n \in \omega_m$, then the set of possible values for p_m is

$$\delta_m(n) = \{q_{\min}(V), \ldots, \min\{\ell, \ N - n + 1 - q_{\min}(V)(M - m)\}\}, \quad m = 1, \ldots, M;$$

(3) if $n_m = n$, $n \in \omega_m$, then the set of possible values for n_{m-1} is

$$\gamma_m(n) = \{\max\{n - T_{\max}, 1 + (m-2)q_{\min}(V)\}, \ldots, n - q_{\min}(V)\}, \ m = 2, \ldots, M;$$

(4) if $n_m = n$, $n \in \omega_m$, and $n_{m-1} = j$, $j \in \gamma_m(n)$, then the set of possible values for p_{m-1} is

$$\theta_m(n, j) = \{q_{\min}(V), \ldots, \min\{\ell, n - j\}\}, \ m = 2, \ldots, M.$$

The validity of Lemma 1 is established by summing up the constraints included in the conditions of Problem 1.

We also need the following auxiliary problem. This problem is new and previously unstudied.

Problem 2. *Given*: the positive integers $T_{\max}$, ℓ, M, and the collection $\{q_1, \ldots, q_K\}$ of positive integers such that $\ell \le T_{\max}$, $q_k \le \ell$, $k = 1, \ldots, K$, $M \le \lfloor N/q_{\min} \rfloor$, where $q_{\min} = \min_{k=1,\ldots,K} q_k$, the collection $\{g_k(n,p),\ k = 1, \ldots, K,$ $p = q_k, \ldots, \ell,\ n = 1, \ldots, N - p + 1\}$ of numerical sequences. *Find*: the collection $\{(n_1, p_1, k_1), \ldots, (n_M, p_M, k_M)\}$ of positive integer triples, where $k_m \in \{1, \ldots, K\}$, $n_m \in \{1, \ldots, N\}$, $p_m \in \{q_{k_m}, \ldots, \ell\}$, $m = 1, \ldots, M$, that minimize

$$G((n_1, p_1, k_1), \ldots, (n_M, p_M, k_M)) = \sum_{m=1}^{M} g_{k_m}(n_m, p_m)$$

under the restrictions (2).

This problem is solved using the method of dynamic programming. The recurrent formulas allowing to obtain the optimal solution to this problem in $\mathcal{O}(K^2 T_{\max}^3 NM)$ time are derived. These recurrent formulas are the core of the algorithm solving Problem 1.

The unformal idea of the algorithm is as follows. First of all, for every part $y_n, \ldots, y_{n+p-1}$ of the input sequence, we suppose that this part is a fragment and try every alphabet element as inducing it (see Step 1). In other words, this step aims to find the best conditionally optimal mapping such that the part $y_n, \ldots, y_{n+p-1}$ of the input sequence and the alphabet element $U = (u_1, \ldots, u_{q(U)})$ stretched according to this mapping turn out to be the closest by the least-squares criterion. The recurrent formulas for the implementation of this step are proved in [3]. Then the results of Step 1 are used as the input data for auxiliary Problem 2. In the case of such input, the value of the objective function of Problem 2 coincides with the value of the objective function of Problem 1 (see Step 2), and the optimal arguments of Problem 2 (see Step 3) together with the sequences found at Step 1 allows to obtain the optimal collections in Problem 1 (see Step 4).

Algorithm $\mathcal{A}$.

INPUT: $T_{\max}$, ℓ, M, V, and Y.

STEP 1

(a) Calculate

$$W(n,p,U) = W_{p,q(U)}(n,p,U),\ U \in V,\ n = 1, \ldots, N - q(U) + 1,\quad p = q(U), \ldots, \min\{\ell, N - n + 1\},$$

using the recurrent formulas

$$W_{s,t}(n,p,U) = \begin{cases} u_1^2 - 2y_n u_1, & s = 1,\ t = 1, \\ W_{s-1,1} + u_1^2 + 2y_{n+s-1}u_1, & s = 2, \ldots, p,\ t = 1, \\ +\infty, & s = 1,\ t = 2, \ldots, q(U), \\ \min\{W_{s-1,t}, W_{s-1,t-1}\} + u_t^2 - 2y_{n+s-1}u_t, & ,s = 1, \ldots, p,\ t = 1, \ldots, q(U). \end{cases}$$

(b) Find the sequence $J(n,p,U) = (J_1(n,p,U),\dots,J_p(n,p,U))$ of summing indices by the rule

$$J_p(n,p,U) = q(U);$$

$$J_{i-1}(n,p,U) = \begin{cases} J_i(n,p,U), & \text{if } W_{i-1,J_i(n,p,U)}(n,p,U) \le W_{i-1,J_i(n,p,U)-1}(n,p,U), \\ J_i(n,p,U)-1, & \text{if } W_{i-1,J_i(n,p,U)}(n,p,U) > W_{i-1,J_i(n,p,U)-1}(n,p,U), \end{cases}$$
$$i = p, p-1, \dots, 2.$$

STEP 2

(a) Set an arbitrary order on the elements of V: $V = \{U_1,\dots,U_K\}$.

(b) Put $q_k = q(U_k)$, $k = 1,\dots,K$.

(c) Calculate

$$G_m(n,p,k) = \begin{cases} W(n,p,U_k), \quad m = 1, n \in \omega_1,\ p \in \delta_1(n),\ k \in \mathcal{K}(p), \\ \min\limits_{j\in\gamma_m(n)} \min\limits_{i\in\theta_m(n,j)} \min\limits_{t\in\mathcal{K}(i)} G_{m-1}(j,i,t) + W(n,p,U_k), \\ \qquad m = 2,\dots,M,\ n \in \omega_m,\ p \in \delta_m(n),\ k \in \mathcal{K}(p), \end{cases}$$

where

$$\mathcal{K}(q) = \left\{k \;\middle|\; k \in \{1,\dots,K\}, q_k \le q\right\},\ q = 1,\dots,N - q_{\min} + 1.$$

(d) Calculate

$$F_{\mathcal{A}} = \min_{n\in\omega_M} \min_{p\in\delta_M(n)} \min_{k\in\mathcal{K}(p)} G_M(n,p,k).$$

STEP 3

Calculate

$$n_M = \arg\min_{n\in\omega_M} \left\{ \min_{p\in\delta_M(n)} \min_{k\in\mathcal{K}(p)} G_M(n,p,k) \right\},$$

$$p_M = \arg\min_{p\in\delta_M(n_M)} \left\{ \min_{k\in\mathcal{K}(p)} G_M(n_M,p,k) \right\},$$

$$k_M = \arg\min_{k\in\mathcal{K}(p_M)} G_M(n_M,p_M,k),$$

$$n_{m-1} = \arg\min_{j\in\gamma_m(n_m)} \left\{ \min_{i\in\theta_m(n_m,j)} \min_{t\in\mathcal{K}(i)} G_{m-1}(j,i,t) \right\},$$

$$p_{m-1} = \arg\min_{i\in\theta_m(n_m,n_{m-1})} \left\{ \min_{t\in\mathcal{K}(i)} G_{m-1}(n_{m-1},i,t) \right\},$$

$$k_{m-1} = \arg\min_{k\in\mathcal{K}(p_{m-1})} G_{m-1}(n_{m-1},p_{m-1},k), \quad m = 2,\dots,M.$$

STEP 4

Put $\mathcal{M}_{\mathcal{A}} = \{n_1,\dots,n_M\}$, $\mathcal{P}_{\mathcal{A}} = \{p_1,\dots,p_M\}$, $J^{(m)} = J(n_m,p_m,U_{k_m})$, $m = 1,\dots,M$, and $\mathcal{J}_{\mathcal{A}} = \{J^{(1)},\dots,J^{(M)}\}$, $\mathcal{U}_{\mathcal{A}} = \{U_{k_1},\dots,U_{k_M}\}$.

OUTPUT: $\mathcal{M}_{\mathcal{A}}$, $\mathcal{P}_{\mathcal{A}}$, $\mathcal{J}_{\mathcal{A}}$, $\mathcal{U}_{\mathcal{A}}$, and $F_{\mathcal{A}}$.

The main result of the research is the following theorem.

Theorem 1. *Algorithm $\mathcal{A}$ finds an exact solution to Problem 1 in time $\mathcal{O}(T_{\max}^3 M K^2 N)$.*

The two-stage structure of Algorithm $\mathcal{A}$ is similar to the structure of the algorithm constructed in [1]. So, the proof of the solution optimality is carried out in a similar way with the replacement of all connected with Problem 2 and formulas for its solution, which are the core of the algorithm.

The type of the objective function and the independence of the elements of the collection $\mathcal{J}_M$ allow us to move from the minimization of the double sum (5) over four collections to two-stage minimization. Minimization of a separate term of the internal sum by $J^{(m)}$, with fixed values of n_m, p_m, and U_m, is allocated to a separate problem of finding a conditional minimum. It allows us to reduce further minimization of the objective function to solving auxiliary Problem 2. This auxiliary problem is unique for Problem 1. The fact that the number M of terms in this sum is known is significantly used when formulating this problem and constructing formulas for its solution. The solution to Problem 2 is the key link of the algorithm.

5 Numerical Simulation

The first example visualizes Algorithm $\mathcal{A}$ operation. In this example, the alphabet consists of some sampled ECG pulses. In fact, from the mathematical point of view, it doesn't matter which sequences are included in the alphabet. ECG-like pulses are selected as the alphabet elements in order to illustrate the potential applicability of the algorithm to bio-medical applications as a step to creating noise-resistant algorithms for processing real biomedical data. It should be mentioned that the optimality of the solution obtained by Algorithm $\mathcal{A}$ has been proven by Theorem 1, so the results of numerical simulation results are given only for illustration.

The input data of Algorithm $\mathcal{A}$ are shown in Fig. 1. The top row contains 4 ECG-like pulses which form the alphabet V of reference sequences, and the bottom row presents an example of the observable sequence Y to be processed.

The data presented in Fig. 2 are unobservable and are not a part of the problem input. These data are used to generate the modeled observable sequence Y. The top row contains the sequence $\mathcal{U}_M$ of M randomly selected alphabet elements. The bottom row presents an example of modeled unobservable pulse sequence X. This sequence contains random extensions of the sequences (pulses) from $\mathcal{U}_M$, and the intervals between these extended pulses are randomly selected according to the quasiperiodicity conditions. The sequence Y, shown in the bottom row of Fig. 1, is the modeled sequence X distorted by additive Gaussian noise (white noise).

Figure 3 shows the result of Algorithm $\mathcal{A}$ operation: the collection $\mathcal{U}_\mathcal{A} \in V^M$— discrimination result (the top row) and the sequence $X_\mathcal{A} = X(\mathcal{U}_\mathcal{A}, \mathcal{M}_\mathcal{A}, \mathcal{P}_\mathcal{A}, \mathcal{J}_\mathcal{A})$ recovered using Algorithm $\mathcal{A}$ output (the bottom row). This example is computed for $K = 4$, $q(U_1) = q(U_2) = q(U_3) = 192$, $q(U_4) = 121$, $M = 4$, $T_{\max} = 384$,

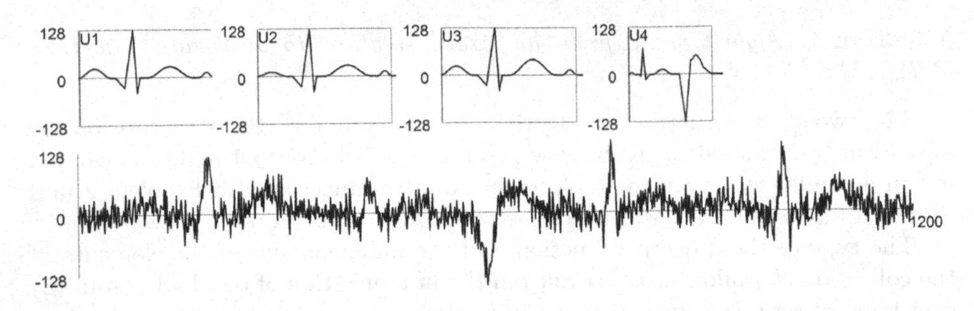

Fig. 1. Example 1. Input data

$\ell = 336$, $N = 1200$, maximal amplitude pulse value is 128, and the noise level $\sigma = 20$.

The second example in this section illustrates the announced importance of using a priori information about the number of fragments (if available) when processing noisy quasiperiodic sequences. For this aim, the same observable sequence is processed by two algorithms: the first one, Algorithm $\mathcal{A}$, uses a priori information about this number, whereas the second one justified in [1] (let's call it Algorithm $\mathcal{A}_1$) don't use it. Recall that the only difference between Problem 1 and the problem studied in [1] is whether the number of fragments is a part of the input or not.

As Figs. 1 and 2, Figs. 4 and 5 contain examples of input data and auxiliary unobservable data. Figure 6 contains sequences $X_{\mathcal{A}}$ (the bottom row) and $X_{\mathcal{A}_1}$ (the top row) restored according to (1) on the base of algorithms $\mathcal{A}$ and $\mathcal{A}_1$ output. In this picture, the inscription above each of the fragments identifies this fragment as an extended alphabet element. The example is calculated for $K = 3$, $q(U_1) = 40$, $q(U_2) = 50$, $q(U_3) = 60$, $M = 5$, $T_{\max} = 300$, $\ell = 240$, $N = 1200$, maximal amplitude pulse value is 140, and the noise level $\sigma = 60$.

You can see, when comparing the unobservable sequence with the sequences in Fig. 6, that Algorithm $\mathcal{A}$ gives better results in terms of restoring the unobservable sequence. This is quite natural since the algorithms $\mathcal{A}$ and $\mathcal{A}_1$ search for the best approximation of the same sequence Y by an element X from $\mathcal{X}_M$ and

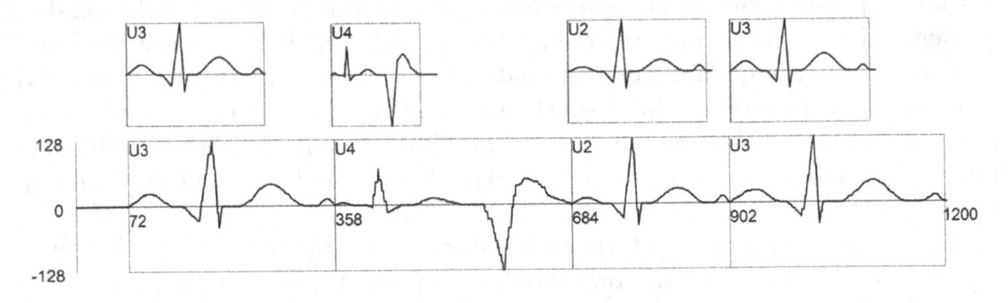

Fig. 2. Example 1. Unobservable data

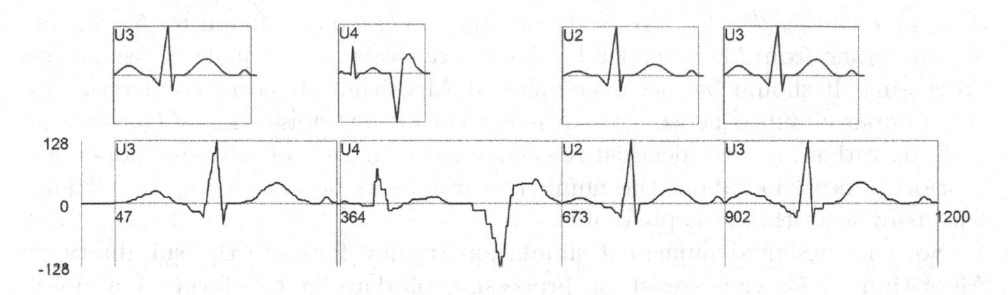

Fig. 3. Example 1. Output data

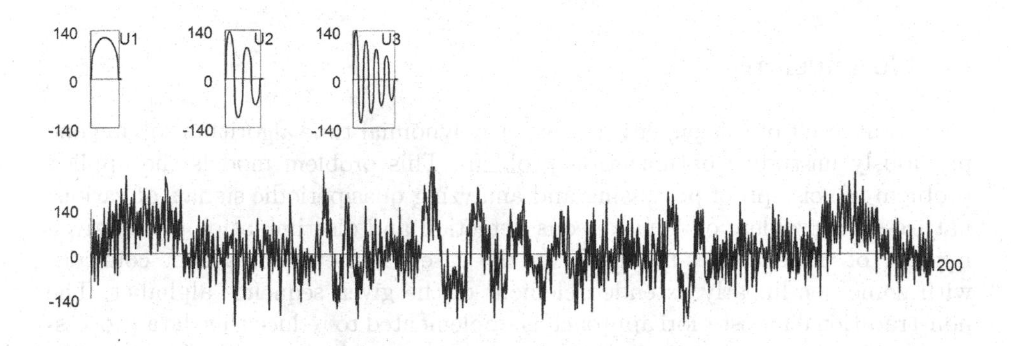

Fig. 4. Example 2. Input data

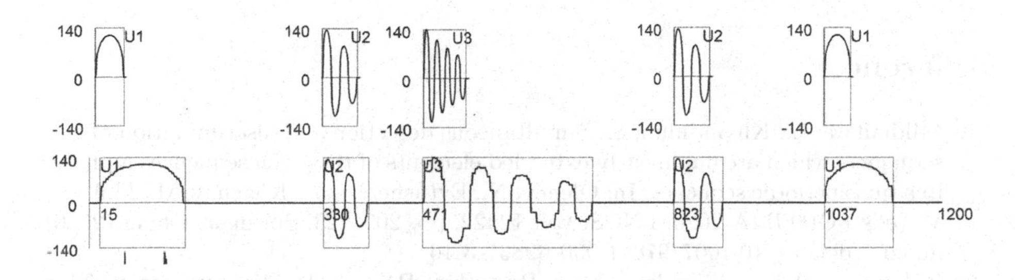

Fig. 5. Example 2. Unobservable data

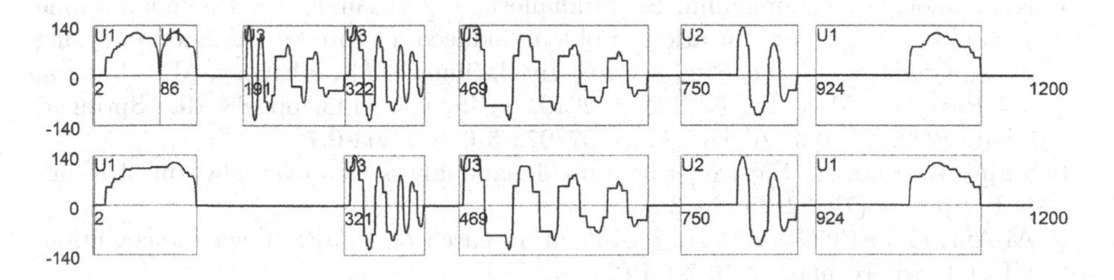

Fig. 6. Example 2. Output data

the wider set $\cup_{M=1}^{\lfloor N/q_{\min}(V)\rfloor} \mathcal{X}_M$. So the number of fragments found by Algorithm $\mathcal{A}_1$ may differ from M, which leads to a deterioration in the quality of sequence processing. It should be mentioned that if Algorithm $\mathcal{A}_1$ correctly determines the number of subsequences, which is typical for low noise, then Algorithm $\mathcal{A}$ and Algorithm $\mathcal{A}_1$ give identical results. But in the case of intensive noise, the a priori information about the number of fragments becomes really important, and using algorithm $\mathcal{A}$ is preferable.

So, the presented numerical simulation results illustrate the suitability of Algorithm $\mathcal{A}$ for noise-resistant processing of data in the form of a noisy quasiperiodic sequence which includes the given number of having the above-described property fragments.

6 Conclusion

The main result of this paper is the exact polynomial-time algorithm solving one previously unstudied optimization problem. This problem models the applied problem of noise-proof processing and analyzing quasiperiodic signals of various nature—the problem of simultaneous detection and discrimination of the given number of subsequences-fragments in the case when every fragment coincides with some non-linearly extended element of the given sequence alphabet. The non-traditional a posteriori approach is implemented to reduce this data processing problem to the new unstudied optimization problem that made it possible to get a result with guaranteed quality estimates. Some numerical simulation results are presented as an illustration of the algorithm operation.

References

1. Mikhailova, L., Khamdullin, S.: Simultaneous detection and discrimination of subsequences which are nonlinearly extended elements of the given sequences alphabet in a quasiperiodic sequence. In: Olenev, N., Evtushenko, Y., Khachay, M., Malkova, V. (eds.) OPTIMA 2020. LNCS, vol. 12422, pp. 209–223. Springer, Cham (2020). https://doi.org/10.1007/978-3-030-62867-3_16
2. Kel'manov, A.V., Mikhailova, L.V., Ruzankin, P.S., et al.: Minimization problem for sum of weighted convolution differences: the case of a given number of elements in the sum. Numer. Anal. Appl. **13**, 103–116 (2020)
3. Kel'manov, A., Khamidullin, S., Mikhailova, L., Ruzankin, P.: Polynomial-time solvability of one optimization problem induced by processing and analyzing quasiperiodic ECG and PPG signals. In: Jaćimović, M., Khachay, M., Malkova, V., Posypkin, M. (eds.) OPTIMA 2019. CCIS, vol. 1145, pp. 88–101. Springer, Cham (2020). https://doi.org/10.1007/978-3-030-38603-0_7
4. Rajni, R., Kaur, I.: Electrocardiogram signal analysis – an overview. Int. J. Comput. Appl. **84**(7), 22–25 (2013)
5. Al-Ani, M.S.: ECG waveform classification based on P-QRS-T wave recognition. UHD J. Sci. Technol. **2**(2), 7–14 (2018)
6. Anderson, B.D., Moore, J.D.: Optimal Filtering. Prentice Hall, Englewood Cliffs (1995)

7. Sparks, T., Chase, G.: Filters and Filtration Handbook. Butterworth-Heinemann, Oxford (2015)
8. Fukunaga, K.: Introduction to Statistical Pattern Recognition, 2nd edn. Academic Press, New York (1990)
9. Duda, R., Hart, P., Stork, D.: Pattern Classification, 2nd edn. Wiley, New York (2000)
10. Polunchenko, A., Tartakovsky, A.: State-of-the-art in sequential change-point detection. Methodol. Comput. Appl. Probab. **14**(3), 649–684 (2012). https://doi.org/10.1007/s11009-011-9256-5
11. Poor, H.V., Hadjiliadis, O.: Quickest Detection. Cambridge University Press, Cambridge (2008)
12. Kel'manov, A.V., Khamidullin, S.A.: A posteriori joint detection and discrimination of a given number of subsequences in a quasiperiodic sequence. Pattern Recogn. Image Anal. **10**(3), 379–388 (2000)
13. Kel'manov, A.V., Okol'nishnikova, L.V.: A posteriori simultaneous detection and discrimination of subsequences in a quasiperiodic sequence. Pattern Recogn. Image Anal. **11**(3), 505–520 (2001)

Network Utility Maximization by Updating Individual Transmission Rates

Dmitry Pasechnyuk(✉) and Andrei M. Raigorodskii

Moscow Institute of Physics and Technology, Moscow, Russia

Abstract. This paper discusses the problem of maximizing the total data transmission utility of the computer network. The total utility is defined as the sum of the individual (corresponding to each node in the network) utilities that are concave functions of the data transmission rate. For the case of non-strongly concave utilities, we propose an approach based on the use of a fast gradient method to optimize a dually smoothed objective function. As an alternative approach, we introduce stochastic oracles for the problem under consideration and interpret them as the messages on the state of some individual node to use randomized switching mirror descent to solve the problem above. We propose interpretations of both described approaches allowing the effective implementation of the protocols of their operation in the real-life computer networks environment, taking into account the distributed information storage and the restricted communication capabilities. The numerical experiments were carried out to compare the proposed approaches on synthetic examples of network architectures.

Keywords: Resource allocation · Computer networks · Utility maximization · Fast gradient method · Primal-dual method · Randomized mirror descent

1 Introduction

Management of the operation of computer communication networks and support of their efficiency, in view of their widespread and massive use, are the crucial tasks today. In particular, the proper management of data transmission in the network should ensure that there are no overloaded, and therefore slow, connections between computers. The network protocol in this setting is not a determining factor; you can fix it and imagine, for example, TCP Internet traffic. At the same time, the task of data flow control is then passed on to the individual computing agents, i.e. network participants. The natural way to control overloading of connections for them is the ability to change their own data transmission rates.

The research of A. Raigorodskii was supported by Russian Science Foundation (project No. 21-71-30005).

N. N. Olenev et al. (Eds.): OPTIMA 2021, CCIS 1514, pp. 184–198, 2021.
https://doi.org/10.1007/978-3-030-92711-0_13

Of course, there are many algorithms and protocols for finding the optimal data transmission rates [1,3,4,6,10]. One of the currently developing approaches proposes to introduce a function of the total utility of the network and consider the optimization problem of maximizing utility in relation to variable data transmission rates [7]. Thus, some intermediary or the network as a whole generate a sequence of values of transmission rates, tending to the minimum of the introduced potential. In such a setting, in order to provide the most resource-efficient procedure for finding the optimal rates, one can use the apparatus and methods developed by the modern theory of convex optimization. The prevalence of this approach in recent years is explained by the fact that many modern methods of convex optimization are directly and effectively applicable to convex problems with linear constraints that naturally arise in it. On the other hand, the variety of optimization methods leads to different modes of operation of these methods on a real-life computer network. In each specific case, one or another protocol may be more beneficial, and in this sense, the convex optimization approach is fruitful: it allows one to form a wide collection of various protocols and provide, in accordance with theoretical considerations, recommendations regarding their appropriate application.

This paper discusses some methods for efficiently optimizing data transmission rates. The first of them follows the idea of setting prices for data transmission over connections [7], so that transmission rates are chosen by each computer for reasons of maximum surplus, i.e. utility minus cost. The second uses a switching scheme to change the rates alternately to more profitable and less loading ones.

However, for the practical application of the schemes proposed in various papers, it is important to represent their implementation in the real-life architecture of a computer network. For both approaches, we describe operating protocols that allow efficiently making updating and storing values distributed, partially parallelized and encapsulated, without losing efficiency. We analyze each method for the convergence rate and indicate, in addition, the dependence of the efficiency on the characteristics of the network and the problem. Thus, the purpose of this work is not so much to propose fundamentally new methods (in this case, we only appropriately apply some recent ideas), but to tightly link the theoretical reasoning with the subject area in which the methods are applied: to clarify the connection with the characteristics of the network; to formulate in more detail the protocol of methods operation in a distributed mode within a real-life computer network, that is, to provide a practical *interpretation* for numerical methods described further.

This paper is organized as follows. Section 2 describes the task of utility maximization in computer networks and formulates the main optimization problem. Further, Sect. 3 using the dual smoothing technique introduce the dual optimization problem and analyses the properties of it. Section 4 describes the primal-dual fast gradient method and analyses its convergence rate for the considered dual problem. We also describe the possibilities of the distributed implementation for this algorithm. Section 5 considers a different type of algorithm, i.e. randomized version of switching mirror descent. We provide the corresponding convergence theorem for the considered problem, and give a description of the possible method operation protocol in a real-life computer network environment.

Finally, Sect. 6 describes experiments on the application of the considered optimization methods to some synthetic computer networks.

2 Problem Statement

By a computer network we mean a structure consisting of a set of vertices of size $n \in \mathbb{N}$, and a set of connections of size $m \in \mathbb{N}$. Each connection in such a model, unlike the edges in a graph, can have a relationship with more than two vertices. The structure of the relationship between connections and vertices is expressed by the matrix $C \in \{0,1\}^{m \times n}$ according to the following rule: $C_{ji} = 1 \Leftrightarrow j$-th connection is in relation to i-th vertex. It turns out to be natural to distinguish separately such statements in which the matrix C is sparse, i.e. the number of its nonzero elements $nnz(C) \ll m \cdot n$.

Data transmission in the network will be characterized from the point of view of each vertex i by the value $x_i \in \mathbb{R}_+$ of its data transmission rate. Quantity x_i determines how much the vertex i loads each of the connections j that are in the relation with it. Below we use the notation $x = (x_1, ..., x_n)^\top \in \mathbb{R}^n_+$. In turn, each of the connections j is characterized by some upper bound on the total load, equal to the throughput $b_j \in \mathbb{R}_+$. Together they form a vector $b = (b_1, ..., b_m)^\top \in \mathbb{R}^m$. One can see that the constraints on the total load of each of the connections is met if and only if the inequality $Cx \leq b$ holds.

The utility achieved by the i-th vertex is characterized by the utility function $u_i(x_i) : \mathbb{R}_+ \to \mathbb{R}$, that depends on the data transmission rate of one i-th vertex. In reasonable formulations of the problem, this function is chosen concave in x_i, for example, $u_i(x_i) = \ln x_i$ (non-strongly concave case) or $u_i(x_i) = a \cdot x_i - b \cdot x_i^2$ ($2b$-strongly concave). By the total utility taken by the entire network, we mean the function $U(x) = \sum_{i=1}^n u_i(x_i)$.

The practical task is to find such values of the data transmission rates for each of the vertices, for which the greatest total utility is achieved. Taking into account the additional details specified above, we can now formulate the main optimization problem [7]:

$$\max_{x : Cx \leq b} \left\{ U(x) = \sum_{i=1}^{n} u_i(x_i) \right\}, \tag{1}$$

where $x \in \mathbb{R}^n_+$, $C \in \mathbb{R}^{m \times n}$, $b \in \mathbb{R}^m$, and u_i is differentiable and concave for all i.

3 Dual Smoothing

Let us move from the formulated main optimization problem with linear constraints to the corresponding dual problem:

$$\min_{\lambda} \left\{ \varphi(\lambda) = \max_{x} \left\{ U(x) + \langle \lambda, b - Cx \rangle \right\} \right\},$$

where $\lambda \in \mathbb{R}^m_+$. Note that the resulting dual factors can be easily interpreted in terms of our subject area. Indeed, by $\lambda = (\lambda_1, ..., \lambda_m)^\top$ we mean the vector of prices for using a unit of bandwidth for each of the connections. Then, rewriting the definition above in the following form:

$$\varphi(\lambda) = \langle \lambda, b \rangle + \sum_{i=1}^{n} \left[u_i(x_i(\lambda)) - \langle \lambda, C_i^\top x_i(\lambda) \rangle \right],$$

using the notation

$$x_i(\lambda) = \arg\max_x \left\{ u_i(x_i) - \langle \lambda, C_i^\top x_i \rangle \right\}, \tag{2}$$

we will obtain a natural interpretation for the dual problem: it thus consists in finding such values of the prices of connections that, provided that each of the vertices is chosen the most profitable for it (in term of utility $u_i(x_i)$ and costs $\langle \lambda, C_i^\top x_i(\lambda) \rangle$) values of data transmission performance, the term associated with violation of the constraints will be the smallest.

We now note the following fact: in the general case, the function U is not strongly concave. At the same time, it is easy to show that in this case the function φ is not necessarily Lipschitz smooth, and this would deprive us of the opportunity to use many of the methods for its optimization, including the fast gradient method. To deal with it, one can use the dual smoothing technique, i.e. instead of the problem (1) consider the optimization problem for the regularized function U_μ with the regularization coefficient $\mu \sim \varepsilon / \|x_* - x_0\|_p^2$, where ε is the required accuracy of solving the primal problem. With this choice of μ, it is known that if

$$\max_x U_\mu(x) - U_\mu(x_*) \leq \frac{\varepsilon}{2},$$

is satisfied for some x_*, then x_* is also a ε-solution of the problem (1).

Let us now move on to a more formal level, namely, we equip the space $\mathbb{R}^n$ with the norm $\|\cdot\|_p$, and the space $\mathbb{R}^m$ with the norm $\|\cdot\|_q$ for some $p, q \in [1, 2]$. We denote the regularized function as U_μ and its dual as follows:

$$U_\mu(x) = U(x) - \frac{\mu}{2}\|x - x_0\|_p^2, \quad \varphi_\mu(\lambda) = \max_x \left\{ U(x) + \langle \lambda, b - Cx \rangle - \frac{\mu}{2}\|x - x_0\|_p^2 \right\}$$

The following lemma describes the properties of the resulting function φ_μ:

Lemma 1. *The function φ_μ has Lipschitz continuous gradient, i.e.* $\forall \lambda_1, \lambda_2 \in \mathbb{R}^m_+$:

$$\|\nabla \varphi_\mu(\lambda_1) - \nabla \varphi_\mu(\lambda_2)\|_q \leq L \|\lambda_1 - \lambda_2\|_q,$$

for $L = \|C\|_{p,q}^2 / \mu$, where

$$\|C\|_{p,q} = \max_{\|x\|_p = 1, \|\lambda\|_q = 1} \langle \lambda, Cx \rangle$$

Proof. Literally coincides with the proof of Theorem 1 from [8].

Assuming at the same time that $\|x_* - x_0\|_p^2 \leq R_p^2$, for an important special case of $p = q = 2$ choosing $\mu = \varepsilon / R_p^2$ we have

$$L = \frac{R_p^2 \lambda_{max}(C^\top C)}{\varepsilon},$$

where $\lambda_{max}(A)$ is the largest eigenvalue of the matrix A. The next lemma provides another, somewhat more intuitive in the framework of the considered subject area, bound for the Lipschitz constant of the φ_μ gradient.

Lemma 2. *When $p = q = 2$, function φ_μ has L-Lipschitz continuous gradient with $L = nnz(C)/\mu$*

Proof. From the first order optimality conditions for (2):

$$\langle \nabla u_i(x_i(\lambda_1)) - \langle \lambda_1, C_i^\top \rangle - \mu(x_i(\lambda_1) - [x_0]_i), x_i(\lambda_1) - x_i(\lambda_2) \rangle \rangle \geq 0,$$

$$\langle \nabla u_i(x_i(\lambda_2)) - \langle \lambda_2, C_i^\top \rangle - \mu(x_i(\lambda_2) - [x_0]_i), x_i(\lambda_2) - x_i(\lambda_1) \rangle \rangle \geq 0.$$

Summing up, we have:

$$\begin{aligned} &\mu \|x_i(\lambda_1) - x_i(\lambda_2)\|_2^2 \\ &\leq \langle \nabla u_i(x_i(\lambda_2)) - \nabla u_i(x_i(\lambda_1)) - \mu(x_i(\lambda_1) - x_i(\lambda_2)), x_i(\lambda_1) - x_i(\lambda_2) \rangle \\ &\leq \langle \langle \lambda_1, C_i^\top \rangle - \langle \lambda_2, C_i^\top \rangle, x_i(\lambda_1) - x_i(\lambda_2) \rangle, \end{aligned}$$

whence

$$\|\nabla \varphi_i(\lambda_1) - \nabla \varphi_i(\lambda_2)\|_2 \leq \|C_i^\top\|_2 \cdot \|x_i(\lambda_1) - x_i(\lambda_2)\|_2 \leq \frac{\|C_i^\top\|_2^2}{\mu} \|\lambda_1 - \lambda_2\|_2.$$

Summing over components and taking into account $C_{ji} \in \{0, 1\}$, we obtain to the expression:

$$\|\nabla \varphi(\lambda_1) - \nabla \varphi(\lambda_2)\|_2 \leq \frac{\sum_{i=1}^n \|C_i^\top\|_2^2}{\mu} \|\lambda_1 - \lambda_2\|_2 = \frac{nnz(C)}{\mu} \|\lambda_1 - \lambda_2\|_2.$$

Thus, the properties of the optimized function directly depend on the sparsity of the matrix C, in other words, in terms of our domain, the fewer vertices on average are in relation to one connection, the less time-consuming the process of finding a solution to the problem is. Further, the nature of this dependence will be refined in the convergence rate bounds for the methods.

4 Fast Gradient Method

Further, we describe the Primal-dual Fast Gradient Method in construction similar to that proposed in [12] for the network utility maximization problem. We provide a slightly modified convergence analysis with improved values of constants and allowing to use arbitrary L_p-norms, and also present in Theorem 1 convergence rates including the non-strongly concave case, for which we apply the dual smoothing technique described above, that extends the class of utility functions for which the considered algorithm is applicable.

4.1 Theoretical Guarantees

Algorithm 1. Primal-dual Fast Gradient Method

Require: λ_0.
1: $\alpha_t = \frac{t+1}{2}$
2: $A_{-1} = 0$, $A_t = A_{t-1} + \alpha_t = \frac{(t+1)(t+2)}{4}$
3: $\tau_t = \frac{\alpha_{t+1}}{A_{t+1}} = \frac{2}{t+3}$
4: **for** $t = 0, 1, \dots, N-1$ **do**
5: Evaluate $\varphi_\mu(\lambda_t)$, $\nabla\varphi_\mu(\lambda_t)$
6: $y_t = \left[\lambda_t - \frac{1}{L}(b - Cx(\lambda_t))\right]_+$
7: $z_t = \left[\lambda_0 - \frac{1}{L}\sum_{k=0}^{t}\alpha_k(b - Cx(\lambda_k))\right]_+$
8: $\lambda_{t+1} = \tau_t z_t + (1-\tau_t) y_t$
9: **end for**
10: **return** λ_N, $\hat{x}_N = \frac{1}{A_N}\sum_{t=0}^{N}\alpha_t x(\lambda_t)$

To analyze the method, we introduce the notation:

$$\psi_t(\lambda) = \sum_{k=0}^{t}\alpha_k\left[\varphi(\lambda_k) + \langle\nabla\varphi(\lambda_k), \lambda - \lambda_k\rangle\right] + \frac{L}{2}\|\lambda - \lambda_0\|_q^2.$$

Lemma 3.

$$A_N\varphi(y_N) \le \min_\lambda \psi_N(\lambda) = \psi_N(z_N). \tag{3}$$

Proof. Let us prove by induction that (3) holds. When $t = 0$, the following holds:

$$\psi_0(z_0) = \min_\lambda\left\{\alpha_0\left[\varphi_\mu(\lambda_0) + \langle\nabla\varphi_\mu(\lambda_0), \lambda - \lambda_0\rangle\right] + \frac{L}{2}\|\lambda - \lambda_0\|_q^2\right\} \ge \alpha_0\varphi_\mu(y_0).$$

Let (3) to be true for t: $A_t\varphi(y_t) \le \psi_t(z_t)$. Let us prove that (3) is true for $t+1$. Indeed, we have

$$\begin{aligned}
\psi_{t+1}(z_{t+1}) &= \min_\lambda\left\{\psi_t(\lambda) + \alpha_{t+1}\left[\varphi(\lambda_{t+1}) + \langle\nabla\varphi(\lambda_{t+1}), \lambda - \lambda_{t+1}\rangle\right]\right\} \\
&\ge \min_\lambda\left\{\psi_t(z_t) + \frac{L}{2}\|\lambda - z_t\|_q^2 + \alpha_{t+1}\left[\varphi(\lambda_{t+1}) + \langle\nabla\varphi_\mu(\lambda_{t+1}), \lambda - \lambda_{t+1}\rangle\right]\right\} \\
&\ge \min_\lambda\left\{A_t\varphi(y_t) + \frac{L}{2}\|\lambda - z_t\|_q^2 + \alpha_{t+1}\left[\varphi_\mu(\lambda_{t+1}) + \langle\nabla\varphi_\mu(\lambda_{t+1}), \lambda - \lambda_{t+1}\rangle\right]\right\} \\
&\ge \min_\lambda\left\{A_t\left(\varphi_\mu(\lambda_{t+1}) + \langle\nabla\varphi_\mu(\lambda_{t+1}), y_t - \lambda_{t+1}\rangle\right) + \frac{L}{2}\|\lambda - z_t\|_q^2 + \alpha_{t+1}[\cdots]\right\}.
\end{aligned} \tag{4}$$

Step $\lambda_{t+1} = \tau_t z_t + (1-\tau_t)y_t$ one can rewrite as the relation $A_{t+1}\lambda_{t+1} = \alpha_{t+1}z_t + A_t y_t$. Using it, we transform:

$$\begin{aligned}
&A_t\langle\nabla\varphi_\mu(\lambda_{t+1}), y_t - \lambda_{t+1}\rangle + \alpha_{t+1}\langle\nabla\varphi_\mu(\lambda_{t+1}), \lambda - \lambda_{t+1}\rangle \\
&\quad = -A_{t+1}\langle\nabla\varphi_\mu(\lambda_{t+1}), \lambda_{t+1}\rangle + \alpha_{t+1}\langle\nabla\varphi_\mu(\lambda_{t+1}), \lambda\rangle + A_t\langle\nabla\varphi_\mu(\lambda_{t+1}), y_t\rangle \\
&\quad = \alpha_{t+1}\langle\nabla\varphi_\mu(\lambda_{t+1}), \lambda - z_t\rangle.
\end{aligned}$$

Hence we have

$$\begin{aligned}
&A_t\left(\varphi_\mu(\lambda_{t+1}) + \langle \nabla\varphi_\mu(\lambda_{t+1}), y_t - \lambda_{t+1}\rangle\right) + \frac{L}{2}\|\lambda - z_t\|_q^2 \\
&+ \alpha_{t+1}\left[\varphi_\mu(\lambda_{t+1}) + \langle \nabla\varphi_\mu(\lambda_{t+1}), \lambda - \lambda_{t+1}\rangle\right] \\
&= A_{t+1}\varphi_\mu(\lambda_{t+1}) + \frac{L}{2}\|\lambda - z_t\|_q^2 + \alpha_{t+1}\langle \nabla\varphi_\mu(\lambda_{t+1}), \lambda - z_t\rangle. \qquad (5)
\end{aligned}$$

After substituting the (5) in the last expression of (4) one can use an extended version of Fenchel's inequality for conjugate functions [9]:

$$\langle g, s\rangle + \frac{\xi}{2}\|s\|_q^2 \geq -\frac{1}{2\xi}\|g\|_{q*}^2.$$

In our case $g = \nabla\varphi(\lambda_{t+1})$, $s = \lambda - z_t$, $\xi = \frac{L}{\alpha_{t+1}}$. Hence,

$$\psi_{t+1}(z_{t+1}) \geq A_{t+1}\varphi_\mu(\lambda_{t+1}) - \frac{\alpha_{t+1}^2}{2L}\|\nabla\varphi_\mu(\lambda_{t+1})\|_q^2. \qquad (6)$$

Further, by Lipschitz smoothness of φ_μ:

$$\begin{aligned}
\varphi(y_{t+1}) &\leq \varphi_\mu(\lambda_{t+1}) + \langle \nabla\varphi_\mu(\lambda_{t+1}), y_{t+1} - \lambda_{t+1}\rangle + \frac{L}{2}\|y_{t+1} - \lambda_{t+1}\|_q^2 \\
&= \min_\lambda\left\{\varphi_\mu(\lambda_{t+1}) + \langle \nabla\varphi_\mu(\lambda_{t+1}), \lambda - \lambda_{t+1}\rangle + \frac{L}{2}\|\lambda - \lambda_{t+1}\|_q^2\right\} \\
&= \varphi_\mu(\lambda_{t+1}) - \frac{1}{2L}\|\nabla\varphi_\mu(\lambda_{t+1})\|_q^2.
\end{aligned}$$

After multiplying both sides of the resulting inequality by A_{t+1}:

$$\begin{aligned}
A_{t+1}\varphi_\mu(y_{t+1}) &\leq A_{t+1}\varphi_\mu(\lambda_{t+1}) - \frac{A_{t+1}}{2L}\|\nabla\varphi_\mu(\lambda_{t+1})\|_q^2 \qquad (7)\\
&\leq A_{t+1}\varphi_\mu(\lambda_{t+1}) - \frac{\alpha_{t+1}^2}{2L}\|\nabla\varphi_\mu(\lambda_{t+1})\|_q^2. \qquad (8)
\end{aligned}$$

Therefore, due to (6) and (8) we have $A_{t+1}\varphi_\mu(y_{t+1}) \leq \psi_{t+1}(z_{t+1})$.

Lemma 4. *It can be assumed that $\|\lambda_*\|_q, \|\lambda_0\|_q \leq R_q$. Then the point $\hat{x}_N$ obtained after N iterations of the Algorithm 1 satisfies the condition*

$$\varphi_\mu(y_N) - U_\mu(\hat{x}_N) + 5R_q\|(C\hat{x}_N - b)_+\|_q \leq \frac{26\,LR_q^2}{A_N}.$$

Proof. The vector of dual factors can be localized using the Slater condition. Considering $\|\lambda_t - \lambda_*\|_q \leq \|\lambda_* - \lambda_0\|_q$ (it was proved in [12], cf. the proof of Lemma 1 from it) we have $\|\lambda_t\|_q \leq \|\lambda_t - \lambda_*\|_q + \|\lambda_* - \lambda_0\|_q + \|\lambda_0\|_q \leq 5R_q$.

From Lemma 3 and considering also $\|\lambda-\lambda_0\|_q^2 \le 2\|\lambda\|_q^2+2\|\lambda_0\|_q^2 \le 2R_q^2+50R_q^2$ we have:

$$A_N\varphi(y_N) \le \min_{\lambda}\left\{\frac{L}{2}\|\lambda-\lambda_0\|_q^2 + \sum_{t=0}^{N}\alpha_t\left[\varphi_\mu(\lambda_t) + \langle\nabla\varphi_\mu(\lambda_t), \lambda-\lambda_t\rangle\right]\right\}$$
$$\le \min_{\|\lambda\|_q\le 5R_q}\left\{\sum_{t=0}^{N}\alpha_t\left[\varphi_\mu(\lambda_t) + \langle\nabla\varphi_\mu(\lambda_t), \lambda-\lambda_t\rangle\right]\right\} + 26\,LR_q^2.$$

Substituting expressions for $\varphi_\mu(\lambda_t)$ and $\nabla\varphi_\mu(\lambda_t)$:

$$\sum_{t=0}^{N}\alpha_t\left[\varphi_\mu(\lambda_t) + \langle\nabla\varphi_\mu(\lambda_t), \lambda-\lambda_t\rangle\right]$$
$$= \sum_{t=0}^{N}\alpha_t\left[\langle\lambda_t, b\rangle + U_\mu(x(\lambda_t)) - \langle\lambda_t, Cx(\lambda_t)\rangle + \langle b - Cx(\lambda_t), \lambda-\lambda_t\rangle\right]$$
$$= \sum_{t=0}^{N}\alpha_t\left[U_\mu(x(\lambda_t)) + \langle\lambda, b - Cx(\lambda_t)\rangle\right] \le A_N\left[U(\hat{x}_N) + \langle\lambda, b - C\hat{x}_N\rangle\right],$$

Which finally leads to

$$A_N\varphi_\mu(y_N) \le A_N U_\mu(\hat{x}_N) + 26\,LR_q^2 + A_N \min_{\|\lambda\|_q\le 5R_q}\langle\lambda, b - C\hat{x}_N\rangle$$
$$= A_N U_\mu(\hat{x}_N) + 26\,LR_q^2 - 5R_q A_N\|(C\hat{x}_N - b)_+\|_q.$$

Theorem 1. *The point $\hat{x}_N$ ($\hat{x}_{N_\mu}$) obtained after N (N_μ) iterations of the Algorithm 1 satisfies the conditions*

$$\max_x U(x) - U(\hat{x}_N) \le \varepsilon, \quad \|(C\hat{x}_N - b)_+\|_q \le \frac{\varepsilon}{4R_q},$$

if the number of iterations satisfies the following inequality (the first for not strongly concave u_i, the second for μ-strongly concave u_i):

$$N \ge \left\lfloor 8\sqrt{13}R_qR_p\cdot\frac{\|C\|_{p,q}}{\varepsilon}\right\rfloor, \quad N_\mu \ge \left\lfloor 2\sqrt{26}\sqrt{R_qR_p}\cdot\frac{\|C\|_{p,q}}{\sqrt{\mu\varepsilon}}\right\rfloor$$

Proof. Due to weak duality, it holds that $\min_\lambda \varphi_\mu(\lambda) \ge \max_x U_\mu(x)$. Due to the fact that $\varphi_\mu(y_N) \ge \min_\lambda \varphi_\mu(\lambda) \ge \max_x U_\mu(x)$, the following estimate follows directly from Lemma 4:

$$\max_x U_\mu(x) - U_\mu(\hat{x}_N) \le \frac{26\,LR_q^2}{A_N} \tag{9}$$

Also, by the properties of duality, we have

$$\max_x U_\mu(x) \geq U_\mu(x) - \langle \lambda_*, (Cx-b)_+ \rangle$$
$$\geq U_\mu(x) - R_q\|(Cx-b)_+\|_q \quad \forall x \in \mathbb{R}^n_+$$

where λ_* is a minimum point of φ_μ. Hence the following estimate follows:

$$\varphi_\mu(y_N) - U_\mu(\hat{x}_N) = (\varphi_\mu(y_N) - \min_\lambda \varphi_\mu(\lambda)) + (\min_\lambda \varphi_\mu(\lambda) - \max_x U_\mu(x))$$
$$+ (\max_x U_\mu(x) - U_\mu(\hat{x}_N)) \geq -R_q\|(C\hat{x}_N - b)_+\|_q.$$

Using Lemma 4 together with the obtained inequality:

$$R_q\|(C\hat{x}_N - b)_+\|_q \leq \frac{13\,LR_q^2}{2A_N} \tag{10}$$

Upper-bounding the right-hand side of (9) by $\frac{\varepsilon}{2}$ (or ε, if the u_i are strongly concave) and taking into account this estimate in (10), we obtain the desired conditions from theorems and condition for N:

$$\frac{104\,LR_q^2}{(N+1)(N+2)} \leq \frac{\varepsilon}{2}$$

Substituting L from Lemma 1 and solving for N, we obtain the inequalities from the theorem.

Lemma 5. *For a μ-strongly concave function U (possibly $\mu = 0$) in a particular case of $p = q = 2$ we have $N \sim \sqrt{nnz(C)}$, i.e.*

$$N = \min\left\{ \left\lceil 8\sqrt{13}R_qR_p \cdot \frac{\lambda_{max}^{1/2}(C^\top C)}{\varepsilon} \right\rceil, \left\lceil 2\sqrt{26}\sqrt{R_qR_p} \cdot \frac{\lambda_{max}^{1/2}(C^\top C)}{\sqrt{\mu\varepsilon}} \right\rceil \right\},$$

whereas in the case of $p = q = 1$ we have $N \sim \max nnz(C_i^\top)$, and more specifically

$$N = \min\left\{ \left\lceil 8\sqrt{13}R_qR_p \cdot \frac{\max_{i=1,\dots,n} \|C_i^\top\|_1}{\varepsilon} \right\rceil, \left\lceil 2\sqrt{26}\sqrt{R_qR_p} \cdot \frac{\max_{i=1,\dots,n} \|C_i^\top\|_1}{\sqrt{\mu\varepsilon}} \right\rceil \right\}.$$

Let us clarify the question of the choice of norms for the problem under consideration. For the setting under consideration, it turns out to be reasonable to preserve freedom only in the choice of $p \in [1, 2]$, while for the dual problem, due to its structure, as a result of the trade off between a decrease in the Lipschitz constant and an increase in R, the most effective choice is $q = 2$ [5]. As you can see from the arguments above, another choice of q also does not bring any benefit in term of dependence on the properties of the matrix C.

4.2 Interpretation

Let us divide the optimizer's responsibility for the primal and dual variables into two natural types of computing agents: connections and vertices. By global process iterations we mean the steps during which each computational agent performs all updates corresponding to the iteration of the Algorithm 1 for the corresponding component of the vectors appearing there. Let us show that the described algorithm can be implemented in a real network architecture at reasonable and insignificant costs of communication between computing agents.

So, let us first consider a procedure performed on some connection j. In the vector y_t it corresponds to the component $[y_t]_j$, updated based on its corresponding previous value $[\lambda_{t-1}]_j$ and some information about $x(\lambda_t)$. We assume that at the iteration t real data transmission rate x_i for each of the vertices i coincides with $x_i(\lambda_t)$ (the procedures described in this section guarantee this). Note now that when multiplying x by C_j the value of the product will be affected only by those x_i for which $C_{ji} = 1$, that is, the vertices directly in relation to the connection j. It is natural then to assume that before updating $[y_t]_j$ the connection j polls all the vertices in relation to it for the values of x_i. $[z_t]_j$ and $[\lambda_{t+1}]_j$ components are updated in the same way. Since all connections perform these procedures in parallel, we can assume that in time independent of the dimension of the problem, all components of the vectors y_t, z_t and λ_{t+1} will be calculated (they will be stored, of course, also distributed).

Now let us move on to considering the procedure performed by the vertex i. Its goal is to calculate the optimal data transmission rate $[\hat{x}_{t+1}]_i$. Note that this value can be obtained, if $x_i(\lambda_{t+1})$ is known, by a simple update in constant time and without additional communications. At the same time, it is easy to see that to solve the auxiliary problem and obtain $x_i(\lambda_{t+1})$ one should know only those components $[\lambda_{t+1}]_j$ for which $C_{ji} = 1$ (due to multiplication by $C_i^\top$). This means that a vertex, before executing the procedure, needs to poll all connections that are in relation to it for the values of $[\lambda_{t+1}]_j$. On the other hand, here it is computationally possible to manage such communication protocol, wherein each connection, calculating a value of $[\lambda_{t+1}]_j$, notifies all related nodes about it, while each node collects incoming messages until it receives up-to-date information from all related connection, and after performing the procedure in a symmetric manner will notify all associated connections about new values of $x_i(\lambda_{t+1})$.

5 Randomized Mirror Descent

5.1 Theoretical Guarantees

Algorithm 2. Stochastic Mirror Descent

Require: x_0.
1: $I = \varnothing, J = \varnothing$
2: **for** $t = 0, 1, \dots, N-1$ **do**
3: **if** $Cx_t - b \le \varepsilon$ **then**
4: $i \sim \mathcal{U}\{1, ..., n\}$
5: $[x_{t+1}]_i = \left[[x_t]_i - \frac{\varepsilon n}{M_U^2}\nabla u_i([x_t]_i)\right]_+$
6: $I = I \cup \{t+1\}$
7: **else**
8: $j_t = \arg\max_{j=1,\dots,m} C_j x_t - b_j$
9: $i \sim \mathcal{U}\{i : C_{j_t i} = 1\}$
10: $[x_{t+1}]_i = \left[[x_t]_i - \frac{\varepsilon n}{\max_{j=1,\dots,m}\|C_j\|_{p*}^2}\right]_+$
11: $J = J \cup \{t+1\}$
12: **end if**
13: **end for**
14: **return** $\hat{x}_N = \frac{1}{|I|}\sum_{t\in I} x_t$

In Algorithm 2 the stochastic oracles of the gradient of the function U and the gradient of constraints $Cx - b$ are used to perform a step of the method, namely, randomized along the vertex i, corresponding to the selected term $u_i(x_t)$ and the component of the updated vector x_t:

$$\mathbb{E}_i[e_i \cdot n\nabla u_i([x_t]_i)] = \nabla U(x_t), \qquad \mathbb{E}_i[e_i \cdot nC_{j_t}] = \nabla(C_{j_t}x_t - b_{j_t}).$$

We assume that the randomized gradient of U used is bounded:

$$|\nabla u_i(x)| \le M_U. \tag{11}$$

Below is a direct consequence of the result on the convergence of the method obtained in [11].

Theorem 2. *Theorem 2 [11] Point $\hat{x}_N$, obtained after N iterations of the Algorithm 2, and $\hat{\lambda}$, chosen so that $[\hat{\lambda}]_j = \frac{1}{|I|}\frac{M_U^2}{\max_{j=1,\dots,m}\|C_j\|_{p*}^2}\sum_{t\in J}\mathbb{1}[j = j_t]$, satisfy the conditions*

$$\mathbb{E}[\max_x U(x) - U(\hat{x})] \le \mathbb{E}[\varphi(\hat{\lambda}) - U(\hat{x})] \le \varepsilon, \quad Cx_t - b \le \varepsilon,$$

if the number of iterations satisfies the following inequality:

$$N \ge \left\lceil \frac{72\max\{M_U, \max_{j=1,\dots,m}\|C_j\|_{p*}\}^2 n^2 R_p^2}{\varepsilon^2} \right\rceil$$

Note that when choosing the norm $p = 1$, we have $\|C_j\|_{p*} = \max_{i=1,\dots,n} C_{ji} \leq 1$, and therefore the estimate ceases depend on the characteristics of the matrix C.

One of the few problems with the presented approach is that the specific functions u used may not satisfy the assumption (11). Note, however, that this problem can be solved — for this one can use adaptive versions of mirror descent, similar, for example, to those described in [2]. Thus, it is possible to obtain estimates similar to those presented above, but including, instead of M_U, the constants of the form $(\frac{1}{N}\sum_{t=1}^{N} M_t^2)^{1/2}$, where M_t is the adaptively selected constant value at iteration t.

5.2 Interpretation

Let us analyze, similarly to the previous considered method, the computational and communication protocol corresponding to the Algorithm 2. In this case, direct duality arises in the described approach not constructively, but only theoretically. This means that in order to calculate the prices of information transmission for each of the connections, it is not necessary to know the values x_i. At the same time, for calculating prices, it becomes necessary to know some data about the structure of the network and the problem, as well as additional information that appears in the course of global iterations of the method. In addition, the very structure of descent with switches requires verification of the fulfillment of all constraints of the problem, which requires information about the operation of all connections (or about the values of the data transmission rates for all vertices, the decision on the preferred verification method is made only for reasons of communication complexity). This gives rise to the need to introduce some kind of decision-making center that aggregates information about the operation of the network and provides it upon request to individual computing agents.

Let's describe the procedures performed by the decision center. At the beginning of the iteration, the center polls the connections or vertices for the fulfillment of the problem's constraints, after which it decides on the type of iteration ($t \in I$ or $t \in J$). If it turns out that some of the connections are overloaded, then the most overloaded of them is selected, which has the number j_t. This completes the center procedure, it is repeated only after one of the vertices notifies it of the completed update of its component $[x_{t+1}]_i$: along with this, the center can also notify with the values $|I|$ and $\{j_t\}_{t=0}^{N}$ all the connections in the network to allow them to calculate their own prices — this, however, is not necessary when the goal is only to find the optimal data transmission rates.

Now let's move on to the procedure performed on the vertex i. At the beginning of a local iteration (which is automatically launched every fixed period of time), the vertex appeals the center to obtain information about the iteration type and value j_t. If the vertex i and the connection j_t are in a relation or the iteration type matches $t \in I$, the vertex accordingly updates the value of the data transmission rate $[x_{t+1}]_i$ and notifies the center (only about the fact of the update, transfer of the new value is not required). If we assume that the number of the vertex i that first requested information from the center is uniform distributed, the resulting protocol will is correspond to the analyzed algorithm.

Note also that the center does not need to poll all connections for overloading every time, since when only one component $[x_{t+1}]_i$ is updated, the vertex can inform the center not only about the fact of overloading, but also about the new value of this components, and then the center can update the values of the constraints more efficiently, without having to go to all the connections of the network every time.

6 Numerical Experiments

Let's move on to the description of numerical experiments. The results presented below were obtained on a macOS 10.15.7 PC running a 3 GHz 6-core Intel Core i5 processor, with a Python 3.8.2 interpreter in a Jupyter Notebook environment, calculations were performed on variables of numpy.float64 type. The source code for setting up experiments is available at https://github.com/dmivilensky/Network-utility-maximization-2021.

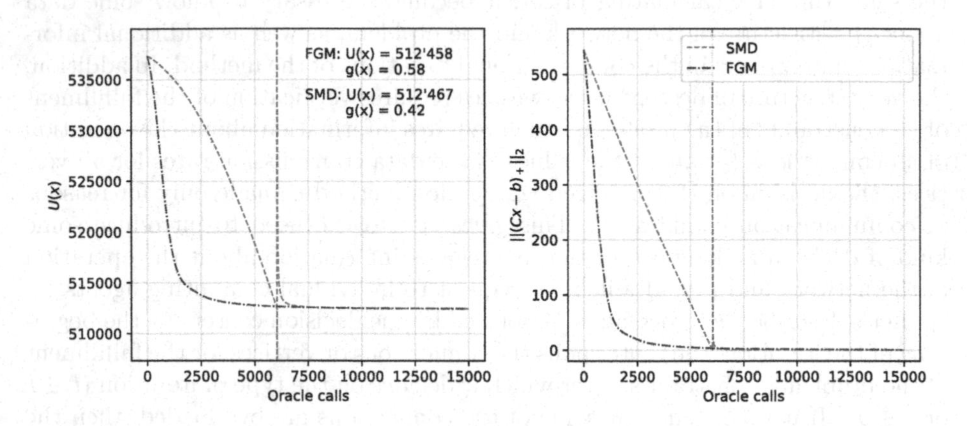

Fig. 1. Algorithm 1 (FGM) and Algorithm 2 (SMD) comparison

Figure 1 shows the dynamics of the values of the total utility function and the constraint residual for the sequences of data transmission rates generated by Algorithm 1 and Algorithm 2 in the course of solving problem (1). For this experiment, we consider a particular formulation of the problem with $p = q = 2$, $m = 40, n = 100$, matrix C with i.i.d. drawn components, such that $nnz(C) \approx 0.001 \cdot m \cdot n$, vector b with i.i.d. components drawn from uniform distribution $\mathcal{U}[0, n]$. We consider utility functions u_i of the form

$$u_i(x) = a_i \cdot x - \frac{\sigma \cdot n}{2} \cdot x^2,$$

where a is vector with i.i.d. components drawn from uniform distribution $\mathcal{U}[1, 50]$, $\sigma = 0.001$. Hence, closed form solution of (2) is

$$x_i(\lambda) = \frac{\left(a - C^\top \cdot \lambda\right)_+}{n \cdot \sigma}.$$

As we can see from the presented figure, there is a distinct moment of the first switching of Algorithm 2 from "constraint correction mode" to the "combined mode", in which the utility function is also being optimized. The abscissa axis of the graphs measures the number of one component oracle calls. In this experiment, it can be seen that Algorithm 2 obtains a slightly better solution than Algorithm 1, using the same number of calls to individual components. The real working time of Algorithm 2 in the simulation exceeds that for Algorithm 1, however, in a real-life environment of a computer network, the main contribution during operation is made by delays in waiting for responses from network participants, which is more objectively reflected by the number of oracle calls.

Optimizing the constraints of the problem at the first stage of the operation of Algorithm 2 can take a long time, determined only by the properties of the problem (that is, arbitrarily long). At the same time, if there is a point in the feasible set, for example, describing the present empirical distribution of data transmission rates in the network, Algorithm 2 immediately starts in "combined" mode. In any case, both proposed methods turn out to be competitive alternatives, and preference can be given to one of the methods mainly for reasons of the protocol for working with a real-life computer network.

7 Conclusion

Following the approach of updating individual data transmission rates and pricing network connections, we propose two methods to maximize the utility of the network. The first is obtained as a generalization of the fast gradient method to the case of not strongly concave utility functions, using the dual smoothing technique. The second is an adaptation of switched stochastic mirror descent for constrained problems. For both proposed methods, a detailed description of the protocols of their operation in the environment of a real-life computer network was presented, taking into account distributed data storage and limited communication capabilities. Numerical experiments on synthetic architectures of computer networks allow us to compare the practical efficiency of the proposed algorithms.

References

1. Arrow, K.J., Hurwicz, L.: Decentralization and computation in resource allocation. Department of Economics, Stanford University (1958)
2. Bayandina, A.: Adaptive stochastic mirror descent for constrained optimization. Constructive Nonsmooth Analysis and Related Topics (dedicated to the memory of VF Demyanov) (CNSA), St. Petersburg, pp. 1–4 (2017)

3. Campbell, D.E., et al.: Resource Allocation Mechanisms. Cambridge University Press, Cambridge (1987)
4. Friedman, E.J., Oren, S.S.: The complexity of resource allocation and price mechanisms under bounded rationality. Econ. Theory **6**(2), 225–250 (1995)
5. Gasnikov, A., Dvurechensky, P., Nesterov, Y.: Stochastic gradient methods with inexact oracle. arXiv preprint arXiv:1411.4218 (2014)
6. Kakhbod, A.: Resource Allocation in Decentralized Systems with Strategic Agents: An Implementation Theory Approach. Springer, New York (2013). https://doi.org/10.1007/978-1-4614-6319-1
7. Kelly, F.P., Maulloo, A.K., Tan, D.K.: Rate control for communication networks: shadow prices, proportional fairness and stability. J. Oper. Res. Soc. **49**(3), 237–252 (1998). https://doi.org/10.1057/palgrave.jors.2600523
8. Nesterov, Y.: Smooth minimization of non-smooth functions. Math. Program. **103**(1), 127–152 (2005)
9. Nesterov, Y.E.: Algorithmic convex optimization. Doctoral (Phys. Math.) Dissertation (Mosc. Phys.-Tech. Inst., Moscow, 2013) (2013)
10. Rokhlin, D.B.: Resource allocation in communication networks with large number of users: the dual stochastic gradient method. Theory Probab. Appl. **66**(1), 105–120 (2021)
11. Tiapkin, D., Stonyakin, F., Gasnikov, A.: Parallel stochastic mirror descent for MDPs. arXiv preprint arXiv:2103.00299 (2021)
12. Vorontsova, E., Gasnikov, A., Dvurechensky, P., Ivanova, A., Pasechnyuk, D.: Numerical methods for the resource allocation problem in a computer network. Comput. Math. Math. Phys. **61**(2), 297–328 (2021)

Applications

A Posteriori Analysis of the Algorithms for Two-Bar Charts Packing Problem

Adil Erzin[1,2(✉)], Georgii Melidi[2], Stepan Nazarenko[2], and Roman Plotnikov[1]

[1] Sobolev Institute of Mathematics, SB RAS, Novosibirsk 630090, Russia
adilerzin@math.nsc.ru
[2] Novosibirsk State University, Novosibirsk 630090, Russia

Abstract. The Two-Bar Charts Packing Problem (2-BCPP) is to pack the bar charts (BCs) of two bars into the horizontal unit-height strip of minimal length. The bars may move vertically within the strip, but it is forbidden to change the order and separate the chart's bars. Recently, for this novel issue, which is a generalization of the Bin Packing Problem (BPP), Strip Packing Problem (SPP), and 2-Dimensional Vector Packing Problem (2-DVPP), several approximation algorithms with guaranteed estimates have been proposed. However, after a preliminary analysis of the solutions constructed by approximation algorithms, we discerned that the guaranteed estimates are inaccurate. This fact inspired us to conduct a numerical experiment in which the approximate solutions are compared to each other and with the optimal ones. We use the Boolean Linear Programming (BLP) formulation of 2-BCPP proposed earlier and apply the CPLEX package to find the optimal solutions or lower bounds for optimum. We also use a database of instances for BPP with known optimal solutions to construct the instances for the 2-BCPP with known minimal packing length. The results of computational experiments comprise the main content of this paper.

Keywords: Bar charts · Strip packing · Approximation algorithms · Simulation

1 Introduction

In [11], we studied the problem of optimizing an investment portfolio in the oil and gas field. Each project is characterized by the annual hydrocarbon production volume, adequately displayed with bar charts. All projects' annual production volume must not exceed a given value (throughput of a pipe). The problem is to complete all projects as early as possible.

This problem is a particular case of the resource-constrained project scheduling problem with one renewable resource [6]. The height of the BC's bar corresponds to the value of the consumed resource. Since the Bar Charts Packing

The study was carried out within the framework of the state contract of the Sobolev Institute of Mathematics (project no. 0314–2019–0014).

N. N. Olenev et al. (Eds.): OPTIMA 2021, CCIS 1514, pp. 201–216, 2021.
https://doi.org/10.1007/978-3-030-92711-0_14

Problem (BCPP) is intractable, we investigate a minor generalization of the Bin Packing Problem (BPP) when each BC consists of two bars. Let us denote further a BC consisting of k bars as k-BC and the problem under consideration as 2-BCPP.

To avoid misunderstanding, we interchangeably use the terms "width" and "length" referring to horizontal dimension. The 2-BCPP can be formulated as follows. We are given a set of 2-BCs. The height of each 2-BC's bar does not exceed 1. In a *feasible* packing, the bars of each 2-BC do not change order and occupy adjacent cells; however, they can move vertically within the strip independently. The 2-BCPP is to find a min-length feasible packing of 2-BCs. If we split the strip into equal unit-width cells of height 1, the packing length is the number of cells with at least one bar.

1.1 Related Results

The 2-BCPP was first formulated in [11] and then examined in [12–14]. Similar problems to the 2-BCPP are the Bin Packing Problem (BPP) [17], the Strip Packing Problem (SPP) [1,8], and the 2-Dimensional Vector Packing Problem (2-DVPP) [19].

In the BPP, a set of items L with given sizes is necessary to pack in the minimal number of unit-size bins. This problem is strongly NP-hard. However, several approximation algorithms are known. One of them is the First Fit Decreasing (FFD). Items are numbered in non-increasing order, and the current item is placed in the first suitable bin. It was proved that the FFD uses no more than $11/9\,OPT(L)+4$ bins [17], where $OPT(L)$ is the minimal number of bins to pack the items from L. Later the additive constant was reduced to 3 [2], then it was reduced to 1 [26], in 1997 to 7/9 [20], and finally, in 2007, the exact value equal to 6/9 of the additive constant was found [9]. For the Modified First Fit Decreasing (MFFD) algorithm, it was shown that $MFFD(L) \leq 71/60\,OPT(L)+31/6$ [18]. This estimate was improved to $71/60\,OPT(L)+1$ [27].

In the SPP, it is necessary to pack (without rotation) a set of rectangles R into the strip of minimal length. The Bottom-Left algorithm arranges rectangles in the descending order of height and yields a 3–approximate solution [1]. Then an algorithm with the ratio of 2.7 was proposed [8]. In [23] an algorithm that builds a 2.5–approximate solution was proposed. Later the ratio was reduced in [22], and [24] to 2. The smallest known estimate for the ratio is $(5/3+\varepsilon)OPT(R)$, for any $\varepsilon > 0$ [16].

2-DVPP is a generalization of BPP and a special case of 2-BCPP. It considers two attributes for each item and bin. The problem is to pack all items in the minimum number of bins, considering both attributes of the bin's capacity limits. In [19] a 2–approximation algorithm for 2-DVPP was presented. [7] presents a survey of approximation algorithms for 2-DVPP. The best algorithm yields a $(3/2+\varepsilon)$–approximate solution, for any $\varepsilon > 0$ [3].

In [12], we proposed an $O(n^2)$–time algorithm that builds a packing for n 2-BCs, which length does not exceed $2OPT + 1$, where OPT is the minimum packing length for the 2-BCPP. When at least one bar of each 2-BC is higher than

1/2 (such 2-BCs we called "big"), an $O(n^3)$–time 3/2–approximation algorithm was proposed. When each 2-BC is big and additionally non-increasing or non-decreasing, the complexity was reduced to $O(n^{2.5})$ preserving the ratio [13]. The paper [14] updates the estimates for the packing length of big 2-BCs, keeping time complexity. In [14], we improve the ratio and give a 5/4–approximation $O(n^{2.5})$–time algorithm for packing big non-increasing or non-decreasing 2-BCs. For the case of big 2-BCs (not necessarily non-increasing or non-decreasing), we proposed a 16/11–approximation $O(n^3)$–time algorithm.

1.2 Our Contribution

The main goal of this paper is a posteriori analysis of the previously developed algorithms. We implement the approximation algorithms and conduct a numerical experiment. We use the CPLEX package for the Boolean Linear Programming (BLP) problem to get an optimum or a lower bound for the packing length. We also use a database of instances for BPP with known optimal solutions for building instances for the 2-BCPP with known optimums.

The remaining parts of the paper are organized as follows. Section 2 provides a statement of the 2-BCPP and the necessary definitions. In Sect. 3, we describe the algorithms under consideration and formulate some properties. In Sect. 4, the numerical experiment results are depicted, and the last section is the paper's conclusion.

2 Formulation of the Problem

Let a semi-infinite unit-height horizontal strip be located on the plane so that its lower boundary coincides with the horizontal axis and its beginning is in origin. For each 2-BC $i \in S$, $|S| = n$, consisting of two unit-width bars, the height of the first (left) bar is $a_i \in (0, 1]$ and of the second (right) is $b_i \in (0, 1]$. Let us split the strip into identical unit-width and unit-height rectangles, which we call the "cells", and number them with naturals starting from the beginning of the strip.

Definition 1. *The* packing *is a function* $p : S \to \mathbb{Z}^+$ *that assigns to each 2-BC* i *a cell number* $p(i)$ *where its first (left) bar falls. The packing is* feasible *if the sum of the bar's heights that fall into each cell does not exceed 1.*

As a result of a packing p, the first bar of ith 2-BC falls into the cell $p(i)$ and the second bar falls into the cell $p(i) + 1$. We will consider only feasible packings; therefore, the word "feasible" will be omitted further.

Definition 2. *The packing* length $L(p)$ *is the number of strip cells in which at least one bar falls.*

In [12], we formulated a 2-BCPP in the form of BLP. However, since we use the CPLEX package for the BLP to get optimal packings in this paper, we replicate its statement below for the reader's convenience. For this purpose, we introduce the variables:

$$x_{ij} = \begin{cases} 1, \text{ if the first bar of the } i\text{th 2-BC is placed into the cell } j; \\ 0, \text{ else.} \end{cases}$$

$$y_j = \begin{cases} 1, \text{ if the cell } j \text{ contains at least one bar;} \\ 0, \text{ else.} \end{cases}$$

Then 2-BCPP in the form of BLP is as follows.

$$\sum_j y_j \to \min_{x_{ij}, y_j \in \{0,1\}}; \tag{1}$$

$$\sum_j x_{ij} = 1, \; i \in S; \tag{2}$$

$$\sum_i a_i x_{ij} + \sum_k b_k x_{kj-1} \leq y_j, \; \forall j. \tag{3}$$

In this formulation, criterion (1) is a minimization of the packing length. Constraints (2) require each 2-BC to be packed into a strip once. Constraints (3) ensure that the sum of the bar's heights in each cell does not exceed 1.

The 2-BCPP (1)–(3) is strongly NP-hard as a generalization of the BPP [17]. Moreover, the problem is $(3/2 - \varepsilon)$-inapproximable for any $\varepsilon > 0$ unless P = NP [25].

3 Algorithms

In [12], we analyzed some approximation algorithms and concluded that the greedy algorithm *GA_LO* with the preliminary lexicographic ordering of 2-BCs in a non-increasing manner significantly outperforms the other examined approximation algorithms. Therefore, in this article, for comparison with new algorithms, we use only one previously analyzed algorithm *GA_LO* [12]. New approximation algorithms were proposed in [13] and [14]. Below is a brief description of the algorithms we analyze in this paper, illustrated by an example.

3.1 Greedy Algorithm *GA_LO*

Algorithm *GA_LO* described in detail (with pseudocode) in [12]. It sorts all 2-BCs lexicographically in non-increasing order of bar's height (Fig. 1*a*). Let P be a list of a lexicographically ordered set of 2-BCs. The first element in P is placed into the first two strip's cells and removed from P. The 2-BCs deleted from P never relocate further. Then until $P \neq \emptyset$, the following procedure is performed. For each 2-BC in P, we search the leftmost position that does not violate the packing feasibility. Among 2-BCs that could be placed into the same leftmost cell, choose a 2-BC with a minimal number, fix its position in the strip and remove it from P. For illustration, the resulting packing of the example in Fig. 1a is presented in Fig. 1b.

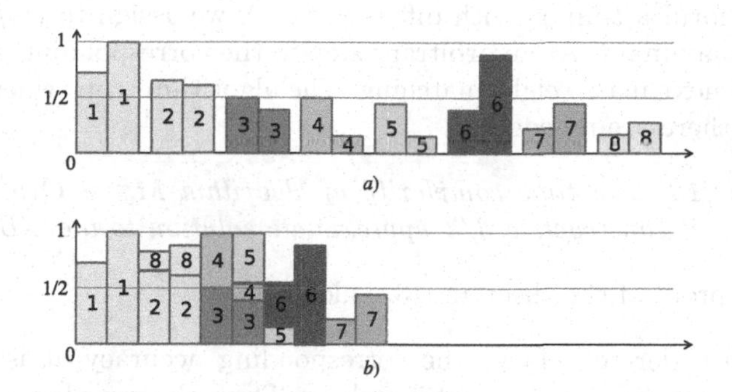

Fig. 1. a) Lexicographically ordered set of 2-BCs; b) Packing built by GA_LO.

In [12], we have proved that algorithm A, which uses GA_LO as a procedure, constructs a packing whose length is at most $2OPT + 1$, where OPT is the problem's optimum. However, a posteriori analysis shows that algorithm GA_LO itself frequently constructs a near-optimal solution.

3.2 Algorithms M_w and $M1_w$

Recall that for any natural k, we denote by k-BC a BC containing k bars.

Definition 3. *Two BCs with an arbitrary number of bars create a t-union if they are packed so that t strip's cells contain bars of both BCs.*

It follows from the definition that if two BCs consist of x bars, their t-union (new BC) has $x - t$ bars. Thus, for example, in Fig. 1b the 2nd and the 8th 2-BCs form a 2-union, and the 5th and the 6th 2-BCs form a 1-union.

Definition 4. *If at least one bar in 2-BC is higher than 1/2, then such 2-BC we call* big. *Consequently, we also call a bar* big *if it is higher than 1/2. Otherwise, let us call a bar* small.

The algorithm M_w was described in detail in [13]. It performs a sequence of steps. At each step, a max-weight matching is built in a specially constructed weighted graph. Initially, using the set S, we build a weighted graph $G_1 = (V_1, E_1)$, in which the vertices are the images of 2-BCs in S ($|V_1| = |S| = n$). The edge $(i, j) \in E_1$ if the ith and jth 2-BCs can form a t-union ($t \in \{1, 2\}$). The edge's weight equals 2 if the ith and jth 2-BCs can form a 2-union. If the ith and jth 2-BCs cannot form a 2-union, but can form a 1-union, then the weight of edge (i, j) equals 1. Then in the graph G_1, a max-weight matching is constructed.

The edges in the matching and their weights indicate the unions of 2-BCs. As a result of these unions, we get a new set of 2- and 3-BCs, which are the prototypes of vertices forming the set V_2 of the next weighted graph $G_2 = (V_2, E_2)$. The edge (i, j) exists in G_2 if the ith and jth BCs can form a union. If the ith and jth BCs can form a t-union with different $t > 0$, we assign to (i, j) the weight equal to the maximal t. At an arbitrary step in the corresponding graph G_k, we construct the next max-weight matching. The algorithm stops when in the next graph G_{k+1}, there are no edges.

Theorem 1. *[13] The time complexity of algorithm M_w is $O(n^4)$, and if all 2-BCs are big, it constructs a 3/2–approximate solution to the 2-BCPP.*

From the proof of the theorem [13] follows the

Remark 1. In order to achieve the corresponding accuracy, it is sufficient to construct only the first max-weight matching. Thus, the complexity of obtaining a 3/2–approximate solution is $O(n^3)$.

Proposition 1. *The algorithm M_w builds only 1- and 2-unions.*

Proof. Suppose that at some step, the algorithm M_w builds a t-union with $t > 2$ using some p-BC and q-BC, $p, q > 2$. Initially, each BC consists of two bars, and only 1- and 2-unions are possible. Moreover, any 2-union of two 2-BCs forms a new 2-BC. Hence, at least one 1-union is needed to create later a BC with more than two bars. Therefore, the 1-unions appear during the formation of these p-BC and q-BC. Let us denote by T, $|T| = t$, the set of cells containing bars of both p-BC and q-BC. In the first cell of T, at least one bar of p-BC and one bar of q-BC are located. However, initially, each BC consists of two bars. Then at least one 2-BC of p-BC and one 2-BC of q-BC fall into the first and second cells of T. Otherwise, these p-BC and q-BC could not be obtained from the initial 2-BCs. Similarly, there exist two 2-BCs that fall into the $(t-1)$th and tth cells. This fact contradicts the M_w performance since these two 2-unions should be formed before the 1-unions. Then at some step, there was a matching of non-maximum weight. Therefore, our assumption is wrong and t-unions with $t > 2$ are impossible. Figure 2*a* shows the example of 3-union when $p, q = 3$. In Fig. 2*b*, the 1st and 3rd 2-BCs should have formed a 2-union at the previous step. The same is true for the 2nd and the 4th 2-BCs. The proposition is proved.

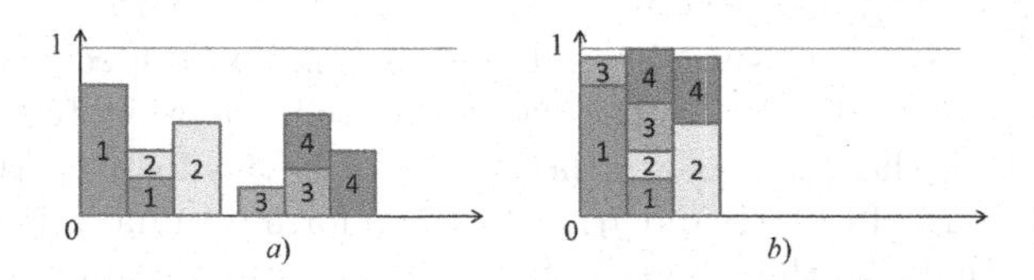

Fig. 2. *a*) p-BC and q-BC for 3-union when $p, q = 3$; *b*) 3-union.

This proposition we use in the implementation of the algorithm M_w. That is, only 1- and 2-unions are built in this algorithm.

In the Computational Experiments section, we also execute the algorithm $M1_w$, which constructs only the first max-weight matching in the graph G_1 (the first stage of algorithm M_w). Both algorithms $M1_w$ and M_w have the ratio at most 3/2, but these algorithms are more accurate on average.

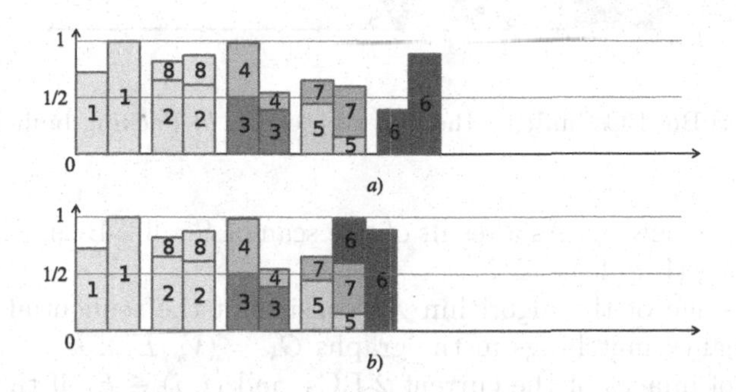

Fig. 3. *a*) Packing built by $M1_w$; *b*) Packing built by M_w.

Figure 3*a* shows the first matching, which is the result of the algorithm $M1_w$. Figure 3*b* shows the packing constructed by the algorithm M_w.

3.3 Algorithms *A*1 and *A*2

In [14] an $O(n^{2.5})$–time 5/4–approximation algorithm for packing non-increasing (or non-decreasing) big 2-BCs is proposed. The algorithm is based on the reduction of 2-BCPP to the Maximum Asymmetric Traveling Salesman Problem with edge's weights 0 or 1 (MaxATSP(0,1)) and using the algorithm proposed in [4,21] for the latter problem. Furthermore, in [14] an $O(n^3)$–time 16/11–approximation algorithm for the packing big (not necessary non-increasing or non-decreasing) 2-BCs is presented. To obtain this estimate, the algorithm for construction a max-cardinality matching [15], and approximation algorithm proposed in [21] is used. We were interested in comparing this algorithm with the previously developed ones in two cases: the 2-BCs are big, and the 2-BCs are arbitrary. If 2-BCs are arbitrary, we first need to construct big 2-BCs to apply the algorithm. We propose two different procedures for constructing big 2-BCs, applying the common part and calling these algorithms *A*1 and *A*2.

In the algorithm *A*1, we apply the first stage of algorithm *A* described in [12]. This procedure is as follows. Set $M = \emptyset$. The 2-BCs are browsing in numerical order. If the current 2-BC is big, then consider the next one. If both bars do not exceed 1/2 and $M = \emptyset$, then put the current 2-BC in M and continue inspecting. If both bars do not exceed 1/2 and $M \neq \emptyset$, then form a 2-union of current 2-BC and 2-BC in M. If the resulting 2-BC is big, then exclude it from M and consider the next 2-BC. If both bars of the resulting 2-BC do not exceed 1/2, leave it in

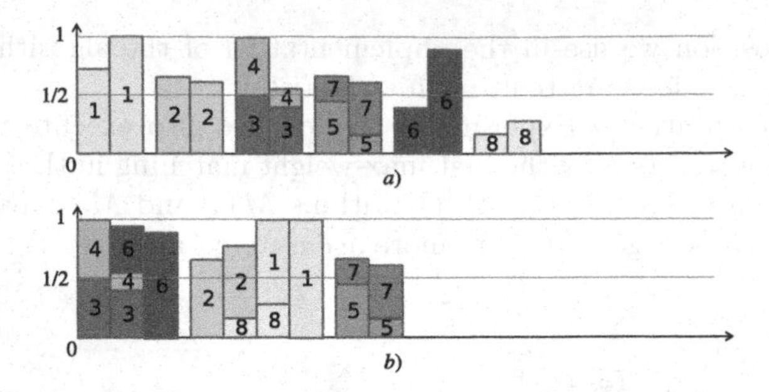

Fig. 4. *a*) Big BCs built by the first step of *A*1; *b*) Packing built by *A*1.

M and continue viewing. As a result of one scan of S, all 2-BCs, except possibly one, become big (Fig. 4*a*).

The first stage of the algorithm $A2$ consists of the sequential constructing of max-cardinality matchings in the graphs $G_k = (V_k, E_k)$, $k = 1, 2, \ldots$, where V_k is the set of images of the current 2-BCs, and $(i, j) \in E_k$ if the ith and jth 2-BCs can form a 2-union (Fig. 5*a*). After each matching, we get new 2-BCs and similarly construct next graph. This stage ends when there are no more 2-unions i.e., $G_k = (V_k, \emptyset)$.

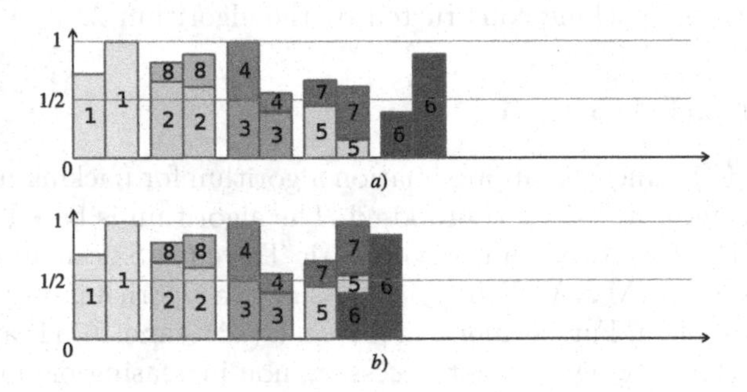

Fig. 5. *a*) Big BCs built by the first step of *A*2; *b*) Packing built by *A*2.

The next common part of both algorithms is the same. The directed graph G_1 is constructed similarly as in the M_w. The vertices of this graph are the images of the big 2-BCs. The arc (i, j) belongs to G_1 if the ith and jth 2-BCs can form a 1-union with ith BC on the left. Then a so-called *admissible* multigraph is constructed [4,21], and its edges are colored with two colors such that the edges in each color class form a collection of node disjoint paths [5]. Then we choose the paths of one color with a maximal number of edges. The number of edges decreases the packing's length compared to the initial length $2\,m$, where $m \leq n$ is the number of big 2-BCs. Figure 4*b* shows the packing constructed by the algorithm $A1$, and Fig. 5*b* shows the packing constructed by the algorithm $A2$.

4 Computational Experiments

In this section, we describe computational experiments that we performed with two groups of test data. The first one consists of the randomly generated data. For the generation of the instances of the second group, we used the existing BPP instances with known optimal solutions. The considered algorithms have been implemented in the Python programming language. The calculations are carried out on the computer Intel Core i7-3770 3.40 GHz 16 Gb RAM.

4.1 Randomly Generated Data

The first group of instances is generated randomly with the different number of 2-BCs $n \in [25, 1000]$. For each n, 50 different instances are generated. To build an optimal solution to BLP (1)–(3) or to find a lower bound for optimum, we use the IBM ILOG CPLEX 12.10 software package (CPLEX) with a limited running time of 300 s.

We consider approximation algorithms $A1$, $A2$, $M1_w$, M_w and GA_LO using three test data sets. The first one consists of arbitrary 2-BCs. For each 2-BC i, $i = 1, \ldots, n$, the heights a_i and b_i take random uniformly distributed values in $(0, 1]$. The second data set consists of big 2-BCs. While generating a 2-BC, one of its bars is chosen randomly as big, and its height takes a random uniformly distributed value in $(0.5, 1]$. Another bar's height takes a random uniformly distributed value in $(0, 1]$. The third data set consists of big non-increasing 2-BCs. Its generation is the same as in the second data set, but additionally, the bars of each generated 2-BC swap places if necessary.

Figure 6 presents numerical experiment results for the randomly generated instances with arbitrary 2-BCs. The value of R is defined in the following way. If we know the optimum, then R is the ratio $R = obj(X)/OPT$, where OPT is the minimum packing length and $obj(X)$ is the length of the packing built by the algorithm $X \in \{A1, A2, M_w, GA\_LO\}$. On the other hand, if CPLEX fails to find the optimum during the allotted time, we set $R = obj(X)/LB$, where LB is the lower bound of the optimum yielded by CPLEX during the allotted time. The figure shows the mean values and standard deviations (vertical segments) of R.

Even for small $n \leq 100$, CPLEX often builds non optimal solution (in 300 s). However, the found by CPLEX approximate packing is near-optimal. In particular, when $n = 25$, CPLEX builds an optimal solution in 34% of cases, and in 6% of cases when $n \in \{50, 75\}$. When $n \geq 500$, CPLEX builds a significantly worse solution than other considered algorithms. Table 1 shows minimum (min), maximum (max), and average (av) values of upper bounds on the *absolute errors* – the differences between the algorithm's objective values and the lower bounds provided by CPLEX. The best values are bold. For example, when $n = 1000$, CPLEX builds the solutions with average absolute error 221.0, while the average absolute errors of the approximation algorithms M_w, $M1_w$, $A1$, $A2$, and GA_LO are 47.1, 47.4, 152.1, 66.3 and 25.2, respectively. As one sees in Table 1 and Fig. 6, the algorithm GA_LO always builds a more accurate solution than

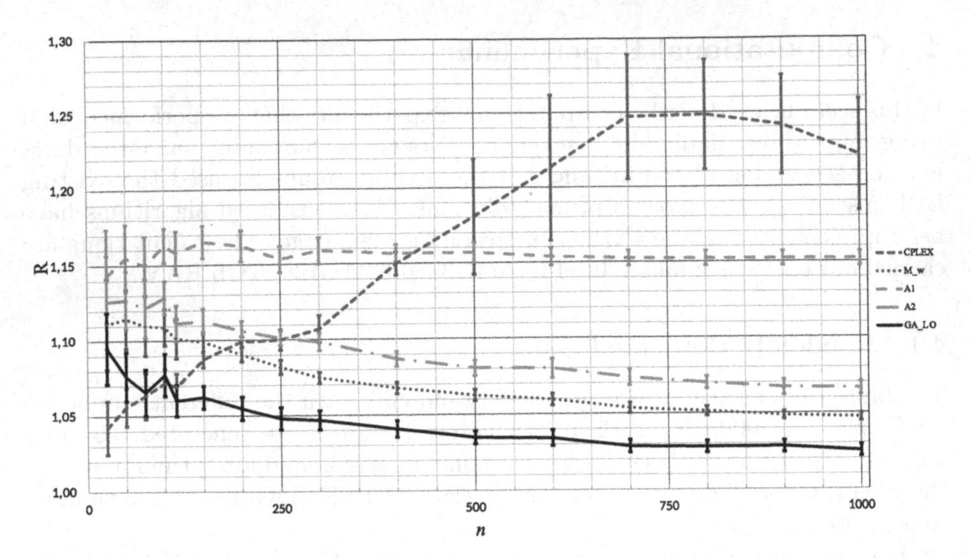

Fig. 6. Comparison of the algorithms on the randomly generated instances with arbitrary 2-BCs.

the other considered algorithms. In most cases, algorithms *A*2 and M_w turn out to be more accurate than the algorithm *A*1. In nearly 95% of cases, the first matching (the result of $M1_w$) turns out to be the only one in the algorithm M_w. Note that all tested approximation algorithms work fast enough for any $n \in [25, 1000]$. The algorithm *GA_LO* appeared to be the fastest one. It terminates in less than 1 s. The algorithms M_w, *A*1 and *A*2 use more running time. For example, when $n = 1000$, the average running time of these algorithms is about one minute.

Tables 2 and 3 present the numerical results for big 2-BCs and big non-increasing 2-BCs, respectively. Since all approximation algorithms yield almost the same values of R in this case, we decide to include only algorithms *GA_LO* and CPLEX into the tables. Starting from $n = 250$, CPLEX builds a worse solution than all other considered algorithms. Therefore, we can conclude that on such a 2-BCs set, all algorithms work well, and we no longer observe such difference in R values as in the case with arbitrary 2-BCs (Fig. 6). For example, when $n \geq 250$, the average value of $R \approx 1.2$ for all considered algorithms.

Based on the numerical experiment results described above, we can conclude that *GA_LO* is preferable among all tested algorithms for the input data with uniformly distributed parameters. Starting from $n = 100$, it builds more accurate solutions than other algorithms. However, considering the particular cases of the 2-BCs (second and third data sets), it is hard to single out the most efficient algorithm. In these cases, the only definite advantage of the algorithm *GA_LO* is its running time, which for all n does not exceed 1 s. For instance, when $n = 1000$, the algorithm *GA_LO* builds solutions in 0.37 s.

Table 1. Absolute errors of the algorithms on the randomly generated instances with arbitrary 2-BCs.

n	CPLEX			M_w			$M1_w$			$A1$			$A2$			GA_LO		
	min	*max*	*av*	*min*	*max*	*av*	*min*	*max*	*av*	*min*	*max*	*av*	*min*	*max*	*av*	*min*	*max*	*av*
25	0	3	**1.0**	0	5	2.8	1	6	3.9	0	6	3.6	1	7	3.2	0	5	2.4
50	0	5	**2.8**	1	9	5.8	3	10	6.5	2	11	7.9	2	10	6.4	0	7	3.9
75	0	9	**4.8**	1	14	8.3	1	14	8.9	5	16	11.4	1	15	9.2	0	12	5.0
100	1	12	**7.2**	2	15	10.9	2	16	11.3	10	23	16.4	9	19	12.9	2	15	7.8
115	1	11	7.9	2	16	11.6	2	18	12.0	11	24	18.2	5	19	12.8	2	15	**6.9**
150	10	18	13.1	12	20	14.8	12	20	15.1	18	31	24.7	12	25	16.9	5	16	**9.3**
200	16	25	19.8	14	23	18.1	14	23	18.2	24	42	32.5	16	27	21.3	5	19	**11.0**
250	18	39	25.0	16	26	20.5	16	27	20.8	27	47	38.4	19	34	25.3	5	20	**12.1**
300	23	51	32.2	18	27	22.5	18	27	22.5	37	61	47.8	23	38	29.7	8	24	**14.1**
400	48	87	60.0	21	33	27.1	21	33	27.3	52	71	62.7	28	50	34.9	8	29	**16.4**
500	61	188	89.9	24	39	31.1	24	39	31.1	60	95	78.1	30	54	40.1	9	31	**17.3**
600	70	255	128.1	28	50	35.7	29	50	36.0	76	108	92.3	32	72	48.1	11	38	**20.5**
700	78	258	173.1	27	49	37.5	28	49	37.7	93	120	106.8	37	66	51.6	12	28	**19.9**
800	92	287	198.6	34	56	41.0	34	56	41.1	102	136	121.7	37	73	56.0	11	39	**22.4**
900	105	335	217.4	34	55	43.7	34	56	43.9	119	159	137.0	41	88	60.4	14	54	**25.4**
1000	112	343	221.0	39	64	47.1	39	64	47.4	129	175	152.1	53	93	66.3	13	45	**25.2**

Table 2. Comparison of the algorithms on the randomly generated instances when all 2-BCs are big (R_{av} is the mean value; R_{sd} is the standard deviation).

n	CPLEX		GA_LO	
	Rav	*Rsd*	*Rav*	*Rsd*
25	**1.003**	0.011	1.013	0.02
50	**1.003**	0.008	1.009	0.01
75	**1.004**	0.007	1.008	0.008
100	**1.003**	0.004	1.006	0.005
250	1.207	0.014	**1.203**	0.015
500	1.242	0.025	**1.203**	0.01
750	1.237	0.01	**1.201**	0.01
1000	1.234	0.01	**1.202**	0.007

As a result of this part of the experiment, we conclude that the algorithm $A2$ turns out to be more beneficial than $A1$. However, the algorithm M_w finds a slightly more accurate solution than $A2$. The difference between R values of M_w and $A2$ is about 0.02 for $n \in \{700, 800, 900, 1000\}$. The difference between R values of $A2$ and GA_LO is about 0.04 for the same n. The running time of the algorithms $A2$ and M_w is almost the same, and on average, for $n = 1000$, they build solutions in 68 and 61 s, respectively.

To make our results reproducible, we have uploaded all the instances to the cloud storage publicly available at the link:
https://disk.yandex.ru/d/sb1IqReFpUu7dg.

Table 3. Comparison of the algorithms on the randomly generated instances when all 2-BCs are big and non-increasing (R_{av} is the mean value; R_{sd} is the standard deviation).

n	CPLEX		*GA_LO*	
	Rav	*Rsd*	*Rav*	*Rsd*
25	**1.005**	0.013	1.022	0.017
50	**1.017**	0.028	1.025	0.025
75	**1.018**	0.02	1.025	0.02
100	**1.022**	0.043	1.026	0.042
250	1.235	0.018	**1.216**	0.014
500	1.26	0.021	**1.21**	0.009
750	1.247	0.012	**1.208**	0.009
1000	1.247	0.008	**1.204**	0.009

4.2 Data Generated from the BPP Benchmarks

Our computational experiment's other test cases are based on the Bin Packing Problem (BPP) instances with known optimal solutions. In [10] the authors provide a review of the most important mathematical models and algorithms developed for BPP. In particular they generate 3840 instances with different number of items (50, 100, 200, 300, 400, 500, 750, 1000) and bins capacities (50, 75, 100, 120, 150, 200, 300, 400, 500, 750, 1000). We use these instances because they are available online as well as their optimal solutions. For each instance of BPP, we generate the instance for 2-BCPP in the following way. Using an optimal solution of the BPP, we sort the bins (cells) in non-decreasing order of the number of items placed in it. Let N be the optimum of BPP, and let n_i, $i = 1, \ldots, N$, be the number of items in the bin i. Then the 2-BCs are generated as follows. The n_1 items in the first and second bins create n_1 2-BCs. To form one 2-BC, we take any one item from the first bin and the second bin. Then the $n_2 - n_1$ items from the second and third bins form the next 2-BCs in the same manner. The remaining $n_3 - (n_2 - n_1)$ items of the third bin create 2-BCs with items from the fourth bin, and so on. The last bin n_N may contain items that were not used in the constructed 2-BCs. Such items are removed, and the optimum for 2-BCPP becomes $N - 1$. Since the size of any bin and item in any BPP's instance may be arbitrary, the height of each bar in 2-BCPP is divided by the bin's capacity. As a result, for each $n \in \{25, 50, 100, 150, 200, 250, 375, 500\}$, we get 480 different instances with known optimal solution.

Let *ratio* be $obj(X)/OPT$, where OPT is the optimum and $obj(X)$ is the objective's value found by the algorithm $X \in \{M_w, A1, A2, GA\_LO\}$. Figure 7 presents the average and standard deviation values of the ratio for the BPP benchmarks-based instances. Again, as in the randomly generated instances, the algorithm *GA_LO* outperforms other algorithms in accuracy and running time. Average ratio is 1.034 for almost each value of n, except 25 and 500,

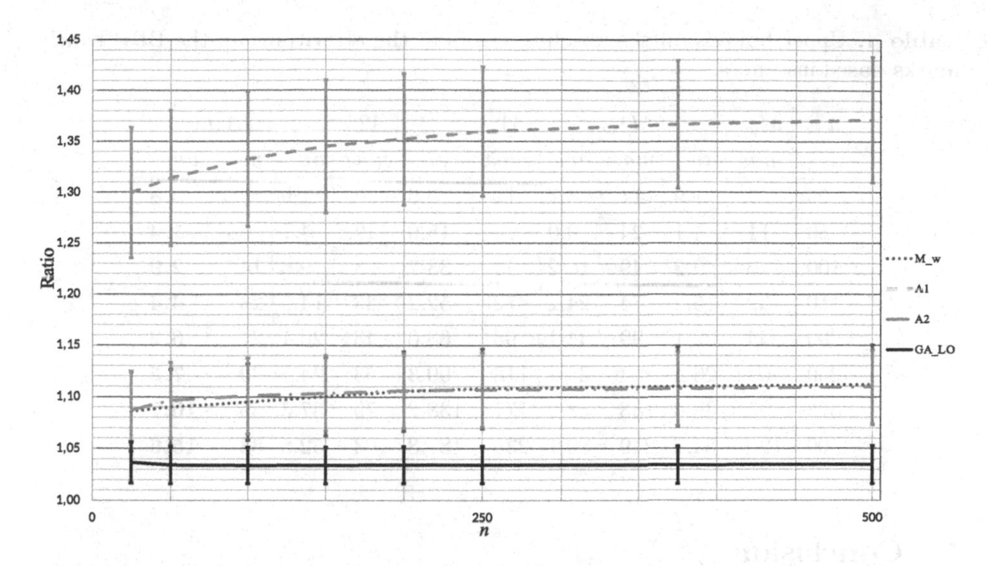

Fig. 7. Comparison of the algorithms on the BPP benchmarks based instances.

where R_{av} equals 1.037 and 1.035, correspondingly. Algorithms M_w and $A2$ build similar solutions. For example, when $n = 500$, the ratio is 1.11 for both of them. Algorithm M_w in 34% of cases performs only one first iteration (builds one matching). $A1$ turns out to be the worst. For example, when $n = 500$ its average ratio is 1.371, while average ratio for $M1_w$ equals 1.181.

Additionally, we analyze the absolute errors. Table 4 presents the difference of objective values yielded by the approximation algorithms and optimum. In the majority of cases, the average absolute error of *GA_LO* appeared to be the smallest. For example, for $n = 500$, the average absolute errors of the approximation algorithms M_w, $M1_w$, $A1$, $A2$, and *GA_LO* equal 54.7, 80.1, 181.8, 52.8 and 15.6, correspondingly. Incidentally, for $n \in \{25, 50, 100\}$, algorithm *GA_LO* builds the optimal solutions in 40%, 26.3% and 14.8% of cases, correspondingly. However, when $n \geq 150$, algorithm M_w builds an optimal solution more frequently than other algorithms. For example, when $n = 500$, M_w builds optimal solution in 12.5% of cases versus 10.6% by *GA_LO* and 11.3% by $A2$.

Consequently, we conclude that all considered algorithms sufficiently solve the test instances obtained using the known BPP instances. Frequently the optimal solutions were built. In the rest of the cases, algorithms yield near-optimal solutions with a ratio close to 1. Like in the previous subsection, we state that algorithm *GA_LO* turns out to be the most beneficial.

Table 4. Upper bounds on the absolute errors of the algorithms on the BPP benchmarks based instances.

n	M_w		$M1_w$		$A1$		$A2$		GA_LO	
	max	av	max	av	max	av	max	av	max	av
25	7	1.9	11	3.7	12	7.0	6	1.9	5	**0.8**
50	11	4.1	24	9.0	24	15.0	12	4.7	6	**1.4**
100	26	9.2	49	16.2	48	33.7	23	9.6	14	**2.9**
150	39	15.2	74	24.2	72	52.2	33	14.5	23	**4.4**
200	54	22.1	99	34.1	98	68.6	43	20.1	27	**6.2**
250	62	26.4	124	38.4	118	90.3	63	24.1	39	**7.5**
375	91	39.6	183	58.2	177	138.2	79	37.3	55	**10.7**
500	120	54.7	249	80.1	235	181.8	103	52.8	83	**15.6**

5 Conclusion

This paper tests several approximation algorithms with known guaranteed estimates to solve the strongly NP-hard problem of packing two-bar charts into the strip of minimal length. This problem is new for the optimization community. It is a generalization of the Bin Packing Problem and 2-Dimensional Vector Packing Problem and a particular case of the resource-constrained project scheduling problem where the jobs consume one renewable resource. For the earlier considered algorithms, we have found the guaranteed accuracy estimates. This paper performs a simulation using the randomly generated instances of different dimensions to compare the approximation algorithms and CPLEX for BLP and instances with known optimum. The numerical experiment shows the high efficiency of the greedy algorithm with the preliminary lexicographic ordering of the 2-BCs (GA_LO) proposed firstly in [12], which significantly outperforms other algorithms in accuracy and runtime.

In this paper, we limited ourselves to approximation algorithms with guaranteed accuracy estimates. Furthermore, we deliberately estimate how much the accuracy of the considered algorithms is higher on average than the guaranteed accuracy. In future research, we plan to test various other heuristics and metaheuristics, for which a priori accuracy estimates are not necessarily known.

References

1. Baker, B.S., Coffman, E.G., Jr., Rivest, R.L.: Orthogonal packing in two dimensions. SIAM J. Comput. **9**(4), 846–855 (1980)
2. Baker, B.S.: A new proof for the first-fit decreasing bin-packing algorithm. J. Algorithms **6**, 49–70 (1985)
3. Bansal N., Eliáš M., Khan A.: Improved approximation for vector bin packing. In: SODA, pp. 1561–1579 (2016)

4. Bläser, M.: A 3/4-approximation algorithm for maximum ATSP with weights zero and one. In: Jansen, K., Khanna, S., Rolim, J.D.P., Ron, D. (eds.) APPROX/RANDOM -2004. LNCS, vol. 3122, pp. 61–71. Springer, Heidelberg (2004). https://doi.org/10.1007/978-3-540-27821-4_6
5. Blaser, M.: An 8/13-approximation algorithm for the maximum asymmetric TSP. J. Algorithms **50**(1), 23–48 (2004)
6. Brucker, P., Knust, S.: Complex Scheduling. Springer, Heidelberg (2006). https://doi.org/10.1007/978-3-642-23929-8
7. Christensen, H.I., Khanb, A., Pokutta, S., Tetali, P.: Approximation and online algorithms for multidimensional bin packing: a survey. Comput. Sci. Rev. **24**, 63–79 (2017)
8. Coffman, E.G., Jr., Garey, M.R., Johnson, D.S., Tarjan, R.E.: Performance bounds for level-oriented two-dimensional packing algorithms. SIAM J. Comput. **9**(4), 808–826 (1980)
9. Dósa, G.: The tight bound of first fit decreasing bin-packing algorithm is $FFD(I) \leq 11/9OPT(I) + 6/9$. In: Chen, B., Paterson, M., Zhang, G. (eds.) ESCAPE 2007. LNCS, vol. 4614, pp. 1–11. Springer, Heidelberg (2007). https://doi.org/10.1007/978-3-540-74450-4_1
10. Delorme, M., Iori, M., Martello, S.: Bin packing and cutting stock problems: mathematical models and exact algorithms. Eur. J. Oper. Res. **225**(1), 1–20 (2016)
11. Erzin, A., Plotnikov, R., Korobkin, A., Melidi, G., Nazarenko, S.: Optimal investment in the development of oil and gas field. In: Kochetov, Y., Bykadorov, I., Gruzdeva, T. (eds.) MOTOR 2020. CCIS, vol. 1275, pp. 336–349. Springer, Cham (2020). https://doi.org/10.1007/978-3-030-58657-7_27
12. Erzin, A., Melidi, G., Nazarenko, S., Plotnikov, R.: Two-bar charts packing problem. Optim. Lett. **15**(6), 1955–1971 (2020). https://doi.org/10.1007/s11590-020-01657-1
13. Erzin, A., Melidi, G., Nazarenko, S., Plotnikov, R.: A 3/2-approximation for big two-bar charts packing. J. Comb. Optim. **42**(1), 71–84 (2021). https://doi.org/10.1007/s10878-021-00741-1
14. Erzin A., Shenmaier V.: An Improved Approximation for Packing Big Two-Bar Charts. http://arxiv.org/abs/2101.00470 (2021)
15. Gabow H.: An efficient reduction technique for degree-constrained subgraph and bidirected network flow problems. In: STOC, pp. 448–456 (1983)
16. Harren, R., Jansen, K., Pradel, L., van Stee, R.: A (5/3 + epsilon)-approximation for strip packing. Comput. Geom. **47**(2), 248–267 (2014)
17. Johnson, D.S.: Near-optimal bin packing algorithms. Massachusetts Institute of Technology. Ph.D. thesis (1973)
18. Johnson, D.S., Garey, M.R.: A 71/60 theorem for bin packing. J. Complex. **1**(1), 65–106 (1985)
19. Kellerer, H., Kotov, V.: An approximation algorithm with absolute worst-case performance ratio 2 for two-dimensional vector packing. Oper. Res. Lett. **31**, 35–41 (2003)
20. Li, R., Yue, M.: The proof of $FFD(L) \leq 11/9OPT(L) + 7/9$. Chin. Sci. Bull. **42**(15), 1262–1265 (1997)
21. Paluch, K.: Maximum ATSP with weights zero and one via half-edges. Theory Comput. Syst. **62**(2), 319–336 (2017). https://doi.org/10.1007/s00224-017-9818-1
22. Schiermeyer, I.: Reverse-fit: a 2-optimal algorithm for packing rectangles. In ESA: Proceedings of 2nd European Symposium on Algorithms, pp. 290–299 (1994)
23. Sleator, D.D.: A 2.5 times optimal algorithm for packing in two dimensions. Inf. Process. Lett. **10**(1), 37–40 (1980)

24. Steinberg, A.: A strip-packing algorithm with absolute performance bound 2. SIAM J. Comput. **26**(2), 401–409 (1997)
25. Vazirani, V.V.: Open problems. In: Approximation Algorithms, pp. 334–343. Springer, Heidelberg (2003). https://doi.org/10.1007/978-3-662-04565-7_30
26. Yue, M.: A simple proof of the inequality $FFD(L) \leq 11/9OPT(L)+1, \forall L$, for the FFD bin-packing algorithm. Acta Math. Appl. Sin. **7**(4), 321–331 (1991)
27. Yue, M., Zhang, L.: A simple proof of the inequality $MFFD(L) \leq 71/60OPT(L)+1, \forall L$, for the MFFD bin-packing algorithm. Acta Math. Appl. Sin. **11**(3), 318–330 (1995)

Pareto Frontier in Multicriteria Optimization of Chemical Processes Based on a Kinetic Model

Kamila Koledina[1,2(✉)], Sergey Koledin[2], and Irek Gubaydullin[1,2]

[1] Institute of Petrochemistry and Catalysis of RAS, Ufa, Russia
[2] Ufa State Petroleum Technological University, Ufa, Russia

Abstract. Modeling laboratory study of catalytic reactions followed by introduction into existing manufacturing and industrial processes require determination of optimal conditions. The formulation of the problem of optimizing the conditions for carrying out a chemical process presupposes the presence of optimality criteria, variable parameters and limitations. In the presence of several conflicting optimality criteria, the problem of multi-criteria optimization of the reaction conditions arises. The solution of optimization problems is based on an adequate kinetic model, as on fundamental knowledge about the chemical transformations that occur in the reaction. The development of a kinetic model is carried out on the basis of a kinetic experiment, which makes it possible to analyze in detail the regularities of the process.

The paper presents a mathematical formulation in the problem of multicriteria optimization of the conditions for carrying out catalytic reactions. Optimality criteria for metal complex catalysis reactions have been developed, which allow determining the optimal conditions for various processes using kinetic models. Physicochemical, ecological and economic criteria of optimality have been formulated: the yield of the target product, the yield of a by-product (for example, toxic), conversion, selectivity, process intensity, productivity, profit, profitability.

A complex catalytic reaction for the synthesis of benzyl butyl ether is considered. On the basis of a detailed kinetic model, the problem of multicriteria optimization of the conditions for conducting in the form of Pareto approximation, using the NSGA-II algorithm, was formulated and solved. Recommendations for conducting the process are made.

Keywords: Multicriteria optimization · Detailed kinetic model · Catalytic reaction conditions · Pareto approximation · Catalytic reaction of benzyl butyl ether synthesis

---

The work has been performed under the theme “Novel theoretical approaches and software for modeling complex chemical processes and compounds with tunable physicochemical properties” (registration number AAAAA19-119022290011-6).

N. N. Olenev et al. (Eds.): OPTIMA 2021, CCIS 1514, pp. 217–229, 2021.
https://doi.org/10.1007/978-3-030-92711-0_15

1 Introduction

Detailed kinetic model in chemical reaction reflects the fundamental properties of the process and includes circuitry chemical reactions, kinetic equation, the calculated values of the parameters, regularity of the change of concentrations. The kinetic model is the basis for the problem of optimizing the conditions for carrying out a chemical process. Together with the criteria of optimality, the problem of single-criterion or multi-criteria optimization of the conditions for carrying out complex chemical processes is determined.

Each chemical reaction is characterized by its own optimality criteria, depending on the nature of the reaction, the objectives of the study and the kinetic model. Optimality criteria are mathematically formulated: target product yield, by-product yield, conversion, selectivity, process intensity, productivity, profit, profitability. The study of one or more optimality criteria determines a single-criterion or multi-criteria optimization (MCO) problem, respectively [1,2].

2 Statement of the Problem of Multicriteria Optimization in Catalytic Reaction Conditions

Formulation and solution of optimization problems require the definition of optimality criteria. It is also necessary to define the variable parameters and their limitations. Process in the optimization problem is a chemical reaction. In chemical technology, such parameters can be: reaction temperature, pressure, initial concentrations of reagents, type of catalyst, reaction time. Restrictions on variable parameters are determined by the nature of the process and technical capabilities.

The mathematical model of chemical kinetics has the form of a system of nonlinear differential equations for a change in the concentration of reaction substance [3,4]. The change is in time with the known values of the kinetic parameters: preexponential factors and activation energies of the rates of the stages (1)–(3).

$$\frac{dy_i}{dt} = \Phi_k \sum_{j=1}^{J} \nu_{ij} w_j, i = 1, ...I; \tag{1}$$

$$w_j = k_j \prod_{i=1}^{I} y_i^{|\alpha_{ij}|} - k_{-j} \prod_{i=1}^{I} y_i^{\beta_{ij}}; \tag{2}$$

$$k_j = k_j^0 exp(-\frac{E_j}{RT}); \tag{3}$$

with initial conditions: $t = 0, y_i(0) = y_i^0$; $t \in [0, t^*]$.

Here y_i - concentration of reaction reagents, mol/L or fraction; t - time, min; Φ_k - determines the amount of catalyst in the reaction, mol/L or mol/kg; J - number of stages; I - number of substances; ν_{ij} - stoichiometric matrix coefficients; w_j - rate j-th stage, 1/min or mol/(kg*min)); k_j, k_{-j} - rate constants of direct and reverse reactions (reduced), 1/min; α_{ij} - negative matrix elements ν_{ij}, β_{ij} - positive elements ν_{ij}, k_j^0,k_{-j}^0 - preexponential factors, 1/min; E_j^+, E_j^- - activation energies of direct and reverse reactions, kcal/mol; R - gas constant, 2 cal/(mol*K); T - temperature, K, t^* - reaction time, min. To determine the optimal conditions for the reaction, it is necessary to solve the problem of multicriteria optimization according to the specified optimality criteria [5].

The mathematical formulation of the MCO problem for the chemical process conditions for the kinetic model is [5,6]: Vector variable parameters

$$X = (x_1, x_2, x_3, x_4, x_5, ...). \tag{4}$$

Here x_1 - reaction temperature; x_2 - initial concentrations of reagents; x_3 - reaction time; x_4 - catalyst type; x_5 - catalyst feed, etc.

Direct constraints on variable parameters $x_1 \in [x_1^-, x_1^+]; x_2 \in [x_2^-, x_2^+]; x_3 \in [x_3^-, x_3^+]; x_4 \in [x_4^-, x_4^+]; x_5 \in [x_5^-, x_5^+]$.

Vector function optimality criteria

$$F(X) = (f_1(X), f_2(X), f_3(X), ...). \tag{5}$$

$F(X)$ with values in target space $F = R^{|F|}$ defined in the area $D_X \subset X = R^{|X|}$. Then maximization (minimization is similar, with the sign -)optimality criteria in D_X can be written as

$$\max_{X \in D_X} F(X) = F(X^*) = F^*. \tag{6}$$

Then MCO problem for conditions of catalytic reaction is a variable determining the parameter values (4), in order to achieve optimality criteria extrema (5) according to (6).

3 Criteria of Optimality Conditions for Carrying Out Catalytic Reactions

Debugging and launching a chemical process into production requires, along with laboratory research, its mathematical modeling. Preliminary modeling and calculation of parameters are necessary to determine the optimal conditions for the chemical process. Theoretical optimization is mainly carried out in terms of physicochemical indicators - product yield, reagent conversion, reaction selectivity. This does not take into account the economic and technological evaluations of the process. The conditions calculated in this way during the real process have to be significantly adjusted [7,8].

Optimality criteria for the studied objects are different and depend on the goals of the chemical process. But in the general case, the optimality criterion is a function, the arguments or parameters of which are the conditions of the reaction, the initial concentrations of the components:

$$f(X) = f(t^*, T, y^0) \rightarrow \max_{X \in D_X} \tag{7}$$

To determine the optimal conditions for the reaction, the following criteria have been developed:

a) Target product yield (prod). It depends on the reaction time t^*, temperature T, initial concentrations of reagents y_0, [g] or [mol / L] or [fraction]:

$$f(X) = y_{prod}(t^*, T, y^0) \rightarrow \max \tag{8}$$

b) By-product yield (by-prod). It depends on the reaction time t^*, temperature T, initial concentrations of reagents y_0, [g] or [mol / L] or [fraction]:

$$f(X) = y_{by-prod}(t^*, T, y^0) \rightarrow \min \tag{9}$$

By-products determine, among other things, the environmental requirements for the process, i.e. the content of toxic, harmful reagents or their maximum permissible concentration.
c) Conversion - part supplied reagent (source), reacted, [fraction]

$$f(X) = \frac{y^0_{source} - y_{source}}{y^0_{source}} \rightarrow \max \tag{10}$$

d) Selectivity - part converted reactant entered into the target response [fraction]

$$f(X) = \frac{y_{prod} M_{Y_{prod}}}{y^0_{source} - y_{source} M_{Y_{prod}}} \rightarrow \max, \tag{11}$$

where $M_{Y_{prod}}$ - molar mass of the target product [g/mol].
e) Intensity - the amount of the target product produced per unit of time per unit volume of the reactor, [g/day].

$$f(X) = \frac{N \cdot y^0_i \cdot \xi_{Y_i}(t^*, T) \cdot M_{Y_i}}{V_P} \rightarrow \max, \tag{12}$$

where V_P - reactor volume, m^3; N - number of cycles per day $[day^{-1}]$; ξ_{Y_i} - conversion of the starting reagent.
f) Productivity - the amount of products produced per unit of time, [g/(L*day)]. For selective reaction, when the amount of product is directly proportional to the conversion of starting reagent, the performance is evaluated by conversion of the starting reagent [9]:

$$f(X) = N(t^*) \cdot y^0_i \cdot \xi_{Y_i}(t^*, T) \cdot M_{Y_i} \rightarrow \max, \tag{13}$$

$$N(t^*) = \frac{1440}{t^* + t_{id}} \rightarrow \max, \tag{14}$$

where t_{id} - idle time between cycles.

Moreover, the increase in productivity is not directly proportional to the increase in conversion. Since the conversion directly depends on t*, and increase t* involves reducing the number of cycles.

Authors [9] suggest that the most objective optimization problem formulation is expressed by the criteria in the form of economic evaluation.

For catalytic processes, the main criterion is the yield of the target product. It is also possible to set it in the form of reduced reaction time for a fixed output. When these criteria are combined in one problem, competition of criteria arises. In this case, it is advisable to formulate the problem taking into account economic criteria - profit, profitability, etc.
g) profit - the difference between the price of the target product and the cost. It is defined as [conventional units] or [normalized]. Cost is determined by the reaction reagent fixed and variable costs of the reaction.

$$f(X) = \sum_{prod=1}^{Pr} y_{prod}(t^*, T, y^0)\eta_{prod} - \sum_{source=1}^{Sr} y_{source}(t^*, T, y^0)\eta_{source} - \mu(t*, T) - A \rightarrow max \quad (15)$$

where y_{prod} - concentration of reaction products; y_{source} - concentration of starting reagents; η_{prod}, η_{source} - vector of unit price weights of components (normalized to the sum of component prices and costs); μ - variable costs (normalized); A - fixed costs (normalized); Pr - number of products; Sr - number of starting reagents.
h) profitability is the ratio of income to investment. (Relative profitability - price values are determined relative to total costs. Absolute profitability - absolute price values), [conventional units] or [normalized]:

$$f(X) = \frac{\sum_{prod=1}^{Pr} y_{prod}(t^*, T, y^0)\eta_{prod}}{\sum_{source=1}^{Sr} y_{source}(t^*, T, y^0)\eta_{source} + \mu(t*, T) + A} \rightarrow max. \quad (16)$$

The criterion of profitability is non-negative, but with an increase in variable costs, the value of profitability passes through a maximum.

In investigating the optimum conditions of the catalytic reactions in the problem multicriteria optimization can be applied physicochemical (8)–(11) (environmental - (9)), processing (12)–(14) and economic optimality criteria (15) and (16). The specified criteria of optimality depend on changes in the concentrations of reagents and can be determined based on the kinetic model of the catalytic reaction. Optimality criteria allow realizing the variability of the process assessment.

4 The Yield of Target and By-Products During Multicriteria Optimization Conditions for the Catalytic Reaction for the Synthesis of Benzylalkyl Ethers

The subject of this paper is benzyl butyl ether catalytic synthesis reaction carried out through the dehydration of benzyl alcohol with butanol-1. Benzyl butyl ether is a bulk industrial product that is widely used for aromatizing perfume, cosmetics and food [10]. In papers [9, 11–14] it was experimentally defined that the intermolecular dehydration of benzyl alcohol with aliphatic alcohols with the yield of ethers can act as a catalyst for copper compounds, the best of which is $CuBr_2$.

In [15], a kinetic model developed benzyl butyl synthesis reaction in the presence of a metal complex catalyst. The scheme of chemical transformations and kinetic equations are shown in Table 1.

Table 1. Stages of chemical transformations in the catalytic reaction for the synthesis of benzyl butyl ether.

N	Stages of chemical transformations	Kinetic equations
1.	$PhCH_2OH(Y_1) + CuBr_2(Y_2) \rightarrow$ $[PhCH_2]^+[CuBr_2(OH)]^-(Y_3)$	$w_1 = k_1y_1y_2$
2.	$[PhCH_2]^+[CuBr_2(OH)]^-(Y_3) + BuOH(Y_4) \rightarrow$ $[PhCH_2OBu]H^+[CuBr_2(OH)]^-(Y_5)$	$w_2 = k_2y_3y_4$
3.	$[PhCH_2OBu]H^+[CuBr_2(OH)]^-(Y_5) \rightarrow$ $PhCH_2OBu(Y_6) + H_2O(Y_7) + CuBr_2(Y_2)$	$w_3 = k_3y_5$
4.	$[PhCH_2]^+[CuBr_2(OH)]^-(Y_3) + PhCH_2OH(Y_1) \rightarrow$ $[PhCH_2OHCH_2Ph]^+[CuBr_2(OH)]^-(Y_8)$	$w_4 = k_4y_3y_1$
5.	$[PhCH_2OHCH_2Ph]^+[CuBr_2(OH)]^-(Y_8) \rightarrow$ $PhCH_2OCH_2Ph(Y_9) + H_2O(Y_7) + CuBr_2(Y_2)$	$w_5 = k_5y_8$
6.	$BuOH(Y_4) + CuBr_2(Y_2) \rightarrow [Bu]^+[CuBr_2(OH)]^-(Y_{10})$	$w_6 = k_6y_2y_4$
7.	$[Bu]^+[CuBr_2(OH)]^-(Y_{10}) + BuOH(Y_4) \rightarrow$ $[BuOHBu]^+[CuBr_2(OH)]^-(Y_{11})$	$w_7 = k_7y_{10}y_4$
8.	$[BuOHBu]^+[CuBr_2(OH)]^-(Y_{11}) \rightarrow$ $BuOBu(Y_{12}) + H_2O(Y_7) + CuBr_2(Y_2)$	$w_8 = k_8y_{11}$
9.	$[Bu]^+[CuBr_2(OH)]^-(Y_{10}) + PhCH_2OH(Y_1) \rightarrow$ $[PhCH_2OBu]H^+[CuBr_2(OH)]^-(Y_5)$	$w_9 = k_9y_1y_{10}$

The values of kinetic parameters for the presented reaction scheme (table 2) were calculated by methods of solving the inverse kinetic problem [16–19].

Table 2. Values of kinetic parameters for the catalytic synthesis of benzyl butyl ether($[k_j]$ = L/(mol*min), j = 1, 2, 4, 6, 7, 9; $[k_j]$ = 1/min, j = 3, 5, 8; $[E_j]$ = kcal/mol).

N	k_j ($T = 140\,°C$)	k_j ($T = 160\,°C$)	k_j ($T = 175\,°C$)	E_j	lnk_j^0
1.	1.701	2.469	2.8	5.40	7.00
2.	2.056	3.911	6.54	12.2	15.4
3.	0.052	0.081	0.1411	10.3	9.50
4.	1.922	5.101	7.02	14.0	17.6
5.	0.113	0.491	0.85	21.7	24.1
6.	0.0006	0.00081	0.0028	15.0	10.7
7.	0.117	0.6031	0.62	18.5	20.4
8.	1.95e−004	0.0021	0.005	35.1	34.1
9.	0.15	0.1631	0.5	11.9	12.4

The reaction proceeds through several catalytic cycles, with the formation of the target product in two possible ways - $PhCH_2OBu$ (Y_6) benzyl butyl ether and by-products - $PhCH_2OCH_2Ph$ (Y_9) dibenzyl ether, BuOBu (Y_{12}) dibutyl ether. The introduction of the process into production requires the determination of the optimal conditions for the reaction with the maximum yield of the target product and the minimum yield of by-products. Chemical experiments [12,13] were carried out at several temperatures from 140 to 175 °C and an initial ratio of reagents. Since molar ratios investigated $[CuBr_2]$:$[BnOH]$:$[n-BuOH]$ = $[Y_2]$:$[Y_1]$:$[Y_4]$ = 1:100:150, 1:100:200, 1:100:300, 1:100:400. In the considered range of temperatures and molar ratios, it is necessary to determine the values for reaching extrema according to the criteria of reaction optimality. Then the variable parameters are: reaction temperature, ratios of reagents and starting time of the reaction with the appropriate physicochemical constraints. In the synthesis reaction of benzyl butyl ether in the presence of a metal complex catalyst, the products are formed: $PhCH_2OBu$ (Y_6), $PhCH_2OCH_2Ph$ (Y_9), $BuOBu$ (Y_{12}). Then the MCO problem for the reaction conditions for the synthesis of benzyl butyl ether has the form:

- Vector variable parameters $X = (x_1, x_2, x_3)$ by (4), where x_1 – reaction temperature, T, $°C$, $x_1 \in [160, 175]$; x_2 – molar ratio of reagents butyl alcohol to benzyl alcohol $N = [Y_4] : [Y_1]$, $x_2 \in [2, 4]$; x_3 – reaction time, $t*$, min, $x_3 \in [400, 800]$.
- Vector function optimality criteria $F(X) = (f_1(X), f_2(X), f_3(X))$: $f_1(X) = y_{PhCH_2OBu(Y_6)}(t^*, T, N) \rightarrow$ max; $f_2(X) = y_{PhCH_2OCH_2Ph(Y_9)}(t^*, T, N) \rightarrow$ min; $f_3(X) = y_{BuOBu(Y_{12})}(t^*, T, N) \rightarrow$ min.

MCO problem is to maximize optimality criteria in the field of DX by (7). The solution of the MCO problem was carried out by the Pareto-approximation algorithm NSGA-II [20–22] in Matlab [23].

Authors of [20] consider a method for solving the Pareto approximation problem by the NSGA-II method based on a genetic algorithm. The algorithm makes it possible to provide acceptable accuracy for complex tasks. Authors of [20] consider modifications of the predator-prey algorithm. In [22], on several known tests and practically significant problems of multicriteria optimization, the comparative efficiency of Pareto approximation using a class of evolutionary algorithms with the formation of a fitness function (fitness function for assessing the quality of each population) is shown.

The essence of the NSGA-II method is that at the first stage, according to the fulfillment of the non-dominance condition, the current individuals are ranked. To increase the diversity of the population, the density of the obtained solutions is estimated. One non-dominated individual is selected from solutions that are close to each other. Further, based on the assessment of crowding and the rank of individuals, crossing and mutation of individuals occurs to obtain offspring. The best solutions are selected from the parents and the resulting offspring (Pareto front). In Fig. 1, Fig. 2, Table 3 shows the results of solving a three-criterion problem of optimizing the reaction conditions for the synthesis of benzyl butyl ether in the presence of a metal complex catalyst.

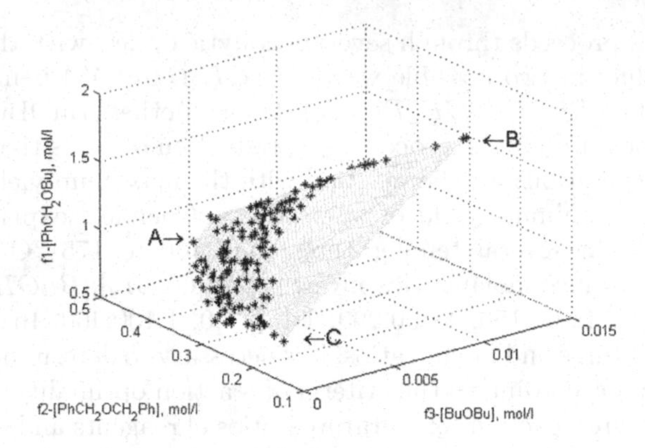

Fig. 1. Approximation Pareto front MCO benzyl butyl problem synthesis reaction at three optimality criteria: maximizing Y_6, Y_9 and Y_{12} minimization

Figure 1 show the set of points of the Pareto front. In Fig. 1, a surface is superimposed on a set of points. The calculated set of points is a fragment of some surface. In Fig. 1 points with extreme values according to one of the criteria are designated A (row 1 of Table 3), B (row 2 of Table 3), C (row 3 of Table 3). Point C corresponds to the minimum values of by-products Y_9 and Y_{12}, but the yield of the target product is also minimal. An increase in the content of dibenzyl ether makes it possible to achieve an increase in the yield of the target benzyl butyl ether - point A. The maximum yield of the target Y6 corresponds to the maximum value of the side dibutyl ether and dibenzyl ether - point B.

Table 3. Approximation value set and Pareto front MCO-benzyl butyl problem synthesis reaction in the presence of a metal complex catalyst.

Points	$x_1 - T, °C$	$x_2 - [Y_4] : [Y_1]$	$x_3 - t^*, min$	$f_1 - Y_6, mol/L$	$f_2 - Y_9, mol/L$	$f_3 - Y_{12}, mol/L$
A	160,8	2,06	720	1,27	0,37	0,001
B	167,6	3,95	784	1,66	0,24	0,011
C	160,0	2,00	400	0,68	0,28	0,0003

In Fig. 2 shows the projections of the three-dimensional visualization of the Pareto front on the axis $f_1 - f_3$.

The calculation of this problem using the NSGA-II algorithm takes about several hours. The NSGA-II parameter values are as follows: PopulationSize - 300, Generations - 100, CrossoverFraction - 0.5, ParetoFraction - 0.5. To compare the results of solving the multi-criteria optimization problem obtained by the NSGA-II algorithm, the problem of linear scalarization of optimality criteria with different coefficients was solved. The criterion is considered

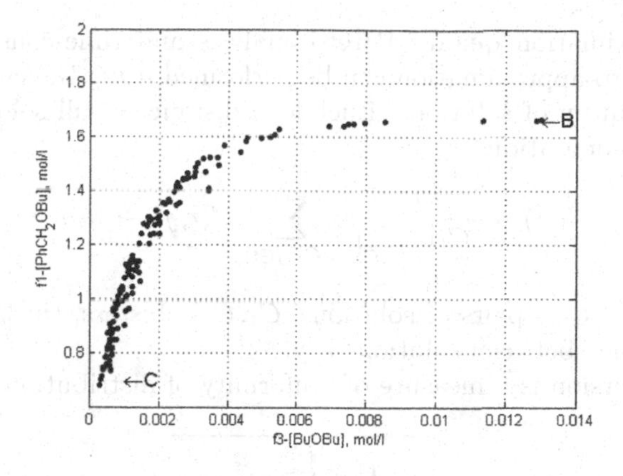

Fig. 2. Projection on the axis of the Pareto front approximation problem MCO-benzyl butyl synthesis reaction at three optimality criteria: maximizing f_1, minimizing f_3.

$$f(X) = \gamma_1 * f_1 - \gamma_2 * f_2 - \gamma_3 * f_3 \rightarrow max. \quad (17)$$

Where $\gamma_1, \gamma_2, \gamma_3$ - weight coefficients of partial optimality criteria.

Table 4 shows the values of the weights of criteria and corresponding values of the reaction conditions. With the same coefficients, the mode corresponding to the point B is determined. In Fig. 1 and in Table 3, providing the maximum yield of the target product.

Since the values of the concentrations of by-products are orders of magnitude less than the target, it is necessary to consider coefficients with a large value for them. At $\gamma_2 = 10$ ($\gamma_1 = \gamma_3 = 1$) and $\gamma_3 = 1000$ ($\gamma_1 = \gamma_2 = 1$) the reaction conditions with the minimum temperature and reaction time are calculated, which is located in the locality of the point C in Fig. 1 and Table 3.

Table 4. Values of weight coefficients and optimality criteria for linear scalarization of the MCO-problem in synthesis of benzyl butyl ether reaction.

γ_1	γ_2	γ_3	$x_1 - T, C$	$x_2 - [Y_4] : [Y_1]$	$x_3 - t^*, min$
1	1	1	167	4	782
1	10	1	140	4	200
1	1	1000	140	1	304

The adequacy of the obtained solutions for the multicriteria optimization in the reaction for the synthesis of benzyl butyl ether in the presence of the $CuBr_2$ metal complex catalyst is based on the agreement with the experimental data and the assessment of the quality of the Pareto approximation. Evaluation of the effectiveness of Pareto approximation is performed based on the resulting approximation quality Pareto analysis and time-consuming. Quality assessment Pareto-approximation can be performed using the characteristics: 1) Uniform distribution of solutions. Euclidean distance of all solution pairs from the obtained condensations.

$$\bar{d}(A) = \frac{1}{|C_1| \cdot |C_2|} \sum_{i \in C_1, j \in C_2} d(i,j) \rightarrow min, \tag{18}$$

where $i \in C_1, j \in C_2$ - pairs of solutions, C_1, C_2- clusters, thickenings, $d(i,j)$ – Euclidean distance between solutions i, j.
2) Average dispersion is a measure of uniformity of distribution of solutions.

$$I_S(A) = \sqrt{\frac{1}{|A|-1} \sum_{j=1}^{|A|} (\bar{d} - d_j)^2} \rightarrow min, \tag{19}$$

where $|A| = |C_1| + |C_2|$ - capacity set of solutions.
3) Capacity of the found set of solutions (the number of elements in the set). Capacity of the set of obtained solutions should tend to the maximum.

$$|A| \rightarrow max, \tag{20}$$

4) The length of the solution. The criterion defines the length of the Pareto-approximation in the space of objective functions across multiple criteria optimization.

$$I_{DE}(A) = \sqrt{\sum_{j=1}^{|F|} \max_{k,l \in [1:(A)]} |f_j(X_k^A) - f_j(X_l^A)|} \rightarrow max, \tag{21}$$

where f – target functions; F – set target functions.

These characteristics are used to assess the Pareto set (genotype) - the optimal values in the set of variable parameters and the Pareto front (phenotype) - the optimal values in the set of objective functions.

The values of quality evaluation criteria Pareto-approximation are given in Table 5. These values correspond to errors in the experimental data and the calculated power solutions allows for the entire range of variable values to determine optimum.

Thus, the values of the Pareto set (variable parameters) and the Pareto front (optimality criteria) are given in Table 4 for the posed MCO problem of the multicyclic catalytic reaction for the synthesis of benzyl butyl ether determine the exhaustive values of the reaction conditions. This allows a decision maker to choose the most optimal temperature for the reaction, the molar ratio of the starting reagents, and the reaction time. This information is necessary when implementing a laboratory-investigated process in the industry.

Table 5. Assessment of the quality of Pareto approximation.

	Phenotype	Genotype
Uniform distribution of solutions:		
Average distance between solutions $\bar{d}(A)$	0,008	0,006
Average dispersion $I_S(A)$	0,153	0,13
Capacity of the solution set $\|A\|$	50	50

5 Conclusion

In the catalytic reaction for the synthesis of benzyl butyl ether, the products are formed $PhCH_2OBu$, $PhCH_2OCH_2P$, $BuOBu$. In the problem of multicriteria optimization, the variable parameters are the reaction temperature, the coefficient of proportionality of butyl alcohol to benzyl alcohol; reaction time. Optimality criteria are the yield of the target product of benzyl butyl ether (maximization), the yield of dibenzyl ether (minimization), and the yield of dibutyl ether (minimization). The calculated values of the reaction conditions for the multicriteria optimization allow a decision maker to choose the most optimal temperature for the reaction, the coefficient of proportionality of the initial reagents, and the reaction time. The proportionality coefficient values for three optimality criteria vary from 2 to 4, due to the fact that at the minimum concentrations of the initial butyl alcohol there is a minimum formation of side dibutyl ether. Accordingly, at maximum values of the proportionality coefficient, the yield of dibutyl ether increases along with an increase in the yield of the target benzyl butyl ether. Carried out three-criterion optimization makes it possible to give technological recommendations for industrial implementation of the process of synthesis benzyl butyl ether in the presence of a metal complex catalyst.

References

1. Vallerio, M., Telen, D., Cabianca, L., Impe, J., Logist, F., Manenti, F.: Robust multi-objective dynamic optimization of chemical processes using the Sigma Point method. Chem. Eng. Sci. **140**, 81–89 (2016). https://doi.org/10.1016/j.ces.2015.09.012
2. Sun, X., Kim, S., Yang, S., Kim, H., Yoon, J.: Multi-objective optimization of a Stairmand cyclone separator using response surface methodology and computational fluid dynamics. Powder. Technol. **320**, 51–65 (2017). https://doi.org/10.1016/j.powtec.2017.06.065
3. Nurislamova, L.F., Gubaydullin, I.M., Koledina, K.F., Safin, R.R.: Kinetic model of the catalytic hydroalumination of olefins with organoaluminum compounds. React. Kinet. Mech. Catal. **117**(1), 1–14 (2015). https://doi.org/10.1007/s11144-015-0927-z
4. Koledina, K., Koledin, S., Schadneva, N., Mayakova, Y., Gubaydullin, I.: Kinetic model of the catalytic reaction of dimethylcarbonate with alcohols in the presence $Co_2(CO)_8$ and $W(CO)_6$. React. Kinet. Mech. Catal. **121**(2), 425–438 (2017). https://doi.org/10.1007/s11144-017-1181-3

5. Koledina, K.F., Koledin, S.N., Karpenko, A.P., Gubaydullin, I.M., Vovdenko, M.K.: Multi-objective optimization of chemical reaction conditions based on a kinetic model. J. Math. Chem. **57**(2), 484–493 (2018). https://doi.org/10.1007/s10910-018-0960-z
6. Rangaiah, G., Andrew, Z., Hoadley, F.: Multi-objective optimization applications in chemical process engineering: tutorial and review. Processes **8**(5), 508 (2020). https://doi.org/10.3390/pr8050508
7. Kreutz, J., Shukhaev, A., Du, W., Druskin, S., Daugulis, O., Ismagilov, R.: Evolution of catalysts directed by genetic algorithms in a plug-based microfluidic device tested with oxidation of methane by oxygen. J. Am. Chem. Soc. **132**(9), 3128–3132 (2010). https://doi.org/10.1021/ja909853x
8. Subbaramaiha, V., Srivastava, V., Mall, I.: Optimization of reaction parameters and kinetic modeling of catalytic wet peroxidation of picoline by Cu/SBA-15. Ind. Eng. Chem. Res. **52**(26), 9021–9029 (2013). https://doi.org/10.1002/aic.14017
9. Tsai, C., Sung, R., Zhuang, B., Sung, K.: TiCl4-activated selective nucleophilic substitutions of tert-butyl alcohol and benzyl alcohols with π-donating substituents. Tetrahedron **66**(34), 6869 (2010). https://doi.org/10.1016/j.tet.2010.06.046
10. Voitkevich, S.: 865 fragrances for perfumes and household chemicals. Food industry, Moscow (1994)
11. Bayguzina, A., Gimaletdinova, L., Khusnutdinov, R.: Synthesis of benzyl alkyl ethers by intermolecular dehydration of benzyl alcohol with aliphatic alcohols under the effect of copper containing catalysts. Russ. J. Org. Chem. **54**(8), 9021–9029 (2018). https://doi.org/10.1134/S1070428018080055
12. Khusnutdinov, R., Bayguzina, A., Gallyamova, L., Dzhemilev, U.: A novel method for synthesis of benzyl alkyl ethers using vanadium-based metal complex catalysts. Pet. Chem. **52**(4), 261–266 (2012). https://doi.org/10.1134/S0965544112040044
13. Khusnutdinov, R., Bayguzina, A., Gimaletdinova, L., Dzhemilev, U.: Intermolecular dehydration of alcohols by the action of copper compounds activated with carbon tetrabromide. Synthesis of ethers. Russ. J. Org. Chem. **48**(9), 1191–1196 (2012). https://doi.org/10.1134/S1070428012090072
14. Yu, J., et al.: Selective synthesis of unsymmetrical ethers from different alcohols catalyzed by sodium bisulfite. Tetrahedron **69**(1), 310–315 (2013). https://doi.org/10.1016/j.tet.2012.10.032
15. Koledina, K.F., Gubaidullin, I.M., Koledin, S.N., Baiguzina, A.R., Gallyamova, L.I., Khusnutdinov, R.I.: Kinetics and mechanism of the synthesis of benzylbutyl ether in the presence of copper-containing catalysts. Russ. J. Phys. Chem. A **93**(11), 2146–2151 (2019). https://doi.org/10.1134/S0036024419110141
16. Koledina, K.F., Koledin, S.N., Gubaydullin, I.M.: Automated system for identification of conditions for homogeneous and heterogeneous reactions in multiobjective optimization problems. Numer. Anal. Appl. **12**(2), 116–125 (2019). https://doi.org/10.1134/S1995423919020022
17. Al'shin, A., Al'shina, E., Kalitkin, N., Koryagina, A.: Rosenbrock schemes with complex coefficients for stiff and differential algebraic systems. Comput. Math. Math. Phys. **46**(8), 1320–1340 (2006). https://doi.org/10.1134/S0965542506080057
18. Reeves, C., Rowe, J.: Genetic Algorithms-Principles and Perspectives. Springer, Boston (2002). https://doi.org/10.1007/b101880
19. Kordabadi, H., Jahanmiri, A.: Optimization of methanol synthesis reactor using genetic algorithms. Chem. Eng. J. **108**(3), 249–255 (2005). https://doi.org/10.1016/j.cej.2005.02.023

20. Kalyanmoy, D., Pratap, A., Agarwal, S., Meyarivan, T.: A fast and elitist multi-objective genetic algorithm: NSGA-II. IEEE Trans. Evol. Comput. **6**(2), 182–197 (2002). https://doi.org/10.1109/4235.996017
21. Hong, H., Ye, K., Jiang, M., Tan, K.C.: Solving large-scale multi-objective optimization via probabilistic prediction model. In: Ishibuchi, H., et al. (eds.) EMO 2021. LNCS, vol. 12654, pp. 605–616. Springer, Cham (2021). https://doi.org/10.1007/978-3-030-72062-9_48
22. Chowdhury, S., Tong, W., Messac, A., Zhang, J.: A mixed-discrete particle swarm optimization algorithm with explicit diversity preservation. Struct. Multidiscipl. Optim. **47**(3), 367–388 (2013). https://doi.org/10.1007/s00158-012-0851-z
23. Heris, M.: NSGA-II in MATLAB. Yarpiz (2015)

The Method of Optimal Geometric Parameters Synthesis of Two Mechanisms in the Rehabilitation System on Account of Relative Position

Dmitry Malyshev[1(✉)], Larisa Rybak[1], Santhakumar Mohan[2], Askhat Diveev[3], Vladislav Cherkasov[1], and Anton Pisarenko[1]

[1] Belgorod State Technological University named after V.G. Shukhov, Belgorod, Russia
[2] Indian Institute of Technology Palakkad, Palakkad, India
[3] Federal Research Center "Computer Science and Control" of Russian Academy of Sciences, Moscow, Russia

Abstract. The paper presents the method of optimal geometric parameters synthesis for a robotic rehabilitation system of the lower limbs as based on a passive orthosis in the form of a serial RRRR mechanism and an active parallel 3-PRRR mechanism. A mathematical model of the kinematics of the RRRR mechanism is formulated. A numerical method is worked out for determining the workspace of the RRRR mechanism. The workspace description is based on the approximation of many solutions of systems of nonlinear inequalities to provide the required rotation angle of rotation in the joints. Effective numerical methods and algorithms are developed and tested. The paves the way for determining the minimum geometric parameters of an active parallel mechanism that ensure movements of a passive orthosis within the workspace under clinical data constrains when simulating walking. In order to implement the developed methods, an effective algorithm, software package and visualization system are synthesized for exported three-dimensional workspaces in STL.

Keywords: Parallel robot · Rehabilitation system · Optimization

1 Introduction

Since the development of electrical engineering in the middle of the 20-th century, robots of various designs have entered into many areas of our life: conveyor production, construction, aerospace industry, medicine and etc. One of the most pressing and complex problems of medicine and neurology is the rehabilitation of patients. Lower limb rehabilitation or treatment has been a hot topic in recent

This work was supported by the Russian Science Foundation, the agreement number 19–19-00692.

N. N. Olenev et al. (Eds.): OPTIMA 2021, CCIS 1514, pp. 230–245, 2021.
https://doi.org/10.1007/978-3-030-92711-0_16

years, since mechanized systems promised effective results and led to significant improvements in the recovery of patients using robotic physiotherapy as pointed out in [1]. Optimizing the parameters of a robotic system is essential in design. In this case, the optimization criterias must consider both the compactness of the structure and its sufficient strength.

2 The Proposed Solution

Figure 1 shows a conceptual design of the proposed system for lower limb rehabilitation. The structure of the rehabilitation system uses an active 3-PRRR parallel robot and a passive orthosis. Each leg of parallel robot consists of a single active prismatic joint (P) and three revolute joints (R). The passive orthosis contains four revolute joints (R), two of which correspond to the hip joint of person, one to the knee and the last to the ankle.

Fig. 1. A conceptual design of the proposed rehabilitation system

The active 3-PRRR parallel mechanism provides the rehabilitation angles required for rehabilitation of all joints of the patient. The passive orthosis is used for supporting the patient's lower limb. The kinematic diagram of the rehabilitation system is shown on Fig. 2. The active mechanism consists of three serial kinematic chains $A_iB_iC_iD_i, i = 3$ and has three degrees of freedom – for translational movements along each of the axes. The position of the rehabilitation platform, which is an equilateral triangle $D_1D_2D_3$ centered at point P and the radius of the circumscribed circle R, is determined by linear displacements $q_1q_2q_3$.). Let us introduce the following notation: a_i is the distance between points A_i and B_i, b_i is between B_i and C_i, c_i is between C_i and $D_i.d_i$ is between B_i and D_i.

A serial RRRR mechanism is used for securing the human foot. with joints corresponding to the patient's. In the joint E, two movements of the hip joint are provided: rotation in the sagittal plane with an angle α and abduction of the legs and the angle ψ between the projection of the EF link onto the XOY plane and the OY axis. In the joint F of the knee joint, the FP link is rotated relative to EF by an angle θ.

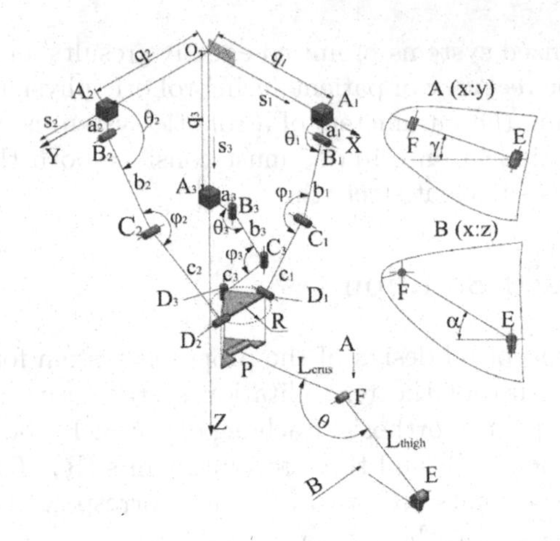

Fig. 2. The kinematic diagram of the rehabilitation system

In process of rehabilitation, a person's leg should take positions according to walking simulation. The extreme positions of the platform for fixing the foot form the shell of the workspace of the passive orthosis. The parallel 3-PRRR mechanism determines the position of the platform. It is required to determine its minimum geometric parameters, which will provide all the extreme positions of the platform. For this, the shell of the workspace of the serial RRRR mechanism used as a passive orthosis can be determined.

3 A Mathematical Model of a Rehabilitation System

Consider the basic kinematic relationships of the mechanisms. The point P is considered as the output point, which is located in the centre of the circumscribed circle of the moving platform of the active parallel robot. It coincides with the centre of the rotational joint of the ankle joint. Restrictions on the geometric parameters of the mechanism can be introduced:

– restrictions on the drive coordinate q:

$$q_i \in [q_{min}, q_{max}] \tag{1}$$

These restrictions take into account the lengths of the guides that determine the minimum and maximum coordinates that the drive hinges A_i can have:

$$q_1 = x_{D1} = x_P + \frac{\sqrt{3}}{2}R, \tag{2}$$

$$q_2 = y_{D2} = y_P + \frac{R}{2}, \tag{3}$$

$$q_3 = z_{D3} = z_P \tag{4}$$

– the distance d between the centers of the joints B_i and D_i

$$d_i \in [0; d_{max}], \tag{5}$$

where $d_{max} = b + c$.

The distances d_i can be defined as the distance between B_i and D_i:

$$d_1 = \sqrt{\left(y_P + \frac{R}{2} - a\right)^2 + (z_P)^2}, \tag{6}$$

$$d_2 = \sqrt{\left(y_P + \frac{\sqrt{3}}{2}R\right)^2 + (z_P - a)^2}, \tag{7}$$

$$d_3 = \sqrt{(x_P - a)^2 + (y_P - R)^2} \tag{8}$$

– restrictions associated with the interference of links.

The interference of the links of the mechanism can be divided into two groups:

– interference at small angles between links connected by rotary joints.
– the interference of links that are not connected to each other.

The first group can be determined, taking into account the restrictions on the angles of rotation in the joints B_i,C_i and D_i[15]:

$$\begin{cases} \theta_i \in [\theta_{min}, \theta_{max}] \\ \phi_i \in [\phi_{min}, \phi_{max}] \\ \psi_i \in [\psi_{min}, \psi_{max}] \end{cases} \tag{9}$$

The second group of interferences is defined using an approach based on determining the minimum distance between the segments drawn between the centers of the joints of each of the links., The approach discussed earlier by the authors in detail in [2] and it is here applied to determine the intersections. In this case, the condition for the absence of interferences of the links can take the form

$$\sqrt{{u_1}^2 + {u_2}^2} > D_{link}, \tag{10}$$

where u_1 is the distance between the axis of the link that does not belong to the plane and the auxiliary plane, u_2 is the distance between the nearest points of the segments connecting the centers of the joints of each of the links when projecting a segment that does not belong to the auxiliary plane onto this plane, D_{link} is the diameter of the links.

System of equations for the connection of the passive mechanism that describes the position of the point P, depending on the rotation angles ψ and α in the hip joint and the rotation angle $\gamma = 180 - \theta$ in the knee joint. The coordinates of the centre of the joint R can be expressed as

$$\begin{cases} x_P = x_E + \sin\psi(L_{thing}\cos\alpha + L_{crus}\cos\beta) \\ y_P = y_E - \cos\psi(L_{thing}\cos\alpha + L_{crus}\cos\beta) \\ z_P = z_E - L_{thing}\cos\alpha - L_{crus}\cos\beta \end{cases}, \tag{11}$$

where $\beta = \alpha + \gamma$, thing – is the length of the link EF, L_{crus} – is the length of the link FP. Let $L_{PE} = L_{thigh} \cos\alpha + L_{crus} \cos\beta$ – the projection of the EP link onto the XOY plane. Replace the variables in the system of Eq. (11)

$$x_{PE} = x_P - x_E, y_{PE} = y_P - y_E, z_{PE} = z_P - z_E. \quad (12)$$

For calculating the optimal dimensions of a parallel robot, it is first necessary to determine the workspace of the RRRR mechanism.

4 Algorithm for Determining the Workspace of the RRRR Mechanism

4.1 Description of the Approximation Method

The following methods are known for determining the workspace: geometric, numerical, and discretization methods [3,4]. In [5,7], interval analysis is a use for approximating the workspace. It is also worth noting the work [8] , devoted to the construction of three-dimensional computational grids in areas of complex shape. The method of non-uniform coverings for approximating the set of solutions of a nonlinear inequalities system was considered in [9] , and the application of this method to determine the workspace of some of the planar robots types was considered in [10,11]. Compared to another methods, the method of non-uniform coverings seems universal, allowing one solving many robotics problems, for example, determining the workspace, solving kinematics problems, taking into account singularity zones, determining positioning errors. The method is based on the approximation of systems of nonlinear inequalities and equations. The tasks posed in this paper have high computational complexity due to the large dimension. However, as will be shown below, the approaches allow effectively solving these problems concerning a serial RRRR mechanism and a parallel 3-PRRR mechanism in one robotic system for the rehabilitation of the lower limbs.

Consider the task of determining the boundaries for the intervals of the RRRR mechanism parameters: the rotation angles in the hip joint in the sagittal plane α, in the knee joint γ and the angle of leg abduction in the hip joint ψ. Interval analysis[[10]] is applied for calculations because variables and conditions for the workspace boundary are presented in interval form, not discrete form. The coverage of the workspace is a set of n-dimensional boxes bounded by the intervals of variable values x_{PE}, y_{PE}, z_{PE}. Ranges of values guaranteed to include the ranges of allowable values

$$\mathbf{Z}_{PE} = [\underline{\mathbf{Z}_{PE}}, \overline{\mathbf{Z}_{PE}}] = \left\{\underline{\mathbf{Z}_{PE}} \le z_{PE} \le \overline{\mathbf{Z}_{PE}}\right\}, \quad (13)$$

$$\mathbf{X}_{PE} = [\underline{\mathbf{X}_{PE}}, \overline{\mathbf{X}_{PE}}] = \left\{\underline{\mathbf{X}_{PE}} \le x_{PE} \le \overline{\mathbf{X}_{PE}}\right\}, \quad (14)$$

$$\mathbf{Y}_{PE} = [\underline{\mathbf{Y}_{PE}}, \overline{\mathbf{Y}_{PE}}] = \left\{\underline{\mathbf{Y}_{PE}} \le y_{PE} \le \overline{\mathbf{Y}_{PE}}\right\}, \quad (15)$$

and the interval $\mathbf{B}$ according to clinical data. For them, the joint intervals $\mathbf{A}', \mathbf{\Gamma}', \mathbf{\Psi}'$ are determined and compared with the clinical data – the intervals

$\mathbf{A}, \mathbf{\Gamma}, \mathbf{\Psi}$. The interval $\mathbf{L}'_{PE}$ for the intervals $\mathbf{X}_{PE}$ and $\mathbf{Y}_{PE}$ and the interval $\mathbf{L}''_{PE}$ for the intervals $\mathbf{Z}_{PE}$ and $\mathbf{B}$ are also determined. In the case of the intersection of the intervals $\mathbf{L}'_{PE}$ and $\mathbf{L}''_{PE}$, as well as the intervals of the calculated and clinically given angles of rotation of the joints, there is a solution to system (11). The lower boundary of the intervals corresponds to the minimum angles required for rehabilitation, and the upper to the maximum, then is: for the angle α; $\mathbf{A} := [\underline{\mathbf{A}}, \overline{\mathbf{A}}] = \{\underline{\mathbf{A}} \leq \alpha \leq \overline{\mathbf{A}}\}$; for the angle γ: $\mathbf{\Gamma} := [\underline{\mathbf{\Gamma}}, \overline{\mathbf{\Gamma}}] = \{\underline{\mathbf{\Gamma}} \leq \gamma \leq \overline{\mathbf{\Gamma}}\}$ for the angle ψ: $\mathbf{\Psi} := [\underline{\mathbf{\Psi}}, \overline{\mathbf{\Psi}}] = \{\underline{\mathbf{\Psi}} \leq \psi \leq \overline{\mathbf{\Psi}}\}$; for the angle β: $\beta = \alpha + \gamma$, that is, $\mathbf{B} := [\underline{\mathbf{B}}, \overline{\mathbf{B}}] = \{\underline{\mathbf{A}} + \underline{\mathbf{\Gamma}} \leq \beta \leq \overline{\mathbf{A}} + \overline{\mathbf{\Gamma}}\}$. For the intervals $\mathbf{X}_{PE}$ and $\mathbf{Y}_{PE}$ let's define the interval $\mathbf{\Psi}'$ of the angle ψ and the interval $\mathbf{L}'_{PE}$, witch is equivalent to the line EP projection segment onto plane XY:

$$\mathbf{\Psi}' := [\underline{\mathbf{\Psi}'}\overline{\mathbf{\Psi}'}], \tag{16}$$

Where $\underline{\mathbf{\Psi}'} = \min\limits_{x_{PE} \in \mathbf{X}_{PE}, y_{PE} \in \mathbf{Y}_{PE}} \left\{ \arctan \frac{x_{PE}}{y_{PE}} \right\}$,
$\overline{\mathbf{\Psi}'} = \max\limits_{x_{PE} \in \mathbf{X}_{PE}, y_{PE} \in \mathbf{Y}_{PE}} \left(\arctan \frac{x_{PE}}{y_{PE}} \right)$

$$\mathbf{L}'_{PE} := [\underline{\mathbf{L}'_{PE}}, \overline{\mathbf{L}'_{PE}}], \tag{17}$$

Where $\underline{\mathbf{L}'_{PE}} = \min\limits_{x_{PE} \in \mathbf{X}_{PE}, y_{PE} \in \mathbf{Y}_{PE}} \left(\sqrt{x_{PE}^2 + y_{PE}^2} \right)$,
$\overline{\mathbf{L}'_{PE}} = \max\limits_{x_{PE} \in \mathbf{X}_{PE}, y_{PE} \in \mathbf{Y}_{PE}} \left(\sqrt{x_{PE}^2 + y_{PE}^2} \right)$.

The angle α expressed taking into account (11): $\alpha = \arcsin \frac{-L_{crus} \sin \beta - z_{PE}}{L_{thingh}}$

The interval $\mathbf{A}'$ of the angle α values defined for the interval $\mathbf{Z}_{PE}$ and the interval $\mathbf{B}$:

$$\mathbf{A}' := [\underline{\mathbf{A}'}, \overline{\mathbf{A}'}], \tag{18}$$

Where $\underline{\mathbf{A}'} = \min\limits_{z_{PE} \in \mathbf{Z}_{PE}, \beta \in \mathbf{B}} \left(\arcsin \frac{-L_{crus} \sin \beta - z_{PE}}{L_{thingh}} \right)$,
$\overline{\mathbf{A}'} = \max\limits_{z_{PE} \in \mathbf{Z}_{PE}, \beta \in \mathbf{B}} \left(\arcsin \frac{-L_{crus} \sin \beta - z_{PE}}{L_{thingh}} \right)$.

For the interval $\mathbf{A}'$ and $\mathbf{B}'$, the interval $\mathbf{L}''_{PE}$ can be expressed as:

$$\mathbf{L}''_{PE} := [\underline{\mathbf{L}''_{PE}}, \overline{\mathbf{L}''_{PE}}], \tag{19}$$

Where $\underline{\mathbf{L}''_{PE}} = \min\limits_{\alpha \in \mathbf{A}', \beta \in \mathbf{B}} (L_{thingh} \cos \alpha + L_{crus} \cos \beta)$,
$\overline{\mathbf{L}''_{PE}} = \max\limits_{\alpha \in \mathbf{A}', \beta \in \mathbf{B}} (L_{thingh} \cos \alpha + L_{crus} \cos \beta)$.

Using equations (16) - (19), system of inequalities is composed. It that takes into account the intersection of intervals in the form

$$\begin{cases} \mathbf{\Psi} \cap \mathbf{\Psi}' \neq \varnothing \\ \mathbf{A} \cap \mathbf{A}' \neq \varnothing \\ \mathbf{\Gamma} \cap \mathbf{\Gamma}' \neq \varnothing \\ \mathbf{L}'_{PE} \cap \mathbf{L}''_{PE} \neq \varnothing \end{cases} . \tag{20}$$

If the system (20) is satisfied, then for the intervals $\mathbf{X}_{PE}$, $\mathbf{Y}_{PE}$, $\mathbf{Z}_{PE}$ and $\mathbf{B}$, there is at least one point entering the workspace. Thus, the volume of the workspace is calculated. Further, for determining the boundary of the workspace, an additional condition in the system (20) is introduced, which takes into account that, at the boundary of the workspace, the intervals of angles $\mathbf{A}, \mathbf{\Gamma}, \mathbf{\Psi}$ cannot simultaneously ultimately include the intervals $\mathbf{A}', \mathbf{\Gamma}', \mathbf{\Psi}'$, respectively.

$$\begin{cases} \underline{\mathbf{\Psi}} \in \mathbf{\Psi}' \vee \overline{\mathbf{\Psi}} \in \mathbf{\Psi}' \vee \underline{\mathbf{A}} \in \mathbf{A}' \vee \overline{\mathbf{A}} \in \mathbf{A}' \vee \underline{\mathbf{\Gamma}} \in \mathbf{\Gamma}' \vee \overline{\mathbf{\Gamma}} \in \mathbf{\Gamma}' \\ \mathbf{\Psi} \cap \mathbf{\Psi}' \neq \varnothing \\ \mathbf{A} \cap \mathbf{A}' \neq \varnothing \\ \mathbf{\Gamma} \cap \mathbf{\Gamma}' \neq \varnothing \\ \mathbf{L}'_{PE} \cap \mathbf{L}''_{PE} \neq \varnothing \end{cases} . \qquad (21)$$

The interval for the constraints of the variables form an n-dimensional box, where n corresponds to the number of variables, i.e. $n = 4$. For determine the shell of the workspace, the initial and subsequent boxes, for which (21) holds, are divided into smaller boxes. For each box B obtained as a result of the partition, the verification is performed similarly. The set $\mathbb{P}_I$ describes the shell of the workspace as boxes, for which the size of the box along the $\mathbf{X}_{PE}$,$\mathbf{Y}_{PE}$ and $\mathbf{Z}_{PE}$ dimensions $d(B) \leq \delta$, where δ is a given positive value that determines the accuracy of the approximation. For the subsequent use of the set of boxes for visualization, the resulting four-dimensional workspace $\mathbb{P}_I$ is projected onto the three-dimensional space XYZ. Projection is to exclude all boxes, except for one, having the same values of the intervals $\mathbf{X}_{PE}$,$\mathbf{Y}_{PE}$ and $\mathbf{Z}_{PE}$,the fourth dimension of the lists $\mathbb{P}$ and $\mathbb{P}_I$ corresponds to the interval $\mathbf{B}$.

4.2 Algorithm Description

The synthesized algorithm for determining the boundaries of the workspace is based on the method of approximating the set of solutions of the resulting system of nonlinear inequalities (21). The algorithm works with two lists of four-dimensional boxes $\mathbb{P}$,$\mathbb{P}_I$ and a list of three-dimensional boxes $\mathbb{P}_J$. Each of the dimensions of the boxes in the list $\mathbb{P}_J$ corresponds to the intervals $\mathbf{X}_{PE}$,$\mathbf{Y}_{PE}$ and $\mathbf{Z}_{PE}$, the fourth dimension of the lists $\mathbb{P}$ and $\mathbb{P}_I$ corresponds to the interval $\mathbf{B}$.

The algorithm works as follows:

1. Set the geometric parameters of the passive orthosis and the approximation accuracy δ. Lists of the internal approximation $\mathbb{P}_I$ and $\mathbb{P}_J$ are empty, the list $\mathbf{P}$ consists of only one box Q, including the intervals $\mathbf{X}_{PE}$,$\mathbf{Y}_{PE}$, $\mathbf{Z}_{PE}$ and $\mathbf{B} : \mathbf{X}_{PE} = \mathbf{Y}_{PE} = \mathbf{Z}_{PE} = [-(L_{thing} + L_{crus}), L_{thing} + L_{crus}]$.
2. Extract box B from list $\mathbb{P}$.
3. If $d(B) \leq \delta$,then add B to the list $\mathbb{P}_I$ and go to step 7.
4. Calculate $\mathbf{A}', \mathbf{\Gamma}', \mathbf{\Psi}', \mathbf{L}'_{PE}$ and $\mathbf{L}''_{PE}$ according to formulas (16) - (19).
5. If the system (21) does not hold, then exclude B and go to step 7.
6. In the remaining cases, divide B by two equal boxes along the edge with the most extended length and add to the end of the list $\mathbb{P}$.

7. If $\mathbb{P} \neq \varnothing$ then go to step 2.
8. Project $\mathbb{P}_I$ in $\mathbb{P}_J$, i.e. sequentially add boxes B from the list $\mathbb{P}_I$ to the list $\mathbb{P}_j$ without the fourth dimension $\mathbf{B}$, with the condition that they should not intersect with each other.

4.3 Simulation Results

A computational example was performed for the following geometric parameters of the mechanism: $\mathbf{A} = [-20°, 10°]$, $\mathbf{\Gamma} = [-60°, 0°]$, $\mathbf{\Psi} = [0°, 25°]$, $\mathbf{B} = [-80°, 10°]$, $L_{thigh} = L_{crus}$=450 mm. These intervals correspond to the size of an adult's foot, as well as the required abduction angle and gait simulation angles. Figure 3 show the projections of the computer shell of the workspace on the plane: a is on the XOY plane, b is on the XOZ plane, c is on the YOZ plane. Figure 3d shows the visualization of the computed full shell of the three-dimensional workspace. The calculation time for an approximation accuracy $\delta = 8$ mm was 19 s.

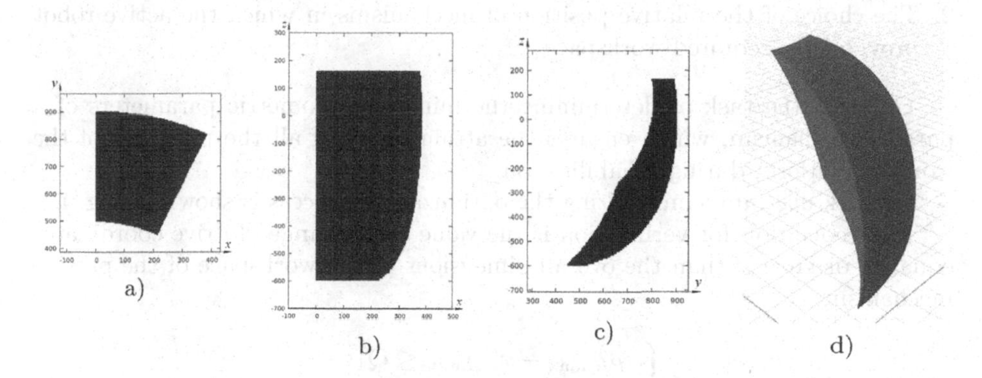

Fig. 3. Workspace: a),b),c) - projections of the shell of the workspace on a plane, d) - 3D workspace

The shell of the workspace represents the many extreme positions of the output link as the platform, which is also the output link for the active 3-PRRR mechanism. Consequently, the resulting set can be used for verifying the achievement of these positions with specific geometric parameters of the active mechanism.

5 Optimization of the 3-PRRR Mechanism Parameters

The choice of optimization criteria is to ensure the compactness of the mechanism and its sufficient strength. A strength calculation perfomed to select a criterion that takes into account strength. It can be concluded that to reduce the forces and moments affecting the strength of the structure, the following optimization criterion should be used:

$$k_1 = b_3 + c_3. \tag{22}$$

The requirement for compactness can be taken into account by adding the sum of the dimensions of the mechanism as an optimization criterion. The a_i and R parameters are constant. As a result, k_2 can be expressed as

$$k_2 = \sum_{i=1}^{2}(b_i + c_i + q_i) + q_3. \tag{23}$$

Two obtained criteria will be used for multi-criteria optimization.

5.1 Description of the Optimization Method

Two optimization problems arise for designing such a hybrid robotic system:

1. Multi-criteria optimization, the parameters of which are nine sizes of the active robot b_i,c_i and q_i, $i \in 1, 3$.
2. The choice of the relative position of mechanisms in which the active robot provides the required workspace.

Consider the task of determining the minimum geometric parameters of a parallel mechanism, which ensures the attainability of all the positions of the working platform during rehabilitation.

A block diagram summarizing the optimization process is shown in Fig. 4.

A precondition for verification is the value of the range of drive coordinates equal to or greater than the overall dimensions of the workspace of the passive mechanism

$$\begin{cases} x_{PE,\max} - x_{PE,\min} \le \mathbf{Q}_1 \\ y_{PE,\max} - y_{PE,\min} \le \mathbf{Q}_2 \\ z_{PE,\max} - z_{PE,\min} \le \mathbf{Q}_3 \end{cases}, \tag{24}$$

Where $\mathbf{Q}_i := [q_{1,\min}, q_{1,\max}] = [\underline{\mathbf{Q}_i}, \overline{\mathbf{Q}_i}] = \{\underline{\mathbf{Q}_i} \le q_i \le \overline{\mathbf{Q}_i}\}$ are the intervals of the driving coordinates of the active mechanism.

Condition (22) can be taken into account by setting the values of $\underline{\mathbf{Q}_i}$

$$\underline{\mathbf{Q_1}} = x_{PE,\max} - x_{PE,\min}, \underline{\mathbf{Q_2}} = y_{PE,\max} - y_{PE,\min}, \tag{25}$$

$$\underline{\mathbf{Q_3}} = z_{PE,\max} - z_{PE,\min}. \tag{26}$$

In this case, the minimum and maximum values of x_{PE},y_{PE} and z_{PE} can be determined using the set of boxes $\mathbb{P}_J$, that describe the shell of the workspace as.

$$x_{PE,\max} = \max_{B \in \mathbb{P}_J} \overline{\mathbf{X}_{PE}}, x_{PE,\min} = \min_{B \in \mathbb{P}_J} \underline{\mathbf{X}_{PE}}, \tag{27}$$

$$y_{PE,\max} = \max_{B \in \mathbb{P}_J} \overline{\mathbf{Y}_{PE}}, y_{PE,\min} = \min_{B \in \mathbb{P}_J} \underline{\mathbf{Y}_{PE}}, \tag{28}$$

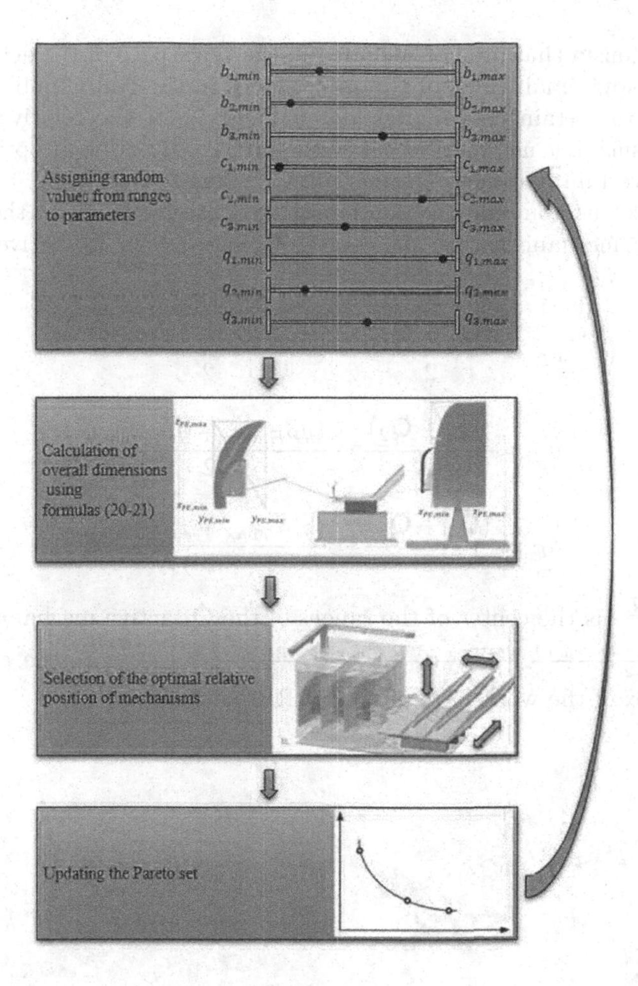

Fig. 4. Optimization process

$$z_{PE,\max} = \max_{B \in \mathbb{P}_J} \overline{\mathbf{Z}_{PE}}, z_{PE,\min} = \min_{B \in \mathbb{P}_J} \underline{\mathbf{Z}_{PE}}. \tag{29}$$

For determining the geometric parameters of a parallel mechanism, it is proposed to use the Hill Climbing technique. It consists of finding a local minimum or maximum by passing between points called nodes, depending on whether the function value is smaller or more extensive compared to the current point. This method is considered in more detail in [13] . In [14] , it was used for constructing an autonomous map and study an unknown environment using mobile robots. It can be used for determining the geometric parameters of other mechanisms depending on the required workspace. The non-uniform coverings method was applied for obtaining the workspace of a passive orthosis. The workspace is a set of small boxes. Hill Climbing technique allows determining the relative position of the mechanisms, in which the number of boxes from the covering set of the

RRRR mechanism that are not included in the workspace of the active 3-PRRR mechanism is minimal. Each of the mesh nodes in the Hill Climbing technique corresponds to certain coordinates x_E, y_E and z_E of the relative position of the mechanisms. Let us describe below the process of finding a optimal relative position for certain parameters using Hill Climbing technique.

Initially let us combine the centers of the described box of the workspace of the RRRR mechanism with the center of ranges of the $\mathbf{Q}_i$ active mechanism Fig. 5a:

$$x_E = \frac{(\overline{\mathbf{Q}_1} + \underline{\mathbf{Q}_1})}{2} - \frac{(x_{PE,\max} + x_{PE,\min})}{2}, \tag{30}$$

$$y_E = \frac{(\overline{\mathbf{Q}_2} + \underline{\mathbf{Q}_2})}{2} - \frac{(y_{PE,\max} + y_{PE,\min})}{2}, \tag{31}$$

$$z_E = \frac{(\overline{\mathbf{Q}_3} + \underline{\mathbf{Q}_3})}{2} - \frac{(z_{PE,\max} + z_{PE,\min})}{2}, \tag{32}$$

where $\frac{(\overline{\mathbf{Q}_i}+\underline{\mathbf{Q}_i})}{2}$ is the center of the ranges of the $\mathbf{Q}_i$ active mechanism, $\left(\frac{(x_{PE,\max}+x_{PE,\min})}{2}, \frac{(y_{PE,\max}+y_{PE,\min})}{2}, \frac{(z_{PE,\max}+z_{PE,\min})}{2}\right)$ is the center of the described box of the workspace of the RRRR mechanism.

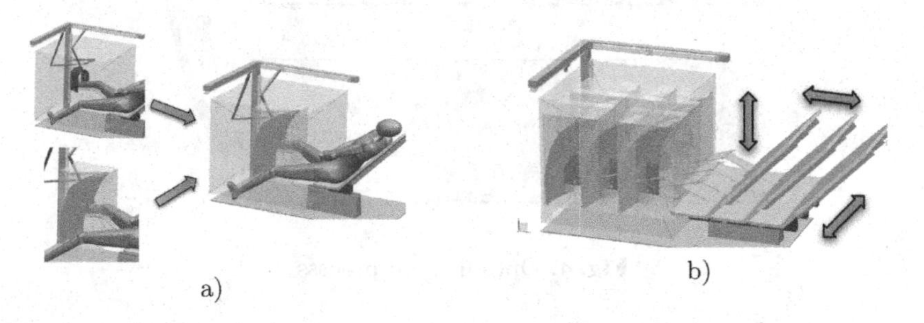

Fig. 5. Relative position of mechanisms: a) Combining of the boxes centers b) Setting of mechanisms relative position.

If the workspace of the active mechanism does not include the workspace of the passive one, then it is necessary to check the other relative position of the mechanisms as showing (Fig. 5b).

Additional offsets along each of the axes can be expressed as

$$x_E = \frac{(\overline{\mathbf{Q}_1} + \underline{\mathbf{Q}_1})}{2} - \frac{(x_{PE,\max} + x_{PE,\min})}{2} + x'_E, \tag{33}$$

$$y_E = \frac{(\overline{\mathbf{Q}_2} + \underline{\mathbf{Q}_2})}{2} - \frac{(y_{PE,\max} + y_{PE,\min})}{2} + y'_E, \tag{34}$$

$$z_E = \frac{(\overline{\mathbf{Q_3}} + \underline{\mathbf{Q_3}})}{2} - \frac{(z_{PE,\max} + z_{PE,\min})}{2} + z'_E, \tag{35}$$

Where x'_E, y'_E, z'_E are additional offsets along each axis. For values $x'_E = y'_E = z'_E = 0$ the center of the described box of the workspace of the RRRR mechanism coincides with the center of the box of the ranges $\mathbf{Q}_i$ of the active mechanism.

When changing the values x_E, y'_E and z'_E, the coordinates of the point E change, therefore, the workspaces of the mechanisms move relative to each other. This fact allows adjusting the relative position of the mechanisms to determine the location at which the workspace of the passive mechanism enters the active workspace.

For each of the boxes of the workspace of the RRRR mechanism for $x'_E = y'_E = z'_E = 0$, it is required to calculate the values of geometric parameters that impose design restrictions on the workspace of the active 3-PRRR mechanism and check the conditions (5), (9) and (10).

Let N denote the number of boxes of the working region of the RRRR mechanism for which conditions (1) or (11) or (10) is not satisfied and which do not enter the working region of the active 3-PRRR mechanism. If $N > 0$, then to determine the minimum value of N when changing$x'_E = y'_E = z'_E$ used Hill Climbing technique. . Starting from the point at $x'_E = y'_E = z'_E = 0$, N is calculated in each of the neighboring nodes. Denote $x'_E = y'_E = z'_E$ for the current point as $x'_{E,0} = y'_{E,0} = z'_{E,0}$, respectively, and the displacements along each axis to nodes as $\Delta x'_E = \Delta y'_E = \Delta z'_E$, which are initially $\pm T_{\max}$ is the maximum distance from the boundary of the workspace of the passive mechanism to the boundaries of the intervals of the driving coordinates of the active mechanism, with $T_{\max} = \max_{i \in 1..6} f_i$, where

$$f_1 = x_{PE,\min} + \frac{(\overline{\mathbf{Q_1}} + \underline{\mathbf{Q_1}})}{2} - \frac{(x_{PE,\max} + x_{PE,\min})}{2} - \underline{\mathbf{Q_1}}, \tag{36}$$

$$f_2 = y_{PE,\min} + \frac{(\overline{\mathbf{Q_2}} + \underline{\mathbf{Q_2}})}{2} - \frac{(y_{PE,\max} + y_{PE,\min})}{2} - \underline{\mathbf{Q_2}}, \tag{37}$$

$$f_3 = z_{PE,\min} + \frac{(\overline{\mathbf{Q_3}} + \underline{\mathbf{Q_3}})}{2} - \frac{(z_{PE,\max} + z_{PE,\min})}{2} - \underline{\mathbf{Q_3}}, \tag{38}$$

$$f_4 = \overline{\mathbf{Q_1}} - \left(x_{PE,\max} + \frac{(\overline{\mathbf{Q_1}} + \underline{\mathbf{Q_1}})}{2} - \frac{(x_{PE,\max} + x_{PE,\min})}{2} \right), \tag{39}$$

$$f_5 = \overline{\mathbf{Q_2}} - \left(y_{PE,\max} + \frac{(\overline{\mathbf{Q_2}} + \underline{\mathbf{Q_2}})}{2} - \frac{(y_{PE,\max} + y_{PE,\min})}{2} \right), \tag{40}$$

$$f_6 = \overline{\mathbf{Q_3}} - \left(z_{PE,\max} + \frac{(\overline{\mathbf{Q_3}} + \underline{\mathbf{Q_3}})}{2} - \frac{(z_{PE,\max} + z_{PE,\min})}{2} \right). \tag{41}$$

Figure 6 shows the search for the minimum value of N, with $N_i \le N_{i-1}$. If N in another node is less than in the current one, then a transition occurs, that is, $x'_{E,0} = x'_E$, $y'_{E,0} = y'_E$, $z'_{E,0} = z'_E$. If at all other nodes N is greater than the current one, then the displacements $\Delta x'_E$, $\Delta y'_E$, $\Delta z'_E$ decrease to find a minimum N closer to the current node. Figure 6 demonstrates coordinate axes X, Y, Z correspond to the displacements x'_E, y'_E, z'_E relative to the starting point of the search.

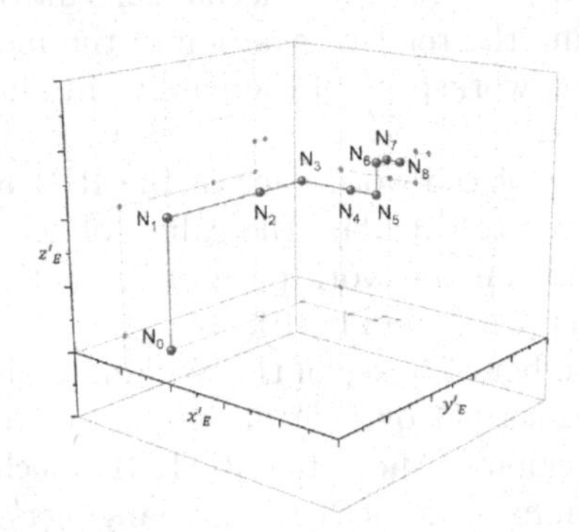

Fig. 6. Search for the minimum value of N

If $N_{\min} > 0$, then we will choose other random optimization parameters b_i, c_i and q_i without changing the Pareto set. If $N_{\min} = 0$, then the Pareto set is updated.

6 Simulation Results

During the simulation, the minimum geometric parameters of the active parallel mechanism were determined for the following sizes: $a = 100$ mm,$R = 115$ mm, $\phi_{min} = \Theta_{min} = \Psi_{min} = 10°$, $\phi_{max} = \Theta_{max} = \Psi_{max} = 170°$,$D_{link} = 20$ mm. Optimization parameters range: $b_i \in [50; 1050], c_i \in [50; 1050], q_1 \in [383; 1383], q_2 \in [450; 1450], q_3 \in [759; 1759]$. As a result, Pareto front was found (Fig. 7). It consist of 5 points.

Let us derive a joint criteria with a coefficient a, taking into account the ratio of two criteria to select a point from the Pareto set $k = ak_1 + k_2$. Using such a representation, the minimum value of the total coefficient for the range of coefficient $a \in [0.5; 7]$ corresponds to the following results: $k_1 = 656, k_2 = 4042, b_1 = 446, c_1 = 450, b_2 = 476, c_2 = 506, b_3 = 319, c_3 = 337, q_1 = 866, q_2 = 466, q_3 = 832$

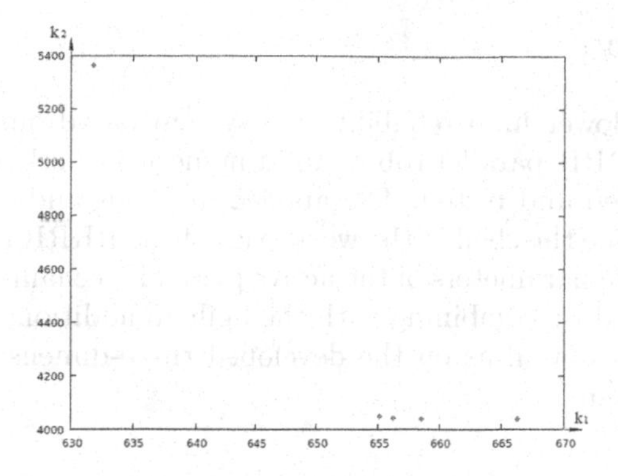

Fig. 7. Pareto front.

For verifying the results, the workspace of the active mechanism was determined following the obtained values of parameters Fig. 8a shows that the workspace of the passive mechanism is fully included in the workspace of the active mechanism. Following the obtained results of the geometric parameters of the mechanisms, a three-dimensional model of the rehabilitation system was developed. The movement of the mechanisms was checked following clinical data. Figure 8b shows the development of the trajectory during rehabilitation. The relative position of the mechanisms and the geometric dimensions of the active 3-PRRR mechanism adequately provide the required movements.

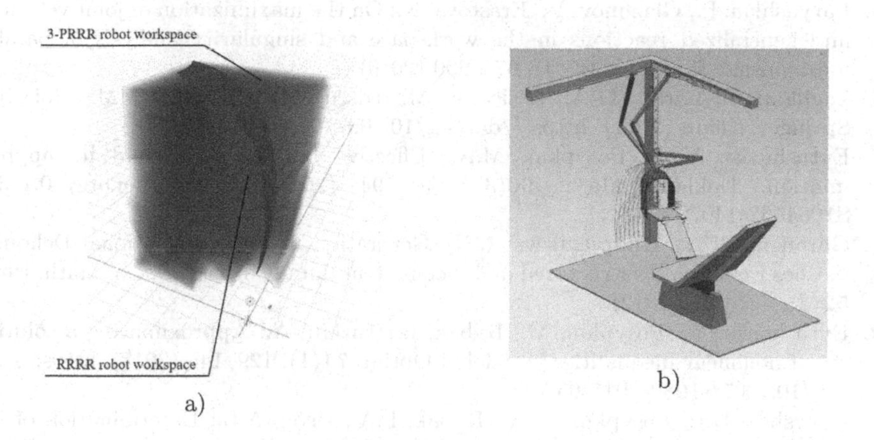

Fig. 8. The workspace of the passive mechanism: a – combined with the workspace of the active mechanism, b – validation on CAD models

7 Conclusion

For the proposed lower limb rehabilitation system based on a passive orthosis and a mobile 3-PRRR parallel robot, valid numerical methods and algorithms have been developed and tested. The applied methods and algorithms made it possible to determine the shell of the workspace of the RRRR mechanism and the minimum geometric parameters of the active parallel mechanism. The simulation results were checked by combining with the built-in additional workspace of the active mechanism, as well as on the developed three-dimensional model of the rehabilitation system.

References

1. Vashisht, N., Puliyel, J.: Polio programme: let us declare victory and move on. Indian J. Med. Ethics **9**(2), 114–117 (2012)
2. Behera, L., Rybak, L., Malyshev, D., Gaponenko, E.: Determination of workspaces and intersections of robot links in a multi-robotic system for trajectory planning. Appl. Sci. **11**(4961) (2021)
3. Evtushenko, Y.G., Posypkin, M.A., Rybak, L.A., Turkin, A.V.: Finding sets of solutions to systems of nonlinear inequalities. Comput. Math. Math. Phys. **57**(8), 1241–1247 (2017)
4. Evtushenko, Y.G., Posypkin, M.A., Rybak, L.A.: The non-uniform covering approach to manipulator workspace assessment. In: IEEE Conference of Russian Young Researchers in Electrical and Electronic Engineering, pp. 386–389. Moscow and St. Petersburg, Russia (2017)
5. Laryushkin, P., Glazunov, V., Erastova, K.: On the maximization of joint velocities and generalized reactions in the workspace and singularity analysis of parallel mechanisms. Robotica, **37**(4), 675–690 (2019)
6. Zeghloul, S., Laribi, M.A., Arsicault, M. (eds.): MEDER 2021. MMS, vol. 103. Springer, Cham (2021). https://doi.org/10.1007/978-3-030-75271-2
7. Evtushenko, Y.G., Posypkin, M.A.: Effective hull of a set and its approximation. Doklady Math. **90**(3), 791–794 (2014). https://doi.org/10.1134/S1064562414070278
8. Garanzha, V.A., Kudryavtseva, L.N.: Generation of three-dimensional Delaunay meshes from weakly structured and inconsistent data. Comput. Math. Math. Phys. **52**(3), 427–447 (2012)
9. Evtushenko, Y., Posypkin, M., Rybak, L., Turkin, A.: Approximating a solution set of nonlinear inequalities. J. Global Optim. **71**(1), 129–145 (2017). https://doi.org/10.1007/s10898-017-0576-z
10. Malyshev, D.I., Posypkin, M.A., Rybak, L.A., Usov, A.L.: Determination of the working area and singularity zones of the 3-RRR robot based on the non-uniform coverings method. Int. J. Open Inf. Technol. **6**(7), 15–20 (2018)
11. Evtushenko, Y.G., Posypkin, M.A., Rybak, L.A., Turkin, A.V.: Int. J. Open Inf. Technol. **4**(12), 1–6 (2016)
12. Jaulin, L., Kieffer, M., Didrit, O., Walter, E.: Applied Interval Analysis: With Examples in Parameter and State Estimation, Robust Control and Robotics. Springer London Ltd (2001)
13. Mitchell, M.: An Introduction to genetic algorithms. The MIT Press (1998)

14. Rocha, R., Ferreira, F., Dias, J.: Multi-robot complete exploration using hill climbing and topological recovery. In: IEEE International Conference on Intelligent Robots and Systems on Proceedings, pp. 1884–1889. Nice, France (2008)
15. Ahmetzhanov, M., Rybak, L., Malyshev, D., Mohan, S.: Determination of the workspace of the system based on the 3-PRRR mechanism for the lower limb rehabilitation. In: Beran, J., Bílek, M., Václavík, M., Žabka, P. (eds.) TMM 2020. MMS, vol. 85, pp. 193–203. Springer, Cham (2022). https://doi.org/10.1007/978-3-030-83594-1_20

Laboratory Analysis of the Social and Psychophysiological Aspects of the Behaviour of Participants in the Lemons Market Game

Olga Menshikova[1,2], Anna Sedush[1], Daria Polyudova[1], Rinat Yaminov[1,3](✉), and Ivan Menshikov[1,3]

[1] Moscow Institute of Physics and Technology, Moscow, Russia
[2] Russian Academy of National Economy and Public Administration, Moscow, Russia
[3] FRC CSC RAS, Moscow, Russia

Abstract. The pair game "Market of Lemons" belongs to the class of signal games. We are talking about the sale of used cars, the quality of which is known only to the seller, who chooses at what price, high or low, to sell the car to the buyer. The buyer can buy a car at the offered price, or he can refuse. The article examines three options for choosing roles and partners and two options for the likelihood of 'bad' cars. These games explore the existence of concealment, revealing, and hybrid perfect Bayesian equilibrium. It is proved that at the level of aggregated behaviour of participants, the experimental results are in good agreement with the concept of Quantal Response Equilibrium (QRE) in the case when random partners meet in each round. To explain the diversity in human behaviour, the psychological Sandra Bem test was used, which allows you to determine the psychological gender of a person. The participants were divided into four groups based on their biological and psychological gender. There was a significant difference in the behaviour of the representatives of these four groups.

Keywords: Experimental Economics · Game theory · Laboratory experiments · Lemon market game · Sandra Bem test

1 Introduction

Under conditions of asymmetric information, the actions of an economic agent acquire the character of a signal about the private information he has. A certain level of strategic thinking is required from the observer to analyse the signal. However, in this kind of socio-economic interactions there is also an ethical side: do you try to create with your signal a false impression of the available private information, trying to make money on it, or send only truthful signals?

Supported by the grant in RFBR 19-01-00296A.

N. N. Olenev et al. (Eds.): OPTIMA 2021, CCIS 1514, pp. 246–257, 2021.
https://doi.org/10.1007/978-3-030-92711-0_17

Successful fraud reduces the level of trust in society, which entails a decrease in the effectiveness of socio-economic interaction.

The ratio of strategic and ethical aspects of decision-making under conditions of asymmetric information can be studied by experimental and behavioural economics on the basis analysis of laboratory signalling games.

In the Laboratory of Experimental Economics of the Moscow Institute of Physics and Technology, a series of experiments was conducted to study the Lemons Market game ("The Market for Lemons" in [1]).

2 The Model and Its Game-Theoretic Analysis

The proposed new version of the classic game is as follows. Two play: the seller (S) and the buyer (B). The common information is that the car can be either good - P (Peach) or bad - L (Lemon). Well known: the probability of L ($0 < a < 1$), the value of the car for the buyer ($V_P = 12$, $V_L = 2$) and for the seller: ($C_P = 10$, $C_L = 1$).

Game scenario. Nature chooses the quality of the car: L with probability a or P with probability $1 - a$. The seller will know the choice of nature, i.e. he knows the type of car: L or P. The seller sets the price p for the car. The buyer sees the price p of the seller but does not know the type of car (P or L). The buyer may agree to the transaction ($Y = 1$) or refuse it ($Y = 0$).

Participants payoffs.

For the seller: $U_S = (p - C_t) \cdot Y$,

for the buyer $U_B = (V_t - p) \cdot Y$.

Here t is the type of car: L or P.

Thus, if there is no agreement, then the gain of both is 0.

The laboratory game consisted of 6 series of 20 independent attempts in accordance with the following choices for roles and partners.

- Option 1. Roles are fixed, partners are permanent.
- Option 2. Roles fixed, partners random.
- Option 3. Roles and partners change randomly.

Each option was played twice: 20 attempts with $a = 10\%$, then 20 attempts with $a = 25\%$. Profits were summed up on attempts and on all six series.

The considered game is a dynamic game with incomplete information. In such games, perfect Bayesian equilibrium (PBE) is investigated, which happens to be pooling, separating or hybrid.

The following statements are true for this game.

1. With any probability of Lemons $0 < a < 1$ in this game there is no "good" separating PBE, in which Peaches are sold.
2. When $a = 10\%$ there is a pooling PBE, when Lemons and Peaches are sold at the same price.
3. When $a = 25\%$ there is no pooling PBE, when Lemons and Peaches are sold at the same price.

4. For any $0 < a < 1$, there is a separating PBE when only Lemons are sold.
5. For any $0 < a < 1$, there is a hybrid PBE, in which the seller of Lemons uses a mixed strategy: with a certain probability to play "fair" (low price), and with an additional probability to try to deceive the buyer (high price).

3 Aggregate Behaviour in Laboratory Experiments

3.1 Quantal Response Equilibrium

In the Laboratory of Experimental Economics 7 experiments on this game were conducted. A total of 84 volunteers from among the students of MIPT took part in experiments.

The Quantal Response Equilibrium, (QRE, [2,3]) was chosen as the behavioural concept of equilibrium for the analysis of the experimental results. This concept is based on replacing the principle of the best Nash response to the given strategies of other players on the principle of the soft best response: the better the response, the greater the probability of its choice. The soft best response depends on the parameter λ that characterizes the player's rationality: with small λ any possible choice is equally likely to be made as an response, when $\lambda \to \infty$ Nash's best answer is obtained. The QRE concept provides ample opportunities for modelling the behaviour of people in game situations and is one of the key tools of experimental economics [3].

Result 1. *As a dynamically stable outcome with an increase in the rationality parameter ($\lambda \to \infty$), the QRE identifies the hybrid PBE that is most beneficial to the seller.*

Result 2. *QRE adequately describes the behaviour of the experiment participants only in scenarios 2 and 3, when partners change by chance.*

Detailed description of the main results.

We first analyse the case of a 25% probability of Lemon. At the aggregated level, it suffices to consider two price ranges:

- low prices (from 1 to 2) and
- high prices (from 10 to 12).

For the sake of simplicity, we will assume for the time being that in the game you can assign only one of two prices: small (1.5) or high (11). Such a game is a standard binary signal game.

In this game there is a separating PBE, when sellers are honest, but the buyer agrees to a deal only at a price of 1.5. It can be verified that this PBE is unstable to small deviations towards the dishonesty of the Lemons seller, and therefore the QRE cannot converge with this PBE with $\lambda \to \infty$.

In this signal game, there is no pooling PBE when sellers of Lemons and Peaches prescribe a price of 11, but there is a hybrid PBE.

In this hybrid PBE, the Lemons seller uses a mixed strategy: with a probability of 2/3, he honestly sets a price of 1.5, and with a probability of 1/3 he tries to deceive the buyer by setting a price of 11.

The buyer always agrees with the price of 1.5 and with a probability of 1/20 agrees with the price 11, i.e., his level of trust in the seller is %.

The seller of Peaches only needs to assign a price of 11.

By "honesty" we mean the probability of a price being set at 1.5 by the Lemons seller, and by "trust", the probability of the buyer's consent to the price of 11.

It is to this hybrid PBE that the QRE converges with $\lambda \to \infty$.

As the rationality parameter increases, QRE describes on the plane (honesty, trust) a certain curve that starts at $(0.5, 0.5)$, when $\lambda = 0$ it ends at $(2/3, 1/20)$, when $\lambda \to \infty$.

Figure 1 shows two curves: the left corresponds to the case when the rationality parameters of the seller and the buyer are the same, and the right one, when the rationality of the buyer is one and a half times higher than the seller of Lemons. At a substantive level, this can be explained by the fact that the buyer has to think more about whether to accept a high price, realizing the possibility of a big loss (9) from the purchase of Lemons. For the Lemon seller, the choice is easier: get a small but guaranteed profit (0.5) or take the risk of cheating with the possibility of encountering an incredulous buyer. From a formal point of view, the ratio of rationality of the buyer to the rationality of the seller at level 1.5 is most consistent with the results of our experiments.

To analyse the behaviour of participants in the experiment, we introduce the following indicators: honesty (calculated for sellers) - this is the proportion of Lemons that the seller offered to buy at a low price; trust (calculated for buyers) - the ratio of the number of offers accepted by the buyer to buy at a high price to the total number of offers to buy at a high price. These values were calculated in each session for each person.

In Fig. 1 aggregated by participants results are represented by dots of different colours, depending on the series. It can be seen that the series 6 points agree best with QRE (the roles and partners are random, and the participants are already the most experienced). Five of the seven points are located near QRE with minimal honesty and adequate trust. At two other points in this series, trust is about the same level, although honesty is noticeably higher than in QRE. Series 4 points (the roles are fixed, and random partners) are a little further from QRE, compared to the points of series 6. Farther from QRE, series 2 points (permanent roles and partners). This is understandable by trying to establish a relationship with a regular partner. In one experiment, the harmony of relationships in series 2 was almost achieved: honesty and trust are more than 80%. In another experiment, the participants got completely "dishonest": in series 2 and 4, there was not a single price fixation of 1.5 sellers of Lemons. But even in this experiment in series 6, honesty increased and the corresponding point on the plane (honesty, trust) approached QRE with minimal honesty. Apparently, the participants realized that with a low level of trust, it is sometimes better to get your 0.5 than with very small chances of chasing a big profit (10).

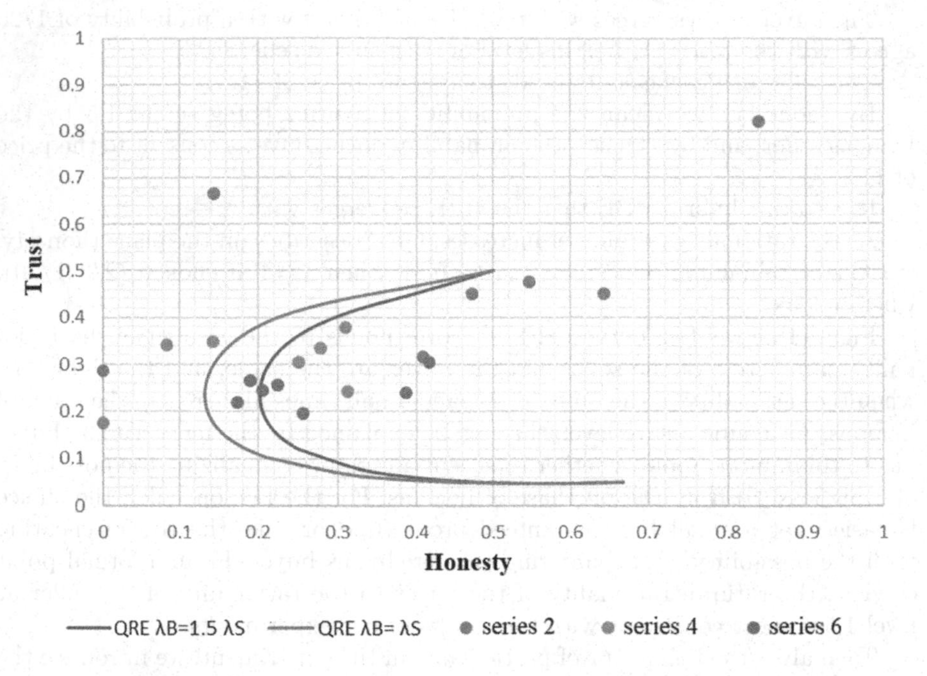

Fig. 1. QRE and results of experiments on the plane (honesty, trust).

The probability of a lemon is 10%. Theoretically, this case turns out to be somewhat more complicated than the 25% case. The fact is that the price $p = 11$ is in a certain sense the point of bifurcation of the model. Turn again to the binary signal game, in which the seller can assign only one of two prices: small (1.5) or large (p) [4].

If $10 < p < 11$, then in the signal game (except for the unstable separating PBE described above), there is a pooling PBE with zero honesty of sellers and full confidence of the buyers. This is due to the low probability of Lemons and the risk-neutrality of buyers, which we assume so far.

When $11 < p < 12$ the pooling PBE disappears, a hybrid PBE appears in mixed strategies.

When $p = 11$ there is an infinite number of PBE, in which the seller is completely dishonest, and the level of customer confidence is any, but not less than 5%. So, at a price of 11, switching from pooling PBE to hybrid occurs.

Let us dwell on the cases $p = 10.5$ and $p = 11.5$. When the high price is 10.5, the coordinates of the PBE on the plane (honesty, trust) are equal $(0, 1)$. As the Fig. 2 shows, it is to this point that QRE (10.5) converges with an increasing level of participants rationality. When the high price is 11.5 instead of pooling

QRE, a hybrid QRE arises with coordinates $(10/19, 1/21)$. QRE (11.5) tends to this point (Fig. 2) when the high price is 11.5.

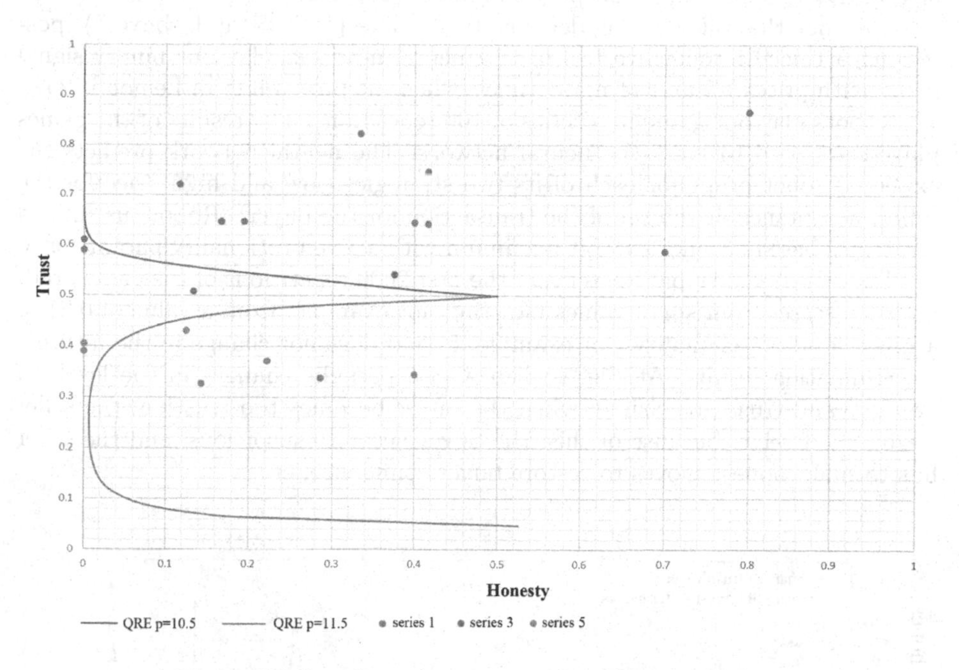

Fig. 2. QRE and experimental results for series 1, 3, 5.

In Fig. 2 dots of different colours represent series 1, 3 and 5. It can be seen that the results of series 1 (regular partners and roles) are not described using QRE, which means that a different model is needed for this case. Series 3 (permanent roles, but casual partners) is well aligned with the QRE branch (10.5). Series 5 (roles and random partners) is more consistent with the QRE branch (11.5).

To clarify these results, we plan to conduct additional control experiments, when the choice of sellers in advance is limited to only two prices.

The main conclusion. Figure 1 and 2 confirm that at the level of aggregated consideration of the behaviour of participants in experiments, variants with random partners are in good agreement with QRE (in series 3–6), which is basically based on the consideration of mixed strategies, as well as the hybrid PBE.

However, a natural question arises: do the participants in the experiments use at least implicitly mixed strategies? The answer to this question, as we shall see, is negative.

3.2 QRE and Reinforcement Learning

As it was shown earlier, series 1 and 2 (with permanent roles and partners) are practically not coordinated with QRE of the original static game seller –

buyer. You can, of course, consider the QRE of a multi-step game, in which the seller and the buyer interact several times. However, in this paper we used an alternative based on the concept of reinforced learning.

Consider the following model. Let two robots ("seller" and "buyer"), possessing artificial intelligence based on a neural network, play our binary signal game with prices 1.5 and 11 many times, when the probability of Lemon is 25%.

Robots play our game in a series of 100 repetitions. The results of the games played are used to train the neural network. The neural network predicts the expected robot profit for each of its two strategies (low and high price for the seller, acceptance or refusal of the transaction for the buyer). Robots use mixed strategies based on taking a soft maximum with a given rationality parameter λ.

The = rationality parameter λ of the first 2000 repetitions of the game gradually increases from small values (starting, for example, from a value of 0.2) to a given level of rationality (for example, 5) and does not change in the future.

Simulation results. At Fig. 3 when $\lambda = 5$ periodic changes in the levels of honesty and trust are visible: when the buyer becomes too trusting, the seller begins to deceive, because of this, the buyer becomes suspicious, and the seller has to make honest moves to restore buyer confidence, etc.

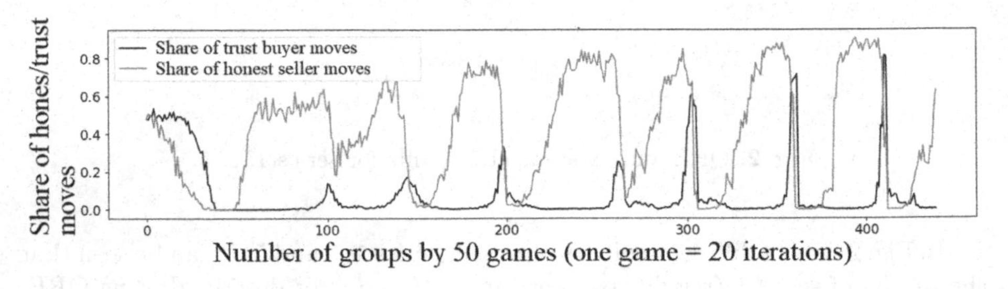

Fig. 3. Dynamics of honesty and trust in a model of learning with reinforcement, $\lambda = 5$.

With an increase in the rationality parameter, the oscillation period first shrinks, and then disappears completely. When the buyer remembers that the seller can cheat and no longer agrees to the transaction at a high price (Fig. 4).

As a result, there is an exit to the separating PBE when only Lemons are sold. Recall that using QRE to get this PBE does not work because of its instability.

If we introduce into the model the inequality aversion, then we can achieve other quasi-stable states of the model. Let the parameter δ is responsible for the seller's "remorse" when selling Lemon. We modify its utility function by subtracting the difference between the seller and the buyer payoffs, multiplied by δ.

In Fig. 5 shows the dynamics of honesty and trust with $\delta = 0.5$ and $\lambda = 5$. Note that without the amendment related to the inequality aversion, with this level of rationality, there is a periodicity, as in Fig. 3. Now honesty weakly

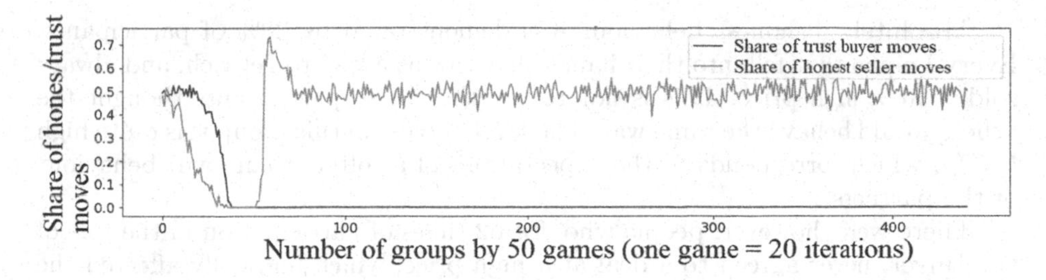

Fig. 4. Dynamics of honesty and trust in a model of learning with reinforcement, $\lambda = 20$.

fluctuates around the level of 0.7, and trust - around the level of 0.5. This mode is more like an approach to a hybrid PBE. In the future, it is planned to learn how to select the parameters of the learning model with reinforcements so as to obtain various behaviors of robots that are close to human behaviour.

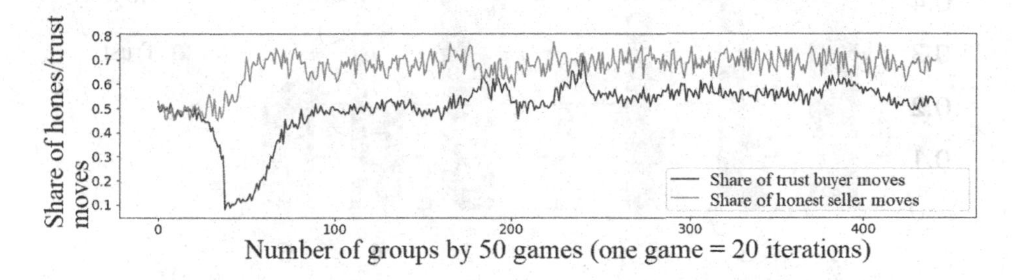

Fig. 5. Dynamics of honesty and trust with $\lambda = 5$, $\delta = 0.5$.

4 Analysis of Individual Behaviour: Psychophysiological Aspects

Each time we observed how the atmosphere of trust that existed in the first series was destroyed by the end of the game. The level of trust, which was measured by the percentage of cars sold at a high price, fell on average two times. In Fig. 6 shows how the levels of honesty and trust changed over the series, and, first, the series are depicted in which the probability of Lemon is 10%, and then the series with the probability of Lemon is 25%.

Only 10% of the participants were honest to the end and did not succumb to the temptation to sell Lemons at the price of Peach. These people not only never deceived themselves, but also believed that others would not do that. They demonstrated a high level of trust of 75% (they agreed to buy offers at a high price in the role of the Buyer), believing that the rest would behave honestly.

Absolutely dishonest behaviour was demonstrated by 28% of participants. Every Lemon that fell into their hands they saw as a way to get rich, and always sold it at a high price. This is not to say that these participants thought the others would behave the same way. The level of trust in this group was quite high (47%), which corresponds to the expectations of a softer (than own) behaviour of the partners.

There were, however, people who, for 62 times of participation in the role of the Buyer, never agreed to a deal at a high price, which markedly affected the decrease in the overall level of trust.

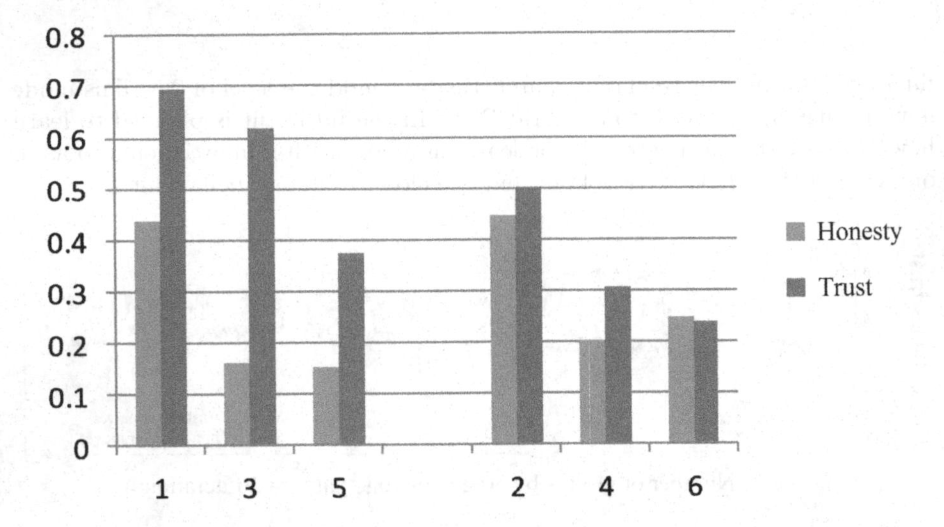

Fig. 6. Changing the level of honesty and trust depending on the series.

Psychological tests were used to explain the diversity in people's behaviour, including the Sandra Bem test [5]. Sandra Bem is an American psychologist who is known for her work on the study of gender roles. She owns a questionnaire, which measures the degree of manifestation in the behaviour of traditionally male or female traits. The questionnaire contains 60 characteristics; you need to answer whether the character trait is inherent or not. 20 qualities in the list reflect masculine traits (independence, assertiveness, dominance, risk appetite, independence, self-confidence, etc.), 20 - feminine (pliability, cordiality, empathy, etc.), 20 do not bear gender load. In the future, we will call feminine a person whose feminine qualities are more pronounced than masculine ones, otherwise we will speak about the masculinity of a person. Table 1 presents the designations of the groups, which will be discussed further.

In our seven experiments, the number of people who passed the Sandra Bem test was 74, and there were an equal number of men and women: 37 people. The feminine and masculine members also shared roughly in half. In Table 2 shows the distribution of participants in four gender groups.

Table 1. Conventions for considered gender groups.

	Masculine	Feminine
Men	mM	fM
Women	mW	fW

Table 2. Distribution of experiment participants by gender groups.

	Masculine	Feminine	People
Men	27%	23%	50%
Women	22%	28%	50%
People	49%	51%	

For each group, the average honesty and trust, the price per Lemon, the profit were calculated (Table 3).

Table 3. Honesty and trust by gender groups.

	Honesty	Trust	Av. price per Lemon	Payoff
mM	0.20	0.37	9.08	55.80
fM	0.37	0.39	7.55	54.03
mW	0.36	0.61	7.58	43.42
fW	0.27	0.50	8.44	52.80

From Table 3 the following main conclusions follow (differences between all groups are statistically significant, $p - value < 0.001$, Wilcoxon-test):

1. The members of the mM group have the least honesty.
2. Great gullibility is inherent in women, especially from the mW group.
3. The lowest average price for a lemon among participants from the mW and fM groups.
4. Men gain more than women. The greatest gain is in the mM group, the smallest in the mW group.

If we consider each of the six series separately, then in all cases, without exception, mW have the greatest gullibility. In four cases, mM has the minimum honesty.

For series 6 difference in the behaviour of gender groups is very pronounced if you look at the share of participants with zero honesty who tried to realize every opportunity to sell Lemon at a high price (Table 4).

Table 4. Participation shares with zero fairness by gender group in series 6.

	mM	fM	mW	fW
Group size	20	17	16	21
Number of totally dishonest	15	9	6	15
The share of totally dishonest players in the gender group	75%	53%	38%	71%

Masculine men turned out to be champions of dishonesty, 75% of whom did not make a single honest move (here you have no mixed strategies). Out of dishonesty, feminine women almost catch up with them (71%). It turns out that gender poles converge here. The minimal share of participants with zero honesty was in the group of masculine women (38%), and this share was almost two times less than in the group of mM.

Predicting the outcome of the game on the distribution of gender types.

All participants in the experiments pass psychological tests before the game. Knowing how many representatives of each of the four gender groups participate in the experiment, one can predict the average honesty and trust of the results of the laboratory game. In the group consisting mainly of masculine men, there will be low honesty and gullibility, and in the group of masculine women both indicators will be significantly higher.

Table 5. Correlation of observed honesty and trust with a share in the group of participants, determined by gender type.

	mM	fM	mW	fW
Honesty	−0.56	0.48	0.29	−0.15
Trust	0.00	0.55	−0.10	−0.54

From Table 5 shows that psychological gender is important in predicting behaviour in our game. In the future, based on the results of preliminary testing, it is assumed to recruit groups of a certain gender composition and to forecast the outcome of the experiment in advance.

The situation is familiar to us. We considered the behaviour of the representatives of the four groups in such games as the Prisoner's Dilemma, the Ultimate Game, The Trust Game, Social dilemma with a common risk. And in all these games, the most prosocial behaviour, which is characterized by cooperativity, fair profit distribution, trust, gratitude, and social responsibility, is significantly high in representatives of two groups: fM and mW.

The four main gender groups also differ in psychophysiological characteristics. The participants in the experiments were seated on stabilographic chairs (see [6, 7]), which are equipped with the Laboratory of Experimental Economics of the Moscow Institute of Physics and Technology. Stabilography makes it possible to identify such a systemic indicator of a person's functional state as the quality of the balance function.

Table 6 shows the main characteristics of the functional state: energy and entropy [6]. These characteristics were measured at rest before the start of the experiment in two versions: eyes open (EO) and eyes closed (EC).

We see that all four gender groups have their own specific features in terms of functional state:

1. Masculine men have the highest energy and lowest entropy with both open and closed eyes.
2. Feminine men have low energy but maximum entropy with their eyes closed.
3. Masculine women have low energy, high entropy with open eyes and low entropy with open eyes.
4. Feminine women have low energy and high entropy.

Table 6. Energy and entropy at rest by gender.

	Energy		Entropy	
	EO	EC	EO	EC
mM	0.04	0.11	9.89	9.79
fM	0.02	0.01	10.05	10.17
mW	0.01	0.02	10.20	9.91
fW	0.01	0.01	10.15	10.14

The individual characteristics of the behaviour of participants in experiments are, of course, determined by many factors. One of the main factors is the biological and psychological gender of the participant.

References

1. Akerlof, G.: The Market for 'Lemons': quality uncertainty and the market mechanism. Q. J. Econ. **84**(3), 488–500 (1970)
2. McKelvey, R.D., Palfrey, T.R.: Quantal response equilibria for extensive form games. Exp. Econ. **1**(1), 9–41 (1998)
3. Goeree, J.K., Holt, C.A., Palfrey, T.P.: Quantal Response Equilibrium. A Stochastic Theory of Games. Princeton University Press, Princeton (2016)
4. Menshikov, I.S.: Lectures on Game Theory and Economic Modelling. 2nd edn. LLC Contact Plus (2010). (in Russian)
5. Bem, S.L.: The measurement of phychological androgyny. J. Consult. Clin. Psychol. **42**(2), 155–162 (1974)
6. Lukyanov, V.I., Maksakova, O.A., Menshikov, I.S., Menshikova, O.R., Chaban, A.N.: Functional state and efficiency of participants in laboratory markets. Proc. Russian Acad. Sci. Theory Control Syst. **6**, 202–219 (2007)
7. Menshikov, I.S.: Analysis of the functional state of participants in laboratory markets. Psychol. J. Higher School Econ. **6**, 125–152 (2009)

Coverage Path Planning for 3D Terrain with Constraints on Trajectory Curvature Based on Second-Order Cone Programming

Timofey Tormagov[1,3](✉) and Lev Rapoport[2,3,4](✉)

[1] Skolkovo Institute of Science and Technology, Moscow, Russia
[2] Topcon Positioning Systems, Moscow, Russia
[3] Moscow Institute of Physics and Technology, Moscow, Russia
tormagov@phystech.edu
[4] V. A. Trapeznikov Institute of Control Sciences of RAS, Moscow, Russia

Abstract. In precision agriculture, the optimal construction of a complete coverage for an uneven field with parallel paths is often considered. As a rule, a coverage can be defined by some initial path relative to which the whole coverage is constructed. For various purposes, such as minimizing the number of turnarounds, some part of the field boundary can be selected as the starting path. The planned paths are used by agricultural machines to perform their job. Assume that a machine following the path has a so called Ackermann steering mechanism. In this case, the vehicle can only use paths whose curvature does not exceed the specified threshold value. For this reason, some paths may be infeasible, which leads to gaps areas in the field's coverage. In this paper we consider the path planning method that uses some specified swaths' overlap for exclusion of gaps areas (unprocessed areas). To approximate the paths, we use uniform cubic B-spline curves. Each elementary spline of this curve is defined by four control points, which are calculated by the algorithm. The target path always lies inside the convex hull of the four adjacent control points. Our method involves imposing conditions on an estimation of a path curvature, which can be represented as second-order cone constraints. In addition, the curvature estimation is included in the optimization criterion, which allows one to straighten the paths when constructing them. As a result, the task of constructing a neighboring path can be represented as a Second-Order Cone Programming problem (SOCP). The algorithm has been tested on a data taken in real fields.

Keywords: Precision agriculture · Coverage path planning · Curvature constraints · Second-order cone programming · SOCP

1 Introduction

Complete Coverage Path Planning problem consists in building such paths (we also use words "routes" and "trajectories") that completely cover some surface.

N. N. Olenev et al. (Eds.): OPTIMA 2021, CCIS 1514, pp. 258–272, 2021.
https://doi.org/10.1007/978-3-030-92711-0_18

In robotics, this task is often applied to precision farming [11,12], object survey by UAV [3], and floor cleaning [4]. In this paper, we will consider the path planning as applied to precision agriculture jobs such as seeding, spraying and lawn mowing. In these cases, the problem is to cover a given area of the field with paths which completely eliminate gaps and minimize an overlapping area. The connection of coverage path planning with the problem of vehicle control and satellite navigation that arise in precision agriculture we consider in our paper [17].

In the work [11] a rapid 2D path planning algorithm is proposed operating as follows. At the first stage the azimuth angle and a reference point in the field are selected. Then the construction of straight paths is carried out. The first path passes through the selected point in selected azimuthal direction. Next, we build the remaining paths parallel to the first one. In literature this method of constructing paths is often called "a boustrophedon" [1] due to its similarity to the corresponding writing technique. The resulting coverage can be estimated using various cost functions. For example, a number of turns at the field boundary can be taken as a cost function. Actually, the target function can take into account factors such as soil erosion, lengths of the turn, and curvature of trajectories, see [12,13]. By changing the initial path, we can obtain the coverage with better value of the selected cost function. However, in 3D case we cannot simply plan the path on the horizontal plane, and then project it on the surface. On an uneven surface, such path planning leads to gaps and overlaps. In the paper [11] a scheme with a construction of cylinders around a trajectory is used in neighboring path design to eliminate this effect. The problem of constructing parallel curves on a parametrically given surface is considered in the work [10], however, curvature constraints are not taken into account there. In our paper we give a rigorous mathematical formulation of the optimal path planning problem on an uneven surface with curvature constraints and least possible overlap between adjacent paths.

After complete coverage problem is solved, we need to schedule routes for vehicles. In many applications, the field is processed by multiple vehicles simultaneously, especially if processing time is an issue. Each path must be covered by one and only one of the vehicles. Each vehicle must be told when and where to go in order to perform a collective and optimal work. As a rule, the task of planning a route is reduced to the one of various vehicle routing problem (VRP). For example, tasks with a resource that is consumed when processing a field, such as spraying, can be represented as a green vehicle routing problem (G-VRP) and electric vehicle routing problem (EVRP). These are routing tasks for machines running on alternative energy sources that require frequent recharging or refueling. The possibilities of solving such problems are discussed, for example, in the papers [7,14]. Various versions of these problems are compared in the survey [6]. In the works [2,20], the routing problem is considered in application to precision agriculture. In these works the simulated annealing metaheuristic algorithm is used to approximate the solution of the vehicle routing problem.

In our paper we consider the complete coverage path planning problem for a vehicle with Ackermann steering mechanism. In this case, we must require that the normal curvature of the target trajectories does not exceed the specified value, which is determined by the maximum possible angle of front wheels rotation and the distance between the front and rear axles. Moreover, large normal curvature values may lead to the "swallowtail" effect, which leads to possible self-intersection of neighboring paths, see [10,18]. We impose constraints on the curvature when constructing a neighboring path. Assume that the gaps in the coverage are prohibited. This condition is true for mowing golf courses. We eliminate skips using small neighboring path overlaps. The problem of optimizing the necessary overlap can be represented as a second order cone problem. In final stage we design routes for vehicles using a simulated annealing algorithm. This part, not reflected in the paper title, is added for completeness.

The paper is organized as follows. The proposed method is introduced in Sect. 2. Subsection 2.1 describes the method used for setting paths and surfaces. Subsection 2.2 introduces our method of neighboring swath calculation which is supplemented by the curvature constraints represented in Subsects. 2.3 and 2.4. In Subsect. 2.5 we introduce a SOCP problem that allows us to calculate a complete coverage for a field. In order to implement these paths, it is necessary to solve the route planning problem, the algorithm for solving which using simulated annealing is presented in Subsect. 2.6. Sections 3 contains a description of the results of an experimental study of the obtained algorithm. Finally, Sect. 4 presents discussion and conclusions.

2 Methods and Results

In this section we include all necessary auxiliary constructions and present the main result.

2.1 Surface and Routes Representation

We use a local Cartesian coordinate system with reference to a certain point of the field. The x-axes and the y-axes are directed East and North respectively. The z-axis coordinate displays the height relative to the reference point. This coordinate systems is called ENU (East, North and Up). When paths planning is performed for a three-dimensional surface, it is necessary to define the height of the point located on the surface, provided (x, y) are known. In other words, a digital elevation model (DEM) must be defined. The heights for some points on a field must be known. They can be obtained, for example, by using navigation measurements taken while the vehicle is driving through the field. Using this data, we need to interpolate a height $z(x, y)$ for an point (x, y). In this work we use bivariate B-spline interpolation for the elevations of surface points.

To define a movement trajectory, we use homogeneous cubic B-splines. A movement trajectory $r(t) \in R^3$ can be fully defined by a set of control points $r_1, r_2, \dots, r_n \in R^3$. It consists of elementary B-splines $r^{(i)}(t) \in R^3$ that can be represented by four neighboring control points $r_{i-1}, r_i, r_{i+1}, r_{i+2} \in R^3$ as follows

$$r^{(i)}(t) = b_0(t)r_{i-1} + b_1(t)r_i + b_2(t)r_{i+1} + b_3(t)r_{i+2}, \tag{1}$$

where t is the spline parameter, $0 \le t \le 1$,

$$\begin{array}{ll} b_0(t) = \frac{(1-t)^3}{6}, & b_1(t) = \frac{4-6t^2+3t^3}{6}, \\ b_2(t) = \frac{1+3t+3t^2-3t^3}{6}, & b_3(t) = \frac{t^3}{6}. \end{array} \tag{2}$$

The set of control points should be supplemented with additional points r_{-1}, r_0, r_{n+1}, r_{n+2} such that $r_{-1} = 3r_1 - 2r_2$, $r_0 = 2r_1 - r_2$, $r_{n+1} = 2r_n - r_{n-1}$, $r_{n+2} = 3r_n - 2r_{n-1}$, see [16]. As a result, the trajectory is C^2-smooth and belongs to the union of $n-1$ convex hulls, each i-th hull being generated by four control points r_{i+2}, r_{i-1}, r_i, and r_{i+1}. For the two-dimensional case ($r(t) \in R^2$) the formulas look similar.

In the case of a movement on an uneven surface by a vehicle with the Ackermann mechanism, the curvature of the trajectory in the tangent plane u, which corresponds to the normal curvature of the curve, must be limited [9]:

$$\|u\| \le \frac{\tan \alpha_{\max}}{L} \tag{3}$$

where $\alpha_{\max}$ is maximum steering angle of the vehicle and L is the distance between the front and rear axles of the vehicle. Define $u_{\max}$ as $u_{\max} = \tan \alpha_{\max}/L$.

2.2 Neighbor Path Design

Suppose we need to calculate the control points of a neighboring boustrophedon path. By $\bar{d}$ we denote the maximum swath width. We require the distance between two neighboring paths to be not less than $\gamma \bar{d}$, $0 < \gamma \le 1$, and not exceeding $\bar{d}$. Thus we allow the overlapping of adjacent paths not larger than $100 \times (1 - \gamma)$ percent of the path width. We assume that the surface can be approximated by an inclined plane within a neighborhood of the size $\bar{d}$.

Let $r(t)$ be the initial swath on the field and $\widetilde{r}(t)$ is a neighboring swath. Let $r_1, r_2, \ldots, r_n \in R^3$ be equidistant points on the initial swath, $d_s = \|r_{i-1} - r_i\| \approx \|r_i - r_{i+1}\|$. Let x_i, y_i, z_i be entries of the vector r_i: $r_i = (x_i, y_i, z_i)$. Let define the normal vector to the spline curve $N_i^{xy} = (N_i^x, N_i^y)$ in the (x, y) plane at the point r_i such that $\|N_i^{xy}\| = \bar{d}$. Define the component N_i^z as $N_i^z = z(x_i + N_i^x, y_i + N_i^y) - z(x_i, y_i)$. Let d_i, $i = 1, 2, \ldots, n$ be a variable. We select a control point $\widetilde{r}_i$, $i = 1, 2, \ldots, n$, of the neighboring spline trajectory on the vector $N_i = (N_i^x, N_i^y, N_i^z)$ such that $\widetilde{r}_i = r_i + d_i N_i$ and $\gamma \bar{d} \le d_i \|N_i\| \le \bar{d}$. Figure 1 shows the neighbouring swath construction. Since a spline may not pass through control points, the distance between the beginning of an elementary spline $r^{(i)}(0)$ and the corresponding control points r_i must be estimated. In the case of the uniform setting of control points, the maximum error in the approximation of the spline curve can be estimated as $d_s^2 k_{\max}/6$, where $k_{\max}$ is the maximum curvature of trajectory (see [8,16]).

2.3 2D Trajectory Curvature Constraints

Let us first consider the two-dimensional field in the plane (x, y). For components of vector r_i we use the notation x_i, y_i such that $r_i = (x_i, y_i)$. The following auxiliary proposition takes place.

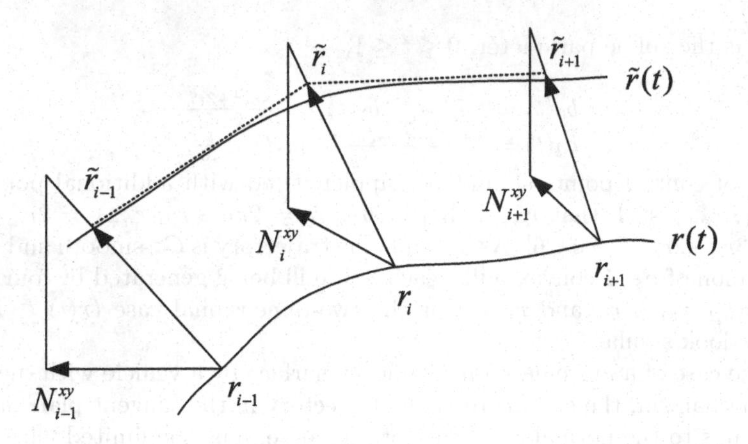

Fig. 1. The neighboring swath construction. The swath $\tilde{r}(t)$ is contracted using the spline trajectory $r(t)$. Dotted line shows the polyline passing through the control points of the new spline.

Proposition 1. *For an elementary cubic B-spline* $r^{(i)}(t)$ *the maximum value of* $\|r^{(i)\prime\prime}(t)\|$, $0 \leq t \leq 1$, *reaches at* $t = 0$ *or at* $t = 1$.

Proof. Since (1) and (2), it follows that

$$r^{(i)\prime\prime}(t) = (r_{i-1} - 2r_i + r_{i+1}) + t\,(-r_{i-1} - 2r_i + r_{i+1} + r_{i+2})\,. \tag{4}$$

Define the constants A, B, C, and D as

$$A = x_{i-1} - 2x_i + x_{i+1}, \tag{5}$$

$$B = -x_{i-1} - 2x_i + x_{i+1} + x_{i+2}, \tag{6}$$

$$C = y_{i-1} - 2y_i + y_{i+1}, \tag{7}$$

$$D = -y_{i-1} - 2y_i + y_{i+1} + y_{i+2}. \tag{8}$$

Using (4), (5), (6), (7), and (8), we obtain

$$\|r^{(i)\prime\prime}(t)\|^2 = t^2\left(B^2 + D^2\right) + 2t(AB + CD) + A^2 + B^2. \tag{9}$$

It means that $\|r^{(i)\prime\prime}(t)\|^2$ is convex function (convex parabola or line). Then the statement of the proposition is straightforward.

□

A norm of a trajectory curvature vector $k^{(i)}(t)$ at the point $r^{(i)}(t) \in R^2$ can be calculated as

$$\|k^{(i)}(t)\| = \frac{\|r^{(i)\prime\prime}(t)\| \sin\varphi(t)}{\left\|r^{(i)\prime}(t)\right\|^2}, \tag{10}$$

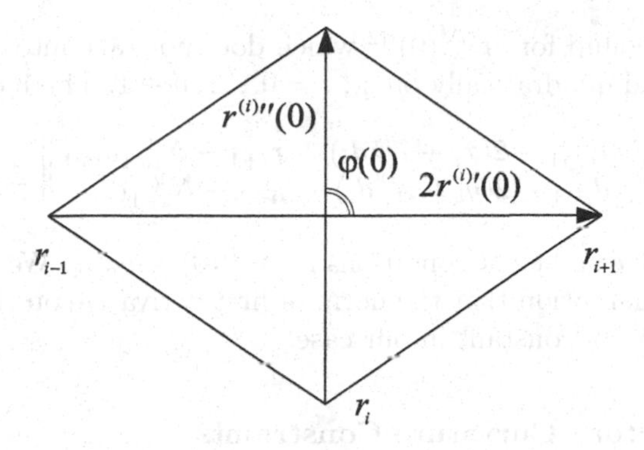

Fig. 2. Equally spaced control points of the B-spline in 2D case and first order derivative of trajectory for $t = 0$.

where $\varphi(t)$ is the acute (or right) angle between the vectors $r^{(i)\prime}(t)$ and $r^{(i)\prime\prime}(t)$. In the case of motion on a two-dimensional surface, the curvature of the B-spline $k^{(i)}(t)$ coincides with the curvature in the tangent plane $u^{(i)}(t)$.

Let us consider the case of $t = 0$ and equally spaced control points $d_s = \|r_{i-1} - r_i\| = \|r_i - r_{i+1}\|$, see Fig. 2. Since the vectors $r_{i-i} - r_i$ and $r_{i+1} - r_i$ form a rhombus, it follows that $\varphi(0) = \pi/2$. Using (1), we obtain

$$r^{(i)\prime}(0) = -0.5r_{i-1} + 0.5r_{i+1} \tag{11}$$

$$r^{(i)\prime\prime}(0) = (r_{i-1} - r_i) + (r_{i+1} - r_i) \tag{12}$$

Using the triangle inequality $\|r_{i+1} - r_{i-1}\| \leq 2d_s$ and (11), we obtain

$$\|r^{(i)\prime}(0)\|^2 \leq d_s^2 \tag{13}$$

Since (10) and (13), it follows that inequalities

$$\left\| \begin{matrix} x_{i-1} + N_{i-1}^x d_{i-1} - 2(x_i + N_i^x d_i) + x_{i+1} + N_{i+1}^x d_{i+1} \\ y_{i-1} + N_{i-1}^y d_{i-1} - 2(y_i + N_i^y d_i) + y_{i+1} + N_{i+1}^y d_{i+1} \end{matrix} \right\| \leq u_{\max} {d_s}^2 \tag{14}$$

for $i = 1, \ldots, n$ give a necessary condition for feasibility of path. Define q_i as $q_i = 1 - d_i$, $i = 0, \ldots, n+1$ and define $\hat{l}_i$, $i = 1, \ldots, n$ as

$$\begin{aligned} 4\hat{l}_i = &\left(x_{i+1} + N_{i+1}^x - x_{i-1} - N_{i-1}^x\right)^2 \\ &+2\left(x_{i+1} + N_{i+1}^x - x_{i-1} - N_{i-1}^x\right)\left(N_{i-1}^x q_{i-1} - N_{i+1}^x q_{i+1}\right) \\ &+\left(y_{i+1} + N_{i+1}^y - y_{i-1} - N_{i-1}^y\right)^2 \\ &+2\left(y_{i+1} + N_{i+1}^y - y_{i-1} - N_{i-1}^y\right)\left(N_{i-1}^y q_{i-1} - N_{i+1}^y q_{i+1}\right). \end{aligned} \tag{15}$$

We obtain

$$\|r^{(i)\prime}(0)\|^2 - \hat{l}_i = \frac{1}{4}\left(\left(N_{i-1}^x q_{i-1} - N_{i+1}^x q_{i+1}\right)^2 + \left(N_{i-1}^y q_{i-1} - N_{i+1}^y q_{i+1}\right)^2\right) \tag{16}$$

Then $\hat{l}_i$ is lower bound for $||r^{(i)'}(0)||^2$ which does not take into account only the terms that depend quadratically on q_i, $i = 0, \ldots, n+1$. The inequalities

$$\left\| \begin{matrix} x_{i-1} + N^x_{i-1} d_{i-1} - 2(x_i + N^x_i d_i) + x_{i+1} + N^x_{i+1} d_{i+1} \\ y_{i-1} + N^y_{i-1} d_{i-1} - 2(y_i + N^y_i d_i) + y_{i+1} + N^y_{i+1} d_{i+1} \end{matrix} \right\| \leq u_{\max} \hat{l}_i \tag{17}$$

for $i = 1, \ldots, n$ give sufficient conditions for $u^{(i)}(0) \leq u_{\max}$. We use the Proposition 1 and the assumption that the norm of first derivative on the spline parameter is approximately constant in our case.

2.4 3D Trajectory Curvature Constraints

In the case of motion over a three-dimensional surface, we need to take into account the constraint on the curvature of the trajectory in the tangent plane (3). Denote $r'(t)$ such as $r(t) = r_i$ by r'_i. As an approximation of the tangent plane at the point $r_i + d_i N_i$ we choose the plane containing the vectors N_i and r'_i. In the inequalities (14) and (17) instead of the coordinates $(x,\ y)$ in the local horizon plane we use the coordinates $(\widetilde{x},\ \widetilde{y})$ in the local horizon plane. The axes $\widetilde{x}$ and $\widetilde{y}$ in each tangent plane are directed along the vectors r'_i and N_i respectively. Let $\langle \cdot, \cdot \rangle$ be a scalar product of two vectors. Similarly to conditions (14) and (17) we have

$$\left\| \begin{matrix} \widetilde{x}_{i-1} + \widetilde{N}^x_{i-1} d_{i-1} + \widetilde{x}_{i+1} + \widetilde{N}^x_{i+1} d_{i+1} \\ \widetilde{y}_{i-1} + \widetilde{N}^y_{i-1} d_{i-1} - 2(\|N_i\| d_i) + \widetilde{y}_{i+1} + \widetilde{N}^y_{i+1} d_{i+1} \end{matrix} \right\| \leq u_{\max} d_s^2, \tag{18}$$

$$\left\| \begin{matrix} \widetilde{x}_{i-1} + \widetilde{N}^x_{i-1} d_{i-1} + \widetilde{x}_{i+1} + \widetilde{N}^x_{i+1} d_{i+1} \\ \widetilde{y}_{i-1} + \widetilde{N}^y_{i-1} d_{i-1} - 2(\|N_i\| d_i) + \widetilde{y}_{i+1} + \widetilde{N}^y_{i+1} d_{i+1} \end{matrix} \right\| \leq u_{\max} \widetilde{l}_i \tag{19}$$

where

$$(\widetilde{x}_{i-1},\ \widetilde{y}_{i-1}) = \left(\langle r_{i-1} - r_i, r'_i \rangle / \|r'_i\|,\ \langle r_{i-i} - r_i, N_i \rangle / \|N_i\|\right), \tag{20}$$

$$(\widetilde{x}_{i+1},\ \widetilde{y}_{i+1}) = \left(\langle r_{i+1} - r_i, r'_i \rangle / \|r'_i\|,\ \langle r_{i+1} - r_i, N_i \rangle / \|N_i\|\right), \tag{21}$$

$$\left(\widetilde{N}^x_{i-1},\ \widetilde{N}^y_{i-1}\right) = \left(\langle N_{i-1} - r_i, r'_i \rangle / \|r'_i\|,\ \langle N_{i-1} - r_i, N_i \rangle / \|N_i\|\right), \tag{22}$$

$$\left(\widetilde{N}^x_{i+1},\ \widetilde{N}^y_{i+1}\right) = \left(\langle N_{i+1} - r_i, r'_i \rangle / \|r'_i\|,\ \langle N_{i+1} - r_i, N_i \rangle / \|N_i\|\right), \tag{23}$$

$$\begin{aligned} 4\widetilde{l}_i = &\left(\widetilde{x}_{i+1} + \widetilde{N}^x_{i+1} - \widetilde{x}_{i-1} - \widetilde{N}^x_{i-1}\right)^2 \\ &+2\left(\widetilde{x}_{i+1} + \widetilde{N}^x_{i+1} - \widetilde{x}_{i-1} - \widetilde{N}^x_{i-1}\right)\left(\widetilde{N}^x_{i-1}(1 - d_{i-1}) - \widetilde{N}^x_{i+1}(1 - d_{i+1})\right) \\ &+\left(\widetilde{y}_{i+1} + \widetilde{N}^y_{i+1} - \widetilde{y}_{i-1} - \widetilde{N}^y_{i-1}\right)^2 \\ &+2\left(\widetilde{y}_{i+1} + \widetilde{N}^y_{i+1} - \widetilde{y}_{i-1} - \widetilde{N}^y_{i-1}\right)\left(\widetilde{N}^y_{i-1}(1 - d_{i-1}) - \widetilde{N}^y_{i+1}(1 - d_{i-1})\right). \end{aligned} \tag{24}$$

2.5 The SOCP Formulation of the Optimal Path Planning

Generation of adjacent paths with trajectory smoothing can be formulated as solution to the SOCP problem

$$\min_{u,d_0,d_1,\ldots d_{n+1}} \beta \|u\| - \sum_{i=0}^{n+1} d_i, \tag{25}$$

$u = (u_1, u_2, \ldots, u_n)$, subjected to constraints

$$\gamma\bar{d} \leq d_i\|N_i\| \leq \bar{d}$$

and

$$\left\| \begin{matrix} \widetilde{x}_{i-1} + \widetilde{N}^x_{i-1}d_{i-1} + \widetilde{x}_{i+1} + \widetilde{N}^x_{i+1}d_{i+1} \\ \widetilde{y}_{i-1} + \widetilde{N}^y_{i-1}d_{i-1} - 2(\|N_i\|d_i) + \widetilde{y}_{i+1} + \widetilde{N}^y_{i+1}d_{i+1} \end{matrix} \right\| \leq u_i d_s^2, \tag{26}$$

(19) where $\widetilde{l}_i$ is defined by (24) and $i = 1, \ldots, n$.

The dimensionless penalty parameter β ($\beta \geq 0$) is responsible for the tradeoff between the paths straightening effect and the overlapping of swaths in the cost function of the optimization problem. After the path is obtained, it is also subjected to re-sampling and extending to the border of the field. The new path is constructed relative to it. The path construction procedure stops when the entire field becomes covered. After that, the paths on the border of the field are cut off.

To solve the SOCP problem, various algorithms can be used, such as [5,15]. However, with certain input data, the formulated SOCP problem may not have feasible solutions. This means that the curvature constraint is violated and we need to choose different β and γ parameters or choose another initial path.

2.6 Route Planning

After the field coverage is constructed, each path must be assigned to one of the multiple vehicles. So, we came to the vehicle route planning problem. Sequential paths, traversed by the same vehicle must be connected by turns. For this purpose, U-turn and Ω-turn near the field boundary are often used, about other maneuvers see [19]. We will be looking for the routes for which the completion time of all machines on the field is minimal possible. Route planning for our field can be represented as vehicle routing problem (VRP). VRP problem is NP-hard, so we need a heuristic method for its solution. In our work, we used simulated annealing (SA) method for two identical vehicles.

The solution to the problem of route planning can be represented as a permutation of the vector p elements that sets the order of swath passage. The example of feasible route planning solution is illustrated in Fig. 3. Let us denote the number of swaths obtained in coverage path planning by M. The value $E(p)$ of the objective function (energy in SA) will correspond to the time from the beginning to the end of all vehicles operation. We assume the constant and identical speed of movement for two vehicles. The SA algorithm for route planning can be written as follows.

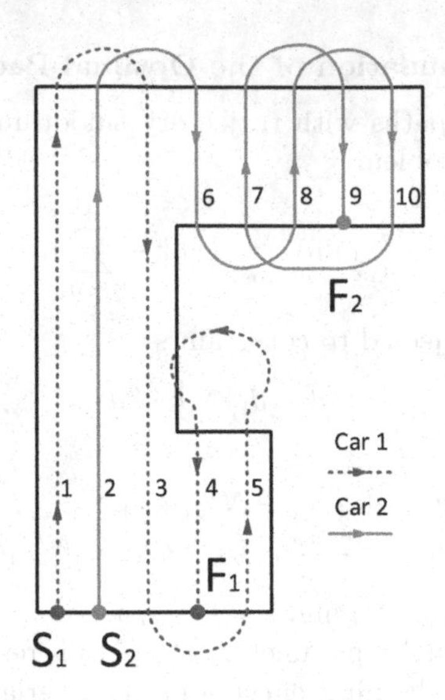

Fig. 3. The example of route planning for two vehicles feasible solution. The solution vector is p = (1, 3, 5, 4, 2, 6, 8, 10, 7, 9). The separator position is $s = 5$. $M = 10$. Switching from row 5 to row 4 is realized using Ω-turn and other switching are implemented using U-turns. Points S_1, S_2 and F_1, F_2 are start and finish points for vehicles 1 and 2 respectively.

1. Set initial solution vector p_0, for example, $p_0 = (1, 2, \ldots, M)$, and initial temperature $T = T_0$ and minimum temperature $T_{\min}$. Determine the matrix of transfer distances between all the starts and all ends of the swaths. Calculate energy for initial statement $E(p_0)$. Choose $iter := 1$.
2. Implement the following SA loop.
   (a) Choose randomly i and j, $i < j$ and obtain new solution $\widetilde{p}$ by inversion the order of vector p_0 components between i and j.
   (b) If $E(p_0) > E(\widetilde{p})$ choose $p_0 = \widetilde{p}$, else choose $p_0 = \widetilde{p}$ with probability $P = \exp(-(E(\widetilde{p}) - E(p_0))/T)$.
   (c) Choose $iter := iter + 1$, $T := 0.1T_{\min}/iter$. If $T \leq T_{\min}$ then exit from the SA loop.
3. Choose the solution with minimum energy that obtained in the SA loop.

We calculate the solution energy $E(p)$ for two vehicles as follows.

1. Choose the initial separator position $s = \lfloor M/2 \rfloor$.
2. Calculate minimum route distances D_1 and D_2 for two cars that can be represented by first s components and last $M - s$ components of vector p.
3. If $D_1 < D_2$ then choose $s := s - 1$ else choose $s := s + 1$.
4. If the previous iteration changed s in the other direction, then choose the solution with the minimum value of $\max(D_1, D_2)$, else go to the step 2.

3 Examples

In this section we present examples illustrating the method described in the previous section. The example 1 illustrates the field experiment and the examples 2, 3, and 4 show the results of model experiments. Results of both paths planning and vehicle routing problems are included for every example.

3.1 Example 1

The first example is pretty simple. We have included it as we used this field in our field experiments with three electric vehicles near the university campus. The surface elevation model is based on the of GNSS land surveying using carrier phase differential (RTK) measurement mode. A section of the field boundary was used as the initial path. The track width is $\bar{d} = 1$ and the minimum turning radius of vehicles is 3 m. We illustrate the resulting full paths coverage and three vehicle routes in the Fig. 4. The picture taken from the quadcopter during the test is shown in the Fig. 5. The lengths of routes are $D_1 = 105.5$ m, $D_2 = 106.6$ m, and $D_3 = 105.1$ m. So, difference between longest and shortest routes is 1.5 m which is 1.4% of the longest route length.

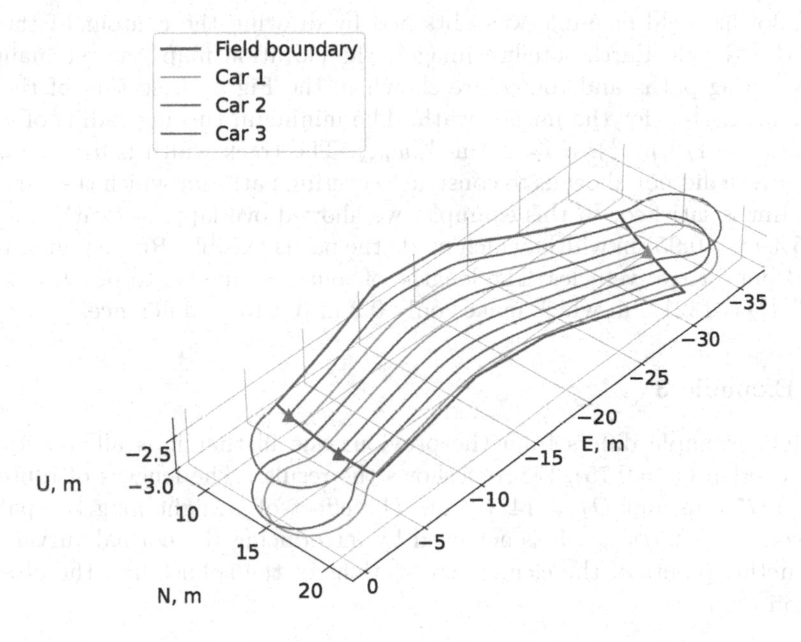

Fig. 4. Example 1. The map of coverage routes.

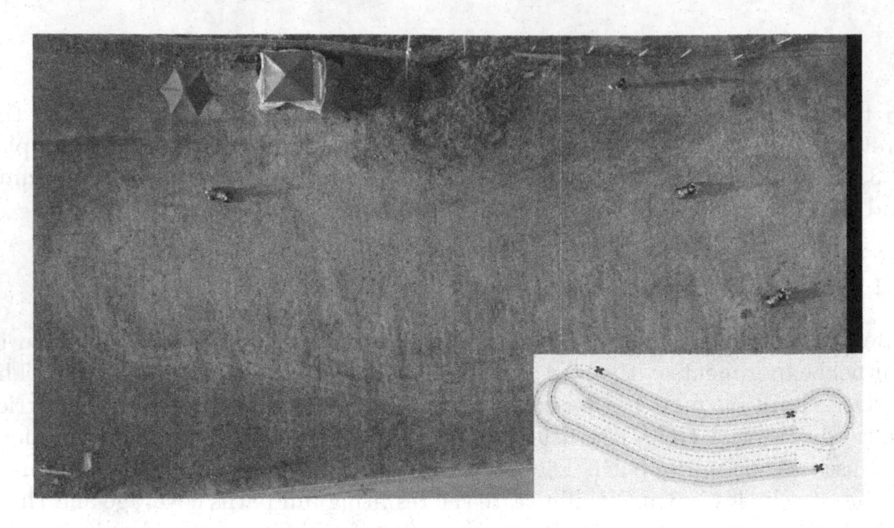

Fig. 5. Example 1. The view from the quadcopter and the operator interface.

3.2 Example 2

The following field example was obtained by drawing the contour of the field using the Google Earth satellite images, the elevation map was set manually. The resulting paths and routes are shown in the Fig. 6. A section of the field border is selected for the initial swath. The minimum turning radius of cars is 3 m ($u_{\max} = 1/3\ \text{m}^{-1}$). $\beta = 100\,\text{m}^{-1}/u_{\max}$. The track width is $\bar{d} = 1$ m. The initial swath did not allow us to construct covering paths for which the curvature constraint is satisfied. In this example, we allowed overlapping swaths no more than 5% ($\gamma = 0.95$) in width, which made the paths feasible. Route planning was carried out for two vehicles. The lengths of routes happened to be $D_1 = 1322.1$ m and $D_2 = 1321.9$ m which makes only 0.2 m (0.015%) difference.

3.3 Example 3

The next example differs from the previous one in that it is allowed to 25% swath overlap ($\gamma = 0.75$). Figure 7 shows the results. The lengths of routes are $D_1 = 1475.7$ m and $D_2 = 1477.2$ m. The effect of straightening the paths is observed, $\beta = 100/u_{\max}$. It is achieved by introducing the normal curvature at the junction points of the elementary sections of the splines into the objective function (25).

3.4 Example 4

For the example 4, it was also impossible to build a coverage for the selected initial path without overlap. The results are provided in Fig. 8. The route lengths are $D_1 = 1159.8$ m and $D_2 = 1158.6$ m. $\beta = 100/u_{\max}$. However, the cultivated

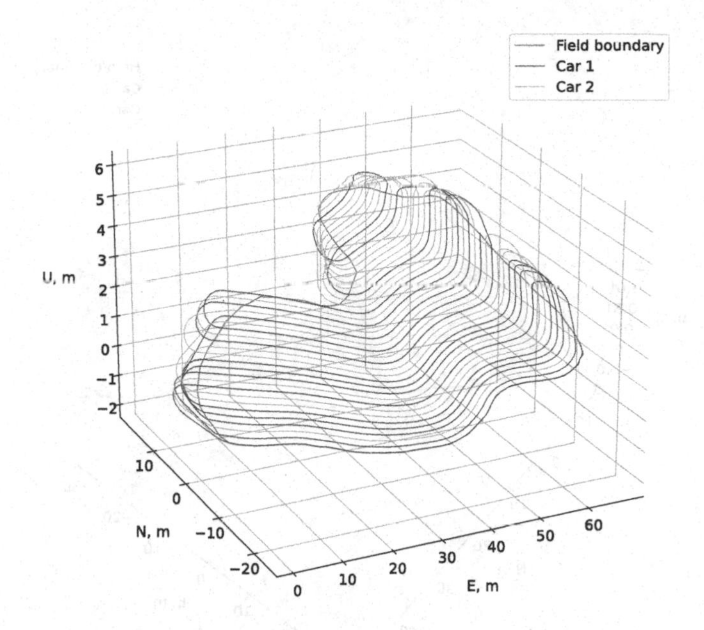

Fig. 6. Example 2.

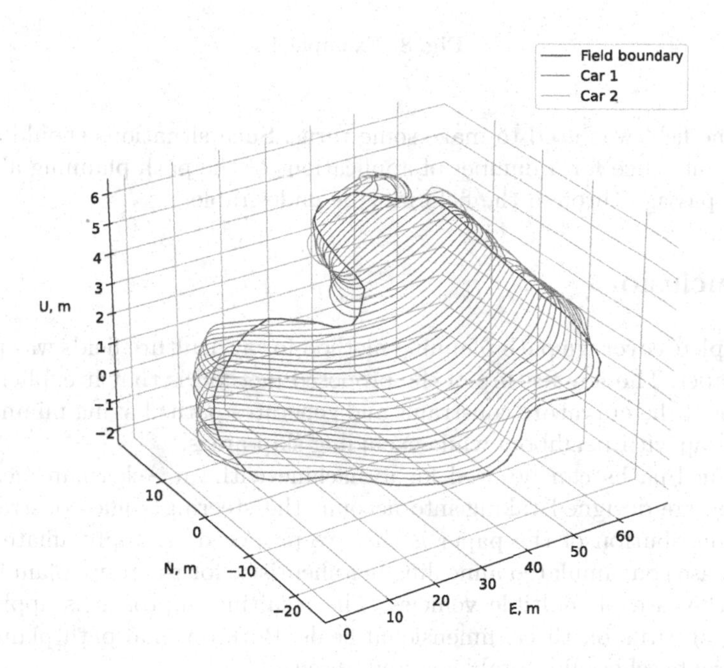

Fig. 7. Example 3.

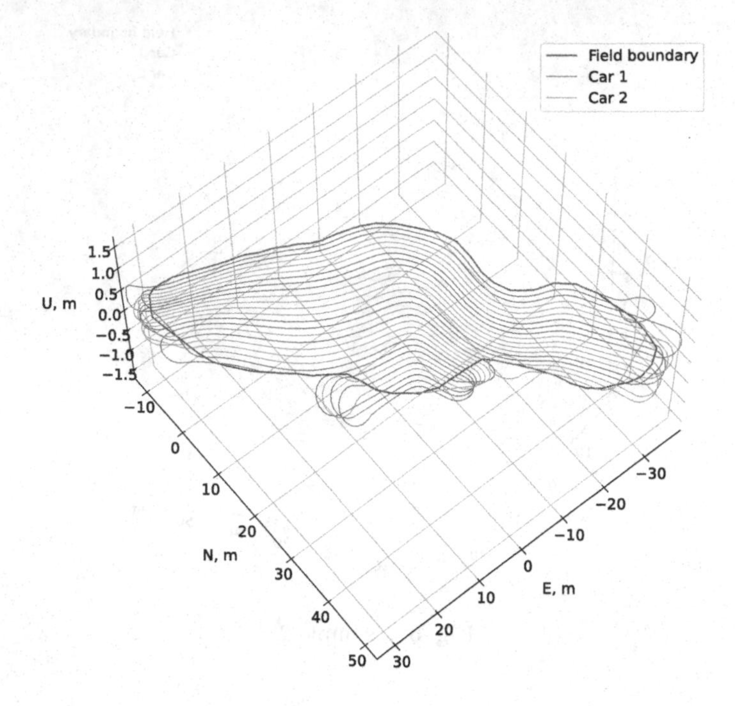

Fig. 8. Example 4.

area of the field was used to make some turns. Such situations should be taken into account, since for a number of applications of the path planning algorithm, repeated passage through the field may be undesirable.

4 Conclusions

The complete coverage path planning method for agriculture fields was proposed in this paper. The key feature of the proposed method is that it explicitly takes into account the curvature constraint and generates paths having minimum possible overlap with neighbors while excluding skip areas.

Obtained paths can be used for a machine with an Ackerman mechanism, since they are designed taking into account the steering angle constraint. The second contribution of the paper is that we proposed an approximate solution method based on simulated annealing metaheuristic for the route planning problem for the case of multiple vehicles. The resulting approach is applicable to planning of paths on three-dimensional fields. Both optimal path planning and optimal route scheduling problems were discussed.

The novelty of the proposed approach is that when considering the first problem the curvature constraint was reduced to the second order cone constraint which in turn allowed for formulation of the optimization problem as the Second-Order Cone Programming Problem which is a convex one. Therefore, it can be effectively solved. In addition, our approach allows to straighten the paths and make them implementable. Thus, in contrast to other works mentioned in the introduction, we give a rigorous mathematical formulation of the problem and a rigorous solution that reduces to the convex optimization.

The natural limitation of the considered approach is that the field surface is assumed to be close to the inclined plane within a neighborhood of the current position. The size of the neighborhood should be comparable to the swath width. Further research is directed to considering an obstacle avoidance in the framework of our approach.

References

1. Choset, H., Pignon, P.: Coverage path planning: the boustrophedon cellular decomposition. In: Zelinsky, A. (ed.) Field and Service Robotics, pp. 203–209. Springer, London (1998). https://doi.org/10.1007/978-1-4471-1273-0_32
2. Conesa-Muñoz, J., Bengochea-Guevara, J.M., Andujar, D., Ribeiro, A.: Route planning for agricultural tasks: a general approach for fleets of autonomous vehicles in site-specific herbicide applications. Comput. Electron. Agric. **127**, 204–220 (2016). https://doi.org/10.1016/j.compag.2016.06.012
3. Coombes, M., Chen, W.H., Liu, C.: Fixed wing UAV survey coverage path planning in wind for improving existing ground control station software. In: 2018 37th Chinese Control Conference (CCC), pp. 9820–9825 (2018). https://doi.org/10.23919/ChiCC.2018.8482722
4. Dakulović, M., Horvatić, S., Petrović, I.: Complete coverage D* algorithm for path planning of a floor-cleaning mobile robot. IFAC Proc. Vol. **44**(1), 5950–5955 (2011). https://doi.org/10.3182/20110828-6-IT-1002.03400. 18th IFAC World Congress
5. Domahidi, A., Chu, E., Boyd, S.: ECOS: an SOCP solver for embedded systems. In: 2013 European Control Conference, ECC 2013, pp. 3071–3076 (2013). https://doi.org/10.23919/ECC.2013.6669541
6. Erdelić, T., Carić, T.: A survey on the electric vehicle routing problem: variants and solution approaches. J. Adv. Transp. **2019** (2019). https://doi.org/10.1155/2019/5075671
7. Erdoğan, S., Miller-Hooks, E.: A green vehicle routing problem. Transp. Res. Part E: Logistics Transp. Rev. **48**(1), 100–114 (2012). https://doi.org/10.1016/j.tre.2011.08.001. Select Papers from the 19th International Symposium on Transportation and Traffic Theory
8. Gilimyanov, R.F., Pesterev, A.V., Rapoport, L.B.: Smoothing curvature of trajectories constructed by noisy measurements in path planning problems for wheeled robots. J. Comput. Syst. Sci. Int. **47**(5), 812–819 (2008)
9. Gilimyanov, R., Pesterev, A., Rapoport, L.: Motion control for a wheeled robot following a curvilinear path. J. Comput. Syst. Sci. Int. **47**(6), 987–994 (2008). https://doi.org/10.1134/S1064230708060129
10. Gálvez, A., Iglesias, A., Puig-Pey, J.: Computing parallel curves on parametric surfaces. Appl. Math. Model. **38**(9), 2398–2413 (2014). https://doi.org/10.1016/j.apm.2013.10.042
11. Hameed, I.A., La Cour-Harbo, A., Osen, O.L.: Side-to-side 3D coverage path planning approach for agricultural robots to minimize skip/overlap areas between swaths. Robot. Auton. Syst. **76**, 36–45 (2016). https://doi.org/10.1016/j.robot.2015.11.009

12. Jin, J., Tang, L.: Coverage path planning on three-dimensional terrain for arable farming. J. Field Rob. **28**(3), 424–440 (2011). https://doi.org/10.1002/rob.20388
13. Jin, J., Tang, L.: Optimal coverage path planning for arable farming on 2D surfaces. Trans. ASABE **53**(1), 283–295 (2010). https://doi.org/10.13031/2013.29488
14. Lin, J., Zhou, W., Wolfson, O.: Electric vehicle routing problem. Transp. Res. Procedia **12**, 508–521 (2016). https://doi.org/10.1016/j.trpro.2016.02.007. Tenth International Conference on City Logistics 17–19 June 2015, Tenerife, Spain
15. O'Donoghue, B., Chu, E., Parikh, N., Boyd, S.: Conic optimization via operator splitting and homogeneous self-dual embedding. J. Optim. Theory Appl. **169**(3), 1042–1068 (2016). https://doi.org/10.1007/s10957-016-0892-3
16. Pesterev, A.V., Rapoport, L.B., Gilimyanov, R.F.: Global energy fairing of B-spline curves in path planning problems. In: 2007 Proceedings of the ASME International Design Engineering Technical Conferences and Computers and Information in Engineering Conference, DETC2007, vol. 8. PART B, pp. 1133–1139 (2008). https://doi.org/10.1115/DETC2007-35306
17. Rapoport, L., Generalov, A., Shavin, M., Tormagov, T.: Navigation and control problems in precision farming. In: 2021 28th Saint Petersburg International Conference on Integrated Navigation Systems (ICINS), pp. 1–8 (2021). https://doi.org/10.23919/ICINS43216.2021.9470810
18. Rodrigues, R.T., Aguiar, A.P., Pascoal, A.: A coverage planner for AUVs using B-splines. In: 2018 IEEE/OES Autonomous Underwater Vehicle Workshop (AUV), pp. 1–6 (2018). https://doi.org/10.1109/AUV.2018.8729760
19. Sabelhaus, D., Röben, F., Meyer zu Helligen, L.P., Schulze Lammers, P.: Using continuous-curvature paths to generate feasible headland turn manoeuvres. Biosyst. Eng. **116**(4), 399–409 (2013). https://doi.org/10.1016/j.biosystemseng.2013.08.012
20. Vahdanjoo, M., Zhou, K., Sørensen, C.A.G.: Route planning for agricultural machines with multiple depots: manure application case study. Agronomy **10**(10), 1608 (2020). https://doi.org/10.3390/agronomy10101608

Determination of Hydrological Model Parameters by Newton Method

Sergey Zasukhin[1] and Elena Zasukhina[2(✉)]

[1] Moscow Institute of Physics and Technology, Dolgoprudny, Russia
[2] Federal Research Center "Computer Science and Control" of the Russian Academy of Sciences, Moscow, Russia

Abstract. The problem of determining parameters of a hydrological model is studied. The model describes vertical water transfer in unsaturated soil and includes some parameters that characterize hydrophysical properties of soil. The problem is formulated as a variational problem with differential constraints in the form of the Richards equation describing a process of water transfer in soil. The objective functional is proportional to the distance squared between measured and modeled values of the soil moisture. Numerical solution of the discretized problem is carried out using Newton method. Gradient and second order derivatives of the objective function are calculated using formulas obtained by fast automatic differentiation techniques.

Keywords: Fast automatic differentiation · Numerical optimization · Newton method

1 Introduction

Various hydrological models have been used in studying the processes of hydrological cycle, the processes of runof formation in the catchment, processes of interaction in the system soil-surface-atmosphere. They are applied in modeling the processes of infiltration, evapotranspiration, surface and subsurface flow, groundwater and river flow, for the planning and management of water resources systems, etc. These models contain parameters that characterize hydro-physical properties of soil, namely, they are included in formulas for calculating hydraulic conductivity and diffusion coefficient. Usually, specialists calculate these hydrophysical characteristics using the formulas of van Genuchten-Mualem model [1]. The parameters included in these formulas are called van Genuchten parameters (VG-parameters).

Successful application of these models depends directly on the exact knowledge of VG-parameters). Their experimental measurement is associated with difficulties. Therefore, the following approach is often applied. The parameters

Supported by Russian Fund of Fundamental Researches, Project No 19-01-00666 and Project No 19-07-00750.

N. N. Olenev et al. (Eds.): OPTIMA 2021, CCIS 1514, pp. 273–285, 2021.
https://doi.org/10.1007/978-3-030-92711-0_19

are determined by comparing the modeled and observed values of some physical quantities. The parameters values are selected in such a way that the difference between the modeled and observed values of mentioned physical quantities is minimal. As a rule, such physical quantities are transient outflow and soil water pressure, transient cumulative infiltration and related data obtained in laboratory transient outflow experiments, for example [2–8].

Problems of this kind continue to be considered by various authors up to the present. They investigate the possibility of determining parameters from various initial data that are available in a particular situation or can be obtained from possible experiments in the current circumstances.

Herewith, arising optimization problem have been solved applying various computational algorithms: gradient-like optimization methods [2,3,6,8], genetic algorithms [4,5,7], methods imitating the behavior of biological populations in conditions of lack of vital resources, algorithms imitating social behavior, see, for example [9–12].

It should be noted that, as a rule, in works in which the numerical solution of the problem is carried out by first-order optimization methods, the gradient of the objective function is approximately calculated using the finite differences, for example [3].

In the present paper, the problem of restoring VG-parameters from accurate soil moisture data is studied. In this connection, the model of vertical water transfer in unsaturated soil is considered. In this model, water movement is described by Richards equations. The problem of the parameters determination is formulated as a variational problem with differential relation in the form of Richards equation. The controlled process is soil moisture, and the objective function is mean-square deviation of the modeled values of soil moisture from its observed values. The numerical solution of the problem is proposed to be carried out using Newton method. Numerical optimization by the steepest descent method did not lead to success due to the large difference between the values of the components of the objective function gradient and its ravine-type shape. This will be discussed in more detail below.

It should be noted that first- and second-order derivatives of the objective function were calculated using exact formulas of fast automatic differentiation method (FAD) [13–17] in contrast to the mentioned works where objective function gradient was calculated approximately. The use of exact formulas for calculating the derivatives of the objective function makes it possible to determine the parameters with high accuracy.

Should be emphasized that only soil moisture data are used to determine the parameters. These initial data are quite accessible.

2 Problem Formulation

2.1 Direct Problem

Suppose that soil is homogeneous isothermal non-deformable porous media. In this case, vertical water transfer in soil is well described by one-dimensional nonlinear second order parabolic partial differential equation. Consider following initial boundary value problem:

$$
\begin{aligned}
&\frac{\partial\theta}{\partial t}=\frac{\partial}{\partial z}\left(D(\theta)\frac{\partial\theta}{\partial z}\right)-\frac{\partial K(\theta)}{\partial z}, \quad (z,t)\in Q,\\
&\theta(z,0)=\varphi(z), \quad z\in(0,\mathrm{L}),\\
&\theta(L,t)=\psi(t), \quad t\in(0,T),\\
&-\left(D(\theta)\frac{\partial\theta}{\partial z}-K(\theta)\right)\bigg|_{z=0}=R(t)-E(t), \quad t\in(0,T),\\
&\theta_{min}\le\theta(0,t)\le\theta_{max}, \quad t\in(0,T),
\end{aligned}
\tag{1}
$$

where z is space variable; t is time; $\theta(z,t)$ is soil moisture at the point (z,t); $Q=(0,L)\times(0,T)$; $\varphi(z)$ and $\psi(t)$ are given functions; $D(\theta)$ and $K(\theta)$ are diffusion coefficient and hydraulic conductivity – the hydrophysical characteristics of the soil; $\theta_{min}=\theta_r+\varepsilon$ and $\theta_{max}=\theta_s-\varepsilon$, where θ_r and θ_s are, respectively, the residual moisture and the saturation moisture depending on the soil type, and ε is a constant such that $0<\varepsilon\ll\theta_r$; $R(t)$ is precipitation; $E(t)$ is evaporation, $0\le E(t)\le M$, $t\in(0,T)$, M is a constant such that $M>0$.

In according to van Genuchten-Mualem model [1], the diffusion coefficient $D(\theta)$ and the hydraulic conductivity $K(\theta)$ are calculated by following formulas:

$$
\begin{aligned}
&K(\theta)=K_0S^{0.5}[1-(1-S^{1/m})^m]^2,\\
&D(\theta)=K_0\frac{1-m}{\alpha m(\theta_s-\theta_r)}S^{0.5-1/m}\\
&\times[(1-S^{1/m})^{-m}+(1-S^{1/m})^m-2],
\end{aligned}
\tag{2}
$$

where $S=\dfrac{\theta-\theta_r}{\theta_s-\theta_r}$; θ_r is given constant; K_0, α, m, θ_s are some parameters. The problem (1)–(2) will be called the direct problem.

2.2 Parameter Identification Problem

Let a function $\hat\theta(z,t)$ be defined on some set $Q_0\subseteq Q$. Further, we call this function $\hat\theta(z,t)$ "experimental data". Denote $[K_0,\alpha,m,\theta_s]^T$ by u. Introduce a set $U=\{u:u\in R^4;\ 0\le a[i]\le u[i]\le b[i],\ i=1,2,3,4\}$. The problem is to find the parameters K_0, α, m and θ_s so that the corresponding solution of the direct problem (1)–(2) would be as close as possible to the function $\hat\theta(z,t)$ on the set Q_0. More precisely, the problem is to determine u^{opt}, $u^{opt}\in U$, and the corresponding solution $\theta^{opt}(z,t)$ of the direct problem (1)–(2) which minimize functional

$$
J=\frac{1}{2}\int_{Q_0}(\theta-\hat\theta)^2dzdt. \tag{3}
$$

2.3 Discretization of the Direct Problem

To approximate the direct problem (1)–(2) by finite differences we divide the intervals $(0,T)$ and $(0,L)$ into N and I equal subintervals with the endpoints

$t^n = \tau n$, $0 \le n \le N$, and $z_i = hi$, $0 \le i \le I$, correspondingly, where $\tau = T/N$ and $h = L/I$.

To approximate the direct problem (1)–(2) the following finite differences scheme was used:

$$\frac{\theta_i^{n+1} - \theta_i^n}{\tau} = \frac{1}{h}\left(D_{i+1/2}^{n+1}\frac{\theta_{i+1}^{n+1} - \theta_i^{n+1}}{h} - K_{i+1/2}^{n+1}\right.$$
$$\left. - D_{i-1/2}^{n+1}\frac{\theta_i^{n+1} - \theta_{i-1}^{n+1}}{h} + K_{i-1/2}^{n+1}\right),$$
$$1 \le i < I; \quad 0 \le n < N,$$
$$\theta_i^0 = \varphi_i, \quad 0 \le i \le I, \quad \theta_I^n = \psi^n, \quad 1 \le n \le N.$$

Here θ_i^n, $D_{i+1/2}^n$, $K_{i-1/2}^n$ are values of the functions $\theta(z,t)$, $D(\theta(z,t))$, $K(\theta(z,t))$ at the points $(ih, n\tau)$, $((i+1/2)h, n\tau)$, $((i-1/2)h, n\tau)$, correspondingly.

The left boundary condition is approximated in the form

$$\frac{\theta_0^{n+1} - \theta_0^n}{\tau} = \frac{2}{h}\left(D_{1/2}^{n+1}\frac{\theta_1^{n+1} - \theta_0^{n+1}}{h} - -K_{1/2}^{n+1} + R^{n+1} - E^{n+1}\right),$$
$$0 \le n < N,$$

where R^{n+1}, E^{n+1} are values of functions $R(t)$ and $E(t)$ at the points $t = (n+1)\tau$.

As a result of this finite-difference approximation of relations (1), we obtain the following systems of equations:

$$\begin{aligned}
&\Phi_0^n = -\left(\frac{1}{\tau} + \frac{2}{h}D_{1/2}^n\right)\theta_0^n + \frac{2}{h}D_{1/2}^n\theta_1^n \\
&+\frac{1}{\tau}\theta_0^{n-1} + \frac{2}{h}\left(-K_{1/2}^n + R^n - E^n\right) = 0, \\
&\theta_{min} \le \theta_0^n \le \theta_{max}, \quad 1 \le n \le N, \\
&\Phi_i^n = \frac{1}{h^2}D_{i-1/2}^n\theta_{i-1}^n - \left\{\frac{1}{\tau} + \frac{1}{h^2}\left(D_{i+1/2}^n + D_{i-1/2}^n\right)\right\}\theta_i^n + \\
&\frac{1}{h^2}D_{i+1/2}^n\theta_{i+1}^n + \left\{\frac{\theta_i^{n-1}}{\tau} + \frac{1}{h}\left(K_{i-1/2}^n - K_{i+1/2}^n\right)\right\} = 0, \\
&1 \le i \le I-1, \quad 1 \le n \le N, \\
&\Phi_I^n = \theta_I^n - \psi^n = 0, \quad 1 \le n \le N, \\
&\theta_i^0 = \varphi_i, \quad 0 \le i \le I.
\end{aligned} \tag{4}$$

The diffusion coefficient and the hydraulic conductivity at the intermediate points are calculated using the formulas

$$D_{i+1/2}^n = \frac{D_i^{n-1} + D_{i+1}^{n-1}}{2}, \quad K_{i+1/2}^n = \frac{K_i^{n-1} + K_{i+1}^{n-1}}{2}, \tag{5}$$
$$1 \le n \le N, \quad 0 \le i < I.$$

2.4 Discrete Parameter Identification Problem

To discretize the functional (3) introduce a set :

$$Q_0 = \{(z,t) : z = ih, t = l\tau, (i,l) \in A\},$$

$$A = \{(i,l) : i = 0, 1, \ldots, I,\ l = 1, \ldots, d\}, 0 < d \leq N,$$

where d is some natural number. The discrete analog of functional (3) has a form

$$W(\theta, u) = \frac{1}{2} \sum_{(j,n) \in A} \left(\theta_j^n - \hat{\theta}_j^n\right)^2 \tau h, \tag{6}$$

where $u \in U$, the set U was defined above.

The problem is to find $u^{opt} \in U$ and corresponding solution $\theta^{opt}(z,t)$ of the direct problem (4)–(5) that minimize the functional $W(\theta, u)$ (6). Thus, as a result of discretization, the continuous problem of identifying the parameters is reduced to a nonlinear programming problem.

3 Solution of the Problem

Previously, the authors tried to identify 3 parameters K_0, α and m using the steepest descent method [18]. But this attempt failed. The complexity of the numerical optimization was caused by the following features of the objective function:

- large difference between values of components of the function gradient;
- a small value of the first component of the objective function gradient;
- ravine-type shape of the objective function.

Numerical results [18] have shown that the deviations of the obtained values from the true values of the parameters strongly depend on the difference between the initial and true values of the parameter K_0. So, a big difference between the initial and true values of the parameter K_0 led to errors up to 44.7% for the parameter α and errors up to 8.2% for the parameter m. As for the parameter K_0, its optimal value was very close to the initial value. Therefore, the numerical solution of the problem of determining 4 parameters K_0, α, m and θ_s was carried out by Newton method.

Newton direction was determined using the first and the second derivatives of the objective function calculated by fast automatic differentiation (FAD) technique. These formulas are accurate, and this is especially important for the problem of identifying the parameters for the reasons indicated above.

3.1 Formulas for Calculating the Derivatives

First, we briefly present the derivation of formulas for the gradient of a function whose variables are linked by functional relations. We use the approach suggested in [13]. Consider twice continuously differentiable mappings

$$W : R^n \times R^r \to R^1, \Phi : R^n \times R^r \to R^n.$$

Let $x \in R^n$ and $u \in R^r$ satisfy the system of n scalar algebraic equations:

$$\Phi(x, u) = 0_n, \tag{7}$$

where 0_n is n-dimensional vector.

In what follows, we will use the following notation. Denote $(n \times m)$-matrix whose (ij)th element equals to $\partial\Phi^j/\partial x^i$, $i = 1, 2, \ldots, n$, $j = 1, 2, \ldots, n$, by $\Phi_x^\top(x, u)$. The matrix transposed to $\Phi_x^\top(x, u)$ will be denoted by $\Phi_{x^\top}(x, u)$.

Let the matrix $\Phi_x^\top(x, u)$ be non-singular. In accordance to the implicit function theorem, the system of algebraic equations (7) defines function $x = x(u) \in C^2(u)$. We will call x phase variable and u – control.

Following the approach from [13,15], one can obtain the formulas for the gradient of the function $W(z(u), u)$:

$$dW(u)/du = W_u(x(u), u) + W_x^\top(x(u), u)p, \tag{8}$$

where the Lagrange multiplier $p \in R^n$ is determined from the solution of next linear system of equations :

$$W_x(x(u), u) + \Phi_x^\top(x(u), u)p = 0_n. \tag{9}$$

Now, using known values of $x(u)$, p and the gradient of the objective function W, one can calculate the second derivatives of W applying an algorithm suggested in [17].

In accordance to this algorithm we determine a matrix Λ as a result of solution matrix equation:

$$\Phi_{x^\top}\Lambda + \Phi_{u^\top} = 0_{nr}, \tag{10}$$

here 0_{nr} is null-matrix of n rows and r columns. The equation (10) is linear with respect to Λ and is split into r subsystems consisted of of n equations. Each of these subsystems has the same basic matrix $\Phi_{x^\top}$ and contains the corresponding column of the matrix Λ as unknowns.

After obtaining the matrix Λ, the second derivatives of the objective function $W(z(u), u)$ (6) can be determined applying following formula:

$$\begin{gathered} d^2W(u)/du^2 = \Lambda^\top(u)L_{xx}(x(u), u, p(u))\Lambda(u) \\ L_{uu}(x(u), u, p(u)) + L_{xu}(x(u), u, p(u))\Lambda(u) \\ \Lambda^\top(u)L_{ux}(x(u), u, p(u)), \end{gathered} \tag{11}$$

here $L(x, u, p) = W(x, u) + p^\top\Phi(x, u)$, $L(x, u, p)$ is Lagrange function.

So, summarizing all of the above, the algorithm for calculating the derivatives of the objective function $W(z(u), u)$ (6) consists of the following sequence of actions:

1. the determination of $x(u)$ as a result of solution of the system (7);
2. the determination of Lagrange multiplier p as a result of solution of the system (9);
3. calculating the gradient of the function $W(z(u), u)$ by the formula (8);
4. the determination of the matrix Λ solving the matrix Eq. (10);
5. calculating the second derivatives of W by the formula (11).

It should be emphasized that the basic matrices of the system (9) and the matrix equation (10) coincide up to transposition.

3.2 Computing the Derivatives of the Objective Function

At first, is necessary to solve the system (4)–(5) with given $u \in U$ to determine $\theta(u)$. The form of the ratios (4)–(5) allows splitting the system (4) into N subsystems. Each of these subsystems corresponds its time layer. It can be seen from the relations (4)–(5) that the system (4) can be solved by sequential solution of these subsystems, starting from the first and ending with the Nth one. The basic matrices of these subsystems are tridiagonal. Therefore, these subsystems can be solved using the tridiagonal matrix algorithm.

At the second stage, to determine Lagrange multiplier p it is necessary to solve the system

$$W_\theta(\theta(u), u) + \Phi_\theta^\top(\theta(u), u)p = 0_n, \tag{12}$$

obtaining as a result of replacing x by θ in (9).

We consider the matrix Φ_θ^T as a $N \times N$-matrix of block-elements of size $(I+1) \times (I+1)$-matrix. This block-matrix Φ_θ^T has the next structure:

$$\Phi_\theta^T = \begin{pmatrix} F_{11} & F_{12} & & & \\ & F_{22} & F_{23} & & \\ & & \ddots & \ddots & \\ & & & F_{kk} & F_{kN} \\ & & & & F_{NN} \end{pmatrix}. \tag{13}$$

here $k = N - 1$. This figure shows all nonzero blocks in the matrix Φ_θ^T.

Each of the block-elements F_{ii}, $i = 1, 2, \ldots, N$, located at the main diagonal of the matrix $\Phi_\theta^\top$, has tridiagonal structure:

$$F_{ii} = \begin{pmatrix} C_0 & B_0 & & & \\ A_1 & C_1 & B_1 & & \\ & \ddots & \ddots & \ddots & \\ & & A_k & C_k & 0 \\ & & & A_N & C_N \end{pmatrix}, \tag{14}$$

$k = N - 1.$

All nonzero elements in each of the block-elements G_{ij}, $j = i + 1$, $i = 1, 2, \ldots, N - 1$, are placed on the main diagonal:

$$F_{ij} = \begin{pmatrix} f & & & & \\ & f & & & \\ & & \ddots & & \\ & & & f & \\ & & & & 0 \end{pmatrix}, \tag{15}$$

where $f = 1/\tau$.

The system (12) can be divided into N subsystems, each of which corresponds to its time layer. The type of system allows you to solve this system, sequentially solving each of these subsystems, starting with the N and ending with the first, i.e. from the bottom to the top. Due to the type of main matrix F_{ii} of the ith subsystem, $i = 1, 2, \ldots, N$, each subsystem can be solved using the tridiagonal matrix algorithm.

After solving the system (12) and determining p, the gradient of the function W can be calculated by formula:

$$dW(u)/du = W_u(\theta(u), u) + W_\theta^\top(\theta(u), u)p. \tag{16}$$

At the fourth stage, the matrix Λ is determined as a result of solution of the matrix equation

$$\Phi_{\theta^\top}\Lambda + \Phi_{u^\top} = 0_{nr}. \tag{17}$$

This matrix equation is linear with respect to Λ and consists of r systems of n equations. Each of these systems has the same main matrix $\Phi_{\theta^\top}$, this matrix is transposed to the matrix $\Phi_\theta^\top$. Therefore, each of these systems can be solved in a similar way to solving the system (12) with the only difference that they are solved from top to bottom.

It is clear that solving the matrix Eq. (17) allows parallelization.

At the fifth stage, after determining $\theta(u)$, Lagrange multiplier p, the gradient of the objective function, the matrix Λ the second derivatives of the objective function $W(\theta(u), u)$ can be calculated using the formula

$$\begin{aligned} d^2W(u)/du^2 = {} & \Lambda^\top(u)L_{\theta\theta}(\theta(u), u, p(u))\Lambda(u) + L_{uu}(\theta(u), u, p(u)) \\ & + L_{\theta u}(\theta(u), u, p(u))\Lambda(u) + \Lambda^\top(u)L_{u\theta}(\theta(u), u, p(u)), \end{aligned} \tag{18}$$

where $L(\theta, u, p) = W(\theta, u) + p^\top\Phi(\theta, u)$, $L(\theta, u, p)$ is Lagrange function.

3.3 Simplification of the Computational Algorithm

We note some features of the problem allowing to simplify the process of calculating the derivatives of the objective problem by FAD formulas.

Currently, there are many computational packages that allow to calculate the gradients of functions the values of which are determined as a result of executing multi-step processes. They allow you to automatically compute the gradients of functions without going into detail about those functions.

But in our case, the authors preferred to calculate the gradient of the objective function using the algorithm described above, rather than applying computational packages of automatic differentiation. The reasons of this decision are:

1. the simplicity of obtaining the main matrix of the system (12) from the main matrix of the system (4);
2. the simplicity of obtaining right sides of the subsystems consisting the system (12);
3. the simplicity of solving the system (12);
4. among all functions included in (4), (6), derivation of derivatives is required only for the functions $D^n_{i+1/2}$ and $K^n_{i+1/2}$. The derivatives of the others are obvious.

Take a closer look at these considerations. Indeed, in order to obtain the main and two adjacent diagonals of the main matrix of the system (12) from corresponding diagonals of the main matrix of the system (4), it is necessary to add to each element of these diagonals terms of the form $(D^n_{i+1/2})_{\theta^n_j}(\theta^n_{i+1}-\theta^n_i)$ or $(K^n_{i+1/2})_{\theta^n_j}$, $j=i,i+1$. The number of such terms varies from 0 to 4.

The derivative of $W(\theta,u)$ with respect to θ is obvious and does not need to be found:

$$\partial W(\theta,u)/\partial\theta^n_j = (\theta^n_j - \hat{\theta}^n_j)\tau h, \quad n=1,\ldots,N, \quad j=1,\ldots,I. \tag{19}$$

Taking into account (15) and (19), we obtain the formula for the right side $H(n)$ of the nth subsystem, $n=1,2,\ldots,N$, of the system (12):

$$\begin{gathered} H(n) = [H^n_0, H^n_1, \ldots, H^n_I]^\top, \\ H^n_j = (\theta^n_j - \hat{\theta}^n_j)\tau h + p^{n+1}/\tau, j=0,1,\ldots,I-1, \quad H^n_I = 0. \end{gathered} \tag{20}$$

The system (12) can be solved by sequentially solving all subsystems from the Nth to the first. Each subsystem is solved using triangular matrix algorithm.

In addition, to facilitate the computational process, we derived all formulas for the first and second order derivatives of $D^n_{i+1/2}$, $K^n_{i+1/2}$ with respect to u and θ applying computer algebra systems.

These features of the problem make it possible to reduce the computation time of the objective function gradient. We tracked the time needed to compute the value of the objective function $W(\theta(u),u)$ as well as the time of computing the objective function gradient. The value of the relative cost (that is, the ratio of the time to compute the gradient to the time to compute the function) turned out to be close to 2.

In [19] the relative costs obtained for several tasks using the automatic differentiation packages ADEPT [19], ADOL-C [20], CppAD [21], Sacado [22] are given. Comparative analysis showed that the fastest of them is the ADEPT package. The best values of relative cost achieved by this package are 2.7–3.8. In the case of the inverse coefficient problem for the heat equation, the best value of the relative cost when applying ADEPT turned out to be 3.6 [23].

The matrix Λ is determined as a result of solution the matrix Eq. (17). This equation consists of r systems:

$$\Phi_{\theta^\top}\Lambda^j + \Phi_{u^j} = 0_n, \quad j = 1, \ldots, r, \tag{21}$$

where Λ^j is jth column of the matrix Λ and u^j is jth component of u. All systems (21) have the same main matrix $\Phi_{\theta^\top}$ that is transposed to the main matrix of the system (12). Due to this factor:

- the matrix $\Phi_{\theta^\top}$ does not need to be determined;
- the systems (21) can be solved in the same way as the system (12), but from top to bottom.

As for the right-hand sides of the systems (21), it is clear from (4) that the determination of Φ_{u^j}, $j = 1, \ldots, r$, is not difficult. Recall that the derivation of the symbolic derivatives of the functions $D^n_{i+1/2}$ and $K^n_{i+1/2}$ was curried out using computer algebra systems.

Calculation of the second derivatives of the objective function by formula (18) allows us to make the following simplifications. The analysis of the relations (4), (5) and (6) shows that there are many duplicate elements in the matrices from (18), some are sparse, some are symmetric. Also, the first term and the third one from (18) are transposed to each other, and

$$W_{\theta\theta} = \tau h E, \quad W_{\theta u} = 0_{rn}, \quad W_{uu} = 0_{rr},$$

where E is unit matrix of size $n \times n$, $n = (I + 1)N$.

Taking into account all these circumstances, we can organize the calculation of the derivatives of the objective function in such a way as to reduce the number of necessary operations a and the laboriousness of the computational process. In addition, a decrease in the number of necessary calculations increases the accuracy of determining the second derivatives of the objective function, which is important for the parameters identification problem.

4 Numerical Results

The described approach was applied in the numerical solution of the parameter identification problem by Newton method.

The problem was considered with the following values of the input parameters:

$$L = 100(\text{cm}), \quad T = 17/96(\text{d}), \quad \theta_r = 0.05(\text{cm}^3/\text{cm}^3), \quad R = 0(cm/d),$$
$$E = 0.0125(cm/d), \quad \varphi(z) = 0.3, \ z \in (0, L), \ \psi(t) = 0.3, \ t \in (0, T),$$
$$a = [0, 0.0005, 0.08, 0.3]^T, \quad b = [300, 0.4, 0.8, 0.8]^T, \quad \varepsilon = 10^{-8}.$$

A grid with $I = 100$ and $N = 17$ was used. Numerical calculations were performed in two stages.

4.1 The First Stage of Calculations

At first, we solved the direct problem (4)–(5) with the parameters $K_0^{true} = 100(\text{cm/d})$, $\alpha^{true} = 0.01$, $m^{true} = 0.2$ and $\theta_s = 0.5$. These parameter values were subsequently declared as true. The found solution was taken as the prescribed function $\hat{\theta}(z,t)$.

4.2 The Second Stage of Calculations

At this stage, the search for a numerical solution of the parameter identification problem was carried out using Newton method. The proposed algorithm was used to determine the first and second order derivatives of the objective function. The step size in Newton direction was determined by minimizing the function obtained by cubic interpolation of the objective function in the selected direction using 20 points The search for a numerical solution of the problem was carried out with various initial approximations. The obtained results are contained in Table 1.

Table 1. Results of solving the problem with various initial approximations

Initial approximation				Errors (%)				Number of iterations
K_0	α	m	θ_s	K_0	α	m	θ_s	
110	0.011	0.22	0.55	0.011	0.004	0.0002	0.0008	590
120	0.015	0.25	0.60	0.009	0.003	0.0002	0.0006	3219
130	0.020	0.30	0.65	0.010	0.004	0.0002	0.0007	1866
90	0.009	0.18	0.45	0.007	0.003	0.0001	0.0005	1159
80	0.005	0.15	0.40	0.007	0.003	0.0002	0.0005	2567
70	0.004	0.10	0.35	0.010	0.004	0.0002	0.0007	4934
60	0.005	0.12	0.35	0.011	0.004	0.0002	0.0007	3863
50	0.005	0.11	0.35	0.007	0.003	0.0001	0.0005	3884
150	0.005	0.11	0.35	0.009	0.003	0.0002	0.0006	5137
160	0.004	0.10	0.35	0.010	0.004	0.0002	0.0007	2628

In this table, the columns from 1 to 4 contain initial values of parameters K_0, α, m and θ_s respectively. The following columns from 5 to 8 contain the absolute values of deviations of obtained (optimal) values of parameters K_0, α, m and θ_s from their true values. Deviations are expressed as a percentage. The last column indicates the number of iterations required to solve the problem.

It should be noted that the solution of the problem with initial approximations occupying rows from 9 to 12 in Table 1 required the use of 40 points instead of 20 points when constructing splines.

As follows from Table 1, the solution of the problem with all considered initial approximations gives good and very close to each other results. So, the error in determining the parameter K_0 for all considered initial approximations varies from 0.007% to 0.011%, for the parameter α this error varies from 0.003% to 0.004%, the error in determining the parameter m varies from 0.0001% to 0.0002% and for the parameter θ_s this error varies from 0.0005% to 0.0008%. As for the number of necessary iterations leading to the solution, this value changes significantly depending on the initial approximation.

5 Conclusions

The analysis of the results obtained leads to the conclusion that the proposed algorithm for calculating the first and second derivatives of the objective function turned out to be effective. Application of Newton method, in which the first and second derivatives of the objective function are calculated using this algorithm, makes it possible to identify the parameters K_0, α, m and θ_s with high accuracy.

So, regardless of the initial approximation, the error in determining the parameter K_0 does not exceed eleven thousandths of a percent. Such an error for parameter α does not exceed four thousandths of a percent, for parameter m - two ten-thousandths of a percent, and for parameter θ_s - eight ten-thousandths of a percent. Note that the case of accurate initial data was studied.

References

1. Van Genuchten, M.T.: A closed form equation for predicting the hydraulic conductivity of unsaturated soils'. Soil Sci. Soc. Am. J. **44**, 892–898 (1980)
2. Van Genuchten, M.T., Leij, F.J., Yates, S.R.: The RETC Code for Quantifying the Hydraulic Functions of Unsaturated Soils. US Salinity Lab, Riverside (1991)
3. Durner, W., Priesack, E., Vogel, H.-J., Zurmuhl, T.: Determination of parameters for flexible hydraulic functions by inverse modeling. In: Proceedings of the International Workshop on Characterization and Measurement of the Hydraulic Properties of Unsaturated Porous Media, pp. 817–829. University of California, Riverside (1999)
4. Takeshita, Y.: Parameter estimation of unsaturated soil hydraulic properties from transient outflow experiments using genetic algorithms. In: Proceedings of the International Workshop on Characterization and Measurement of the Hydraulic Properties of Unsaturated Porous Media, pp. 761–768. University of California, Riverside (1999)
5. Vrugt, J.A., Weerts, A.H., Bouten, W.: Information content of data for identifying soil hydraulic parameters from outflow experiments. Soil Sci. Soc. Am. J. **65**, 19–27 (2001)
6. Ngo, V.V., Latifi, M.A., Simonnot, M.-O.: Estimability analysis and optimisation of soil hydraulic parameters from field lysimeter data. Transp. Porous Media **98**(2), 485–504 (2013)
7. Zhang, K.: Parameter identification for root growth based on soil water potential measurements - an inverse modeling approach. Procedia Environ. Sci. **19**, 574–579 (2013)
8. Filipovic, V., et al.: Inverse estimation of soil hydraulic properties and water repellency following artificially induced drought stress. J. Hydrol. Hydromech. **66**(2), 170–180 (2018)

9. Abbaspour, K.C., Schulin, R., van Genuchten, M.T.: Estimating unsaturated soil hydraulic parameters using ant colony optimization. Adv. Water Resour. **24**, 827–841 (2001)
10. Yang, X., You, X.: Estimating parameters of van Genuchten model for soil water retention curve by intelligent algorithms. Appl. Math. Inf. Sci. **7**(5), 1977–1983 (2013)
11. Chen, G., Jiao, L., Li, X.: Sensitivity analysis and identification of parameters to the van Genuchten equation. J. Chem. (2016) https://doi.org/10.1155/2016/9879537. Article ID 9879537
12. Zhang, J., Wang, Z., Luo, X.: Parameter estimation for soil water retention curve using the SALP swarm algorithm. Water **10**(6), 815 (2018). https://doi.org/10.3390/w10060815
13. Evtushenko, Y.: Automatic differentiation viewed from optimal control theory. In: Griewank. A., Corliss, G.F. (eds.) Automatic Differentiation of Algorithms. Theory, Implementation and Application, vol. III, pp. 25–30. SIAM, Philadelphia (1991)
14. Griewank. A., Corliss, G.F. (eds.): Automatic Differentiation of Algorithms. Theory, Implementation and Application, vol. III. SIAM, Philadelphia (1991)
15. Evtushenko, Yu.: Computation of exact gradients in distributed dynamic systems. Optim. Methods Softw. **9**(1–3), 45–75 (1998)
16. Griewank, A., Walther, A.: Evaluating Derivatives: Principles and Techniques of Algorithmic Differentiation. 2nd edn. SIAM (2008)
17. Evtushenko, Y.G., Zasuhina, E.S., Zubov, V.I.: FAD method to compute second order derivatives. In: Corliss, G., Faure, C., Griewank, A., Hascoët, L., Naumann, U. (eds.) Automatic Differentiation of Algorithms, pp. 327–333. Springer, New York (2002). https://doi.org/10.1007/978-1-4613-0075-5_39
18. Zasukhina, E.S., Zasukhin, S.V.: Identification of parameters of the basic hydrophysical characteristics of soil. In: Evtushenko, Y.G., Khachay, M.Y., Khamisov, O.V., Kochetov, Y.A., Malkova, V.U., Posypkin, M.A. (eds.) Proceedings of the OPTIMA-2017 Conference, Petrovac, Montenegro, 02 October 2017, CEUR Workshop Proceedings, pp. 584–590 (2017)
19. Hogan, R.J.: Fast reverse-mode automatic differentiation using expression templates in C++. ACM Trans. Math. Softw. **40**(4) (2014). Article 26
20. Griewank, A., Juedes, D., Utke, J.: Algorithm 755: ADOL-C: a package for the automatic differentiation of algorithms written in C/C++. ACM Trans. Math. Softw. **22**, 131–167 (1996)
21. Bell, B.: CppAD: a package for C++ algorithmic differentiation (2007). http://www.coin-or.org/CppAD
22. Gay, D.M.: Semiautomatic differentiation for efficient gradient computations. In: Bücker, M., Corliss, G., Naumann, U., Hovland, P., Norris, B. (eds.) Automatic Differentiation: Applications, Theory, and Implementations. LNCSE, vol. 50, pp. 147–158. Springer, Heidelberg (2006). https://doi.org/10.1007/3-540-28438-9_13
23. Gorchakov, A.Y.: About software packages for fast automatic differentiation. Intell. Syst. Technol. **1**, 30–36 (2018). (In Russian)

Author Index